Environmental and Human Health – Risk Management in Developing Countries

T0227541

Environmental and Human Health – Risk Management in Developing Countries

Editors

Eddie N. Laboy-Nieves
Universidad del Turabo, Puerto Rico

Mattheus F. A. Goosen
Alfaisal University, Kingdom of Saudi Arabia

Evens Emmanuel
Université Quisqueya, Haiti

CRC Press
Taylor & Francis Group
Boca Raton London New York

CRC Press is an imprint of the
Taylor & Francis Group, an **informa** business

Cover photo credits: Eddie N. Laboy-Nieves.

http://ciemades.org

CRC Press/Balkema is an imprint of the Taylor & Francis Group,
an informa business

First issued in paperback 2017

© 2010 Taylor & Francis Group, London, UK
Typeset by MPS Ltd (A Macmillan Company) Chennai, India

British Library Cataloguing in Publication Data
A catalogue record for this book is available from the British Library

Library of Congress Cataloging-in-Publication Data

Environmental and human health : risk management in developing countries /
editors, Eddie N. Laboy-Nieves, Evens Emmanuel, Mattheus Goosen.
        p. cm.
    Includes index.
    ISBN 978-0-415-60162-7 (hard cover : alk. paper) — ISBN 978-0-203-83595-1
(e-book)   1. Environmental health—Developing countries.   2. Health risk
assessment—Developing countries.   3. Environmental management—Health
aspects—Developing countries.   4. Sustainable development—Health aspects—
Developing countries. 5. Economic development—Health aspects—Developing
countries.   I. Laboy, Eddie Nelson.   II. Goosen, Mattheus F. A.   III. Emmanuel, Evens.
IV. Title.

    RA566.5.D44E585 2010
    362.109172'4—dc22

                                                          2010028419

Published by: CRC Press/Balkema
            P.O. Box 447, 2300 AK Leiden, The Netherlands
            e-mail: Pub.NL@taylorandfrancis.com
            www.crcpress.com – www.taylorandfrancis.co.uk – www.balkema.nl

ISBN: 978-1-138-11669-6 (Pbk)
ISBN: 978-0-415-60162-7 (Hbk)

Contents

Preface

In the past, the world's ecosystems were able to absorb the ecological damage resulting from extensive industrialization and development. However, the rapid increases in human population, as well as in technological and economic expansion has put increasingly greater demands on the Earth's natural resources, such as fresh water supplies, which in turn has made it more difficult to sustain healthy ecosystems. Many international environmental organizations, such as the World Watch Institute and Green Peace, have reported that global natural resources continue to be over-exploited. They also noted significant problems with environmental pollution, resource depletion, conflicting land uses, and quite often a general indolence of society when it comes to the protection of natural resources. The Earth's biosphere as a sustainable habitat is endangered because of the domino effect of anthropogenic (i.e. man made) activities. Hence, the natural environment as well as human wellbeing is at risk. There is thus a greater need than ever, in the midst of this present planetary crisis, for individuals as well as for institutions to continue their research, education and public outreach in order to better promote environmentally sensible economic growth, with an aim to better understand the interrelationship between human development and nature, and to advocate for a more sustainable world.

The primary focus of this book is the use of a multidisciplinary team approach to assess and interpret the multitude of environmental hazards and risks, and to discuss the strategies that may be employed to manage them. The book expresses the viewpoints and experiences of faculty, scientists and graduate students representing 17 academic and research institutions from 11 countries worldwide. It is hoped that this intellectual exercise will allow the reader to access reliable and updated information in order to better acquaint them with the environmental and human risks facing society, the communication and decision-making processes involved in sustainable economic development, the effective management of surface, ground and wastewater resources, and the need for the conservation not only of natural but also cultural assets. Through the contents of this book, the authors and editors aim to provide natural resource managers, policy makers, researchers, government personnel, and faculty and students, with a comprehensive review of environmental and human health risk management, from the perspective of developing countries.

Through this edited book, the International Center for Environmental Studies and Sustainable Development (http://ciemades.org) would like to contribute in a special way in helping to intensify the process for the reconstruction of the research physical structures in environmental sciences at the *Université Quisqueya* (Haiti), which were severely damaged by the devastating earthquake of January 12, 2010.

The views expressed in this volume are those of the authors and not necessarily of the institutions to which they are affiliated. We hope that this book will contribute to the advancement of research, management and education projects to produce reliable and practical information that will aid in the reduction of environmental and human health risks, especially for emerging nations.

Eddie N. Laboy-Nieves
Universidad del Turabo, Puerto Rico

Mattheus F. A. Goosen
Alfaisal University, Kingdom of Saudi Arabia

Evens Emmanuel
Université Quisqueya, Haiti

2010

List of Authors (∗) and Co-Authors

Agnamey, Patrice
Université de Picardie Jules Verne, Amiens, France. Email: agnamey.patrice@chu-amiens.fr

∗Balthazard-Accou, Ketty
Université de Picardie Jules Verne, Amiens, France. Email: kettybal@gmail.com

∗Bosque, Adalberto
Universidad del Turabo, Gurabo, Puerto Rico. Email: abosque@email.suagm.edu

Brasseur, Phillippe
Institut de Recherche pour le Développement, Dakar, Sénégal. Email: brasseur@ird.sn

∗Cáceres-Acosta, Nancy María
Puerto Rican Association of Women, Science and Technology, San Juan, Puerto Rico. Email: nancymca55@yahoo.com

∗Carro-Anzalotta, Antonio E.
Universidad del Turabo, Gurabo, Puerto Rico. Email: acarro1@email.suagm.edu

Chiappetta Jabbour, Charbel José
University of São Paulo, Brazil. Email: charbel@usp.br

Bernard, Chocat
Lyon University, Lyon, France. Email: bernard.chocat@insa-lyon.fr

∗Cosme-Colón, Iris Nanette
Universidad del Turabo, Gurabo, Puerto Rico. Email: i.cosme@yahoo.com

Cotto-Maldonado, María del Carmen
Universidad del Turabo, Gurabo, Puerto Rico. Email: mcotto48@suagm.edu

∗del Amo-Rodríguez, Silvia
Centro de Investigaciones Tropicales, Universidad Veracruzana, Mexico. Email: sdelamo@uv.mx

∗Dorval, Farah A.
Lyon University, Lyon, France. Email: farah-altagracia.dorval@insa-lyon.fr

Emmanuel, Evens
Université Quisqueya, Port-au-Prince, Haiti. Email: evemm1@yahoo.fr

*Fifi, Urbain
Lyon University, Lyon, France. Email: fiur2000@yahoo.fr

Fischer, Wolfgang
Forschungszentrum Jülich, Institute of Energy Research, *Jülich*, Germany. Email: wo.fischer@fz-juelich.de

*Goosen, Mattheus (Theo) F.A.
Alfaisal University, Riyadh, Kingdom of South Arabia. Email: mgoosen@alfaisal.edu

Grelle-Mpakam, Hernanie
Université de Yaoundé I. Cameroon. Email: mhernaniegrelle@yahoo.fr

Hake, Friedrich-Jürgen
Forschungszentrum Jülich, Institute of Energy Research, *Jülich*, Germany. Email: jfh@fz-juelich.de

*Kouam-Kenmogme, Guy R.
Université de Liège. Belgium. Email: grkouam@yahoo.fr

*Laboy-Nieves, Eddie Nelson
Universidad del Turabo, Gurabo, Puerto Rico. Email: elaboy@suagm.edu

Lipeme-Kouyi, Gislain
Lyon University, Lyon, France. Email: gislain.lipeme-kouyi@insa-lyon.fr

*Malavé-Llamas, Karlo
Universidad del Este, Carolina, Puerto Rico. Email: kmalave@suagm.edu

Nono, Alexandre
Université de Dschang. Cameroon. Email: alex_nono2000@yahoo.fr

Oms, Julio A.
US Geological Survey, San Juan, Puerto Rico. Email: jaoms@usgs.gov

*Phillips, Peter
Winthrop University, Rock Hill, South Carolina, USA. Email: phillipsp@winthrop.edu

Raccurt, Christian
Université de Picardie Jules Verne, Amiens, France. Email: raccurt.christian@chu-amiens.fr

*Ramos-Jusino, Yolanda
Puerto Rico Electric Power Authority, San Juan, Puerto Rico. E-mail: y-ramos@prepa.com

Ramos-Prado, José María
Centro de Investigaciones Tropicales. Universidad Veracruzana, Mexico. Email: jramos@uv.mx

Rosillon, Francis
Université de Liège. Belgium. Email: f.rosillon@ulg.ac.be

Salgado-Herrera, Miriam
Puerto Rican Association of Women, Science and Technology, San Juan, Puerto Rico. Email: miriamsalhe@yahoo.com

San Miguel-Rivera, Lisbeth
Universidad del Turabo,Gurabo, Puerto Rico. Email: lsan@email.suagm.edu

Schaffner, Fred C.
Universidad del Turabo, Gurabo, Puerto Rico. Email: fschaffner@suagm.edu

*Schlör, Holger
Forschungszentrum Jülich, Institute of Energy Research, *Jülich*, Germany. Email: h.schloer@fz-juelich.de

*Silva, Eliciane Maria da
Methodist University, Piracicaba, Brazil. Email: eimsilva@unimep.br

Totet, Anné
Université de Picardie Jules Verne, Amiens, France. Email: totet.anne@chu-amiens.fr

Vergara-Tenorio, María del Carmen
Centro de Investigaciones Tropicales. Universidad Veracruzana, Mexico. Email: cvergara@uv.mx

*Villafañe-Deyack, Eileen C.
Universidad Metropolitana, Río Piedras, Puerto Rico. Email: eileenvillafae@yahoo.com

Winiarski, Thierry
Lyon University, Lyon, France. Email: Thierry.winiarski@entpe.fr

Curriculum Vitae of the Editors

EDDIE N. LABOY-NIEVES works as Associate Professor in the School of Science and Technology, *Universidad del Turabo* (http://ut.pr), Puerto Rico. He has nearly 25 years experience in teaching undergraduate and graduate level courses related to environmental sciences. He is an alumni of Frostburg State University and the Venezuelan Institute for Scientific Research, where he earned his M.S. and Ph.D. degree in Wildlife Management and Ecology, respectively. He worked as Manager of the Jobos Bay National Estuarine Research Reserve (Puerto Rico). He has received competitive grants from the USA EPA, NOAA and NSF to conduct environmental education projects. His research interests are focused on environmental characterization and habitat analysis of tropical terrestrial and marine ecosystems, and the populations dynamics of shallow water sea cucumbers. Dr. Laboy-Nieves has authored, co-authored, edited and peer-reviewed over 35 scientific publications. He is the President of the International Center for Environmental and Sustainable Development Studies: CIEMADeS (http://ciemades.org), Puerto Rico Chapter, the organization that promoted the publishing of this volume. He serves as Scientific Advisor for many community, national, and international organizations.

MATTHEUS (THEO) F. A. GOOSEN is Associate Vice President for Research and Graduate Studies at *Alfaisal University* (www.alfaisal.edu), a new private non-profit institution in Riyadh, Kingdom of Saudi Arabia. Previously he held the position of Campus Dean (CAO) at the New York Institute of Technology in Amman, Jordan. Dr. Goosen has also held academic dean positions at the Universidad del Turabo in Puerto Rico, USA, and at the Sultan Qaboos University in Muscat, Oman. He obtained his doctoral degree in Chemical/Biomedical Engineering from the University of Toronto, Canada, in 1981. After graduation he spent three years as a post doc at Connaught Laboratories in Toronto and then ten years at Queen's University in Kingston, Ontario. He has been on the Board of Directors of two companies. Dr. Goosen has published extensively with over 150 papers, book chapters, books, and patents to his credit.

 **EVENS EMMANUEL** is the Dean of the Faculty of Sciences and Engineering, Director of the Laboratory of Water and Environment Quality, and Head of the Graduate School of Ecotoxicology, Environment and Management of Water of the *Université Quisqueya*, a private nonprofit institution of 2,000 students in Port-au-Prince, Haiti. He holds a Master of Sciences in Sanitary Engineering from *Universidad San Carlos de Guatemala* and a doctoral degree in Environmental Engineering from the National Institute of Applied Sciences of Lyon, France. Dr. Emmanuel has produced more than 50 publications including 25 conference papers, and participated as peer reviewer for 10 international journals. His academic and research field are centered in methodological approaches of human health and ecological risk assessment of water and wastewater. Professor Emmanuel is member of the Caribbean Academy of Sciences (CAS) and the Puerto Rican Academy of Sciences and Arts. He is the President/Coordinator of the Haiti CIEMADeS Chapter, Scientific Director of the Haitian National Committee of UNESCO International Hydrological Program, National Coordinator of UNESCO ISARM Program, and member of the task force "Education and Capacity Building" of UNESCO Ecohydrology Program.

Communicating environmental risks and the decision making process

Tamarindo Beach calcareous sandstone formation, Guánica, Puerto Rico (@Eddie N. Laboy-Nieves)

Teachers examining hypersaline mud flats in Jobos Bay, Puerto Rico (@Eddie N. Laboy-Nieves)

Chapter 1

Environmental and human health risk management: An overview

Mattheus F. A. Goosen, Eddie N. Laboy-Nieves and Evens Emmanuel

SUMMARY

This chapter provides a brief overview of environmental and human health risk management in developing countries with an emphasis on communicating environmental risks and the decision making process, management of surface and groundwater resources, conservation of land, air, water and cultural assets, and risk management as it relates to wastewater reuse. This is important so to improve the interpretation of a wide variety of information related to the holistic management of our environment. The synopsis presented here may hopefully be applied towards a reduction in the many environmental risks faced by society and assist in the betterment of human health.

1.1 INTRODUCTION

Emerging states have often experienced a lack of harmony in coordinating economic development with environmental management (Laboy-Nieves 2009). This has often affected the health of people living in those countries, as well as efforts to reach environmentally sustainable economic expansion (Jabour et al. 2009). This chapter presents and discusses the multidisciplinary aspects of managing environmental risks and their effects on individual wellbeing

Communicating ecological risks to the general populace as well as gaining a better understanding of the decision making process as it relates to sustainable development play major roles in achieving changes in attitudes that contribute to environmental awareness in society. We can argue that the average person often does not fully understand the basis of environmental risks or how to deal with them (Hunt and Auster 1990; Azzone and Bertele 1994). Environmental education is in particular vital for young people who will be potential decision makers.

Management of surface and groundwater resources is a crucial aspect for sustainable growth in emerging nations (OTHU 2009). Compounding the issue of water availability is contamination of water supplies and a lack of wastewater treatment facilities (Hillstrom and Hillstrom 2004). This affects human health. Furthermore, sustainable economic expansion is now considered by many organizations and their stakeholders as being the model to follow. Many firms currently operate and converse based on their triple performance (i.e. economic, environmental and social) (USDHHS 1997).

The aim of this chapter is to provide a brief overview of environmental and human health risk management in developing countries with an emphasis on communicating environmental risks and the decision making process, management of surface and groundwater resources, conservation of land, air, water and cultural assets, and risk management as it relates to wastewater reuse.

1.2 COMMUNICATING ENVIRONMENTAL RISKS AND THE DECISION-MAKING PROCESS

Malavé-Llamas and Cotto-Maldonado, in Chapter 2, reported that communicating risk information to society can be an intimidating challenge especially when environmental factors are involved. They argue that the increased visibility of global change, terrorism, industrialization and catastrophic emergencies underscores the potential for environmental related human tragedy along with economic, social, and political consequences. Therefore, environmental threat communication must be targeted, understandable, and effective without inadvertently provoking hostility, mistrust or panic. In addition, an important component of the risk is perception. Many low-income communities have been exposed to disproportionate environmental risks and disease. Often companies, as well as news and governmental agencies have failed in their efforts to overcome public distrust in these communities. Communication is therefore a powerful and necessary tool to help solve this delema.

Knowing how to communicate information is essential. This process requires a sound structure and organization focusing on social, epidemiological, environmental, behavioral, and administrative elements (USDHHS 1997). Diverse audiences perceive communicated information differently thus the risk assessment process should think about cultural diversity, values, and language and communication skills of the target audience. Face to face communication is generally the most effective form to divulgate the message, especially to small audiences.

One consequence of improved communication in society is the growth of environmental consciousness in companies. This has brought changes in their organizational routine, called environmental management (see Chapter 3; as well as Jabour et al. 2009). The planning and organizational practices include environmental considerations in the decisions-making process, the establishment of ecological policies, and the adoption of green management tools such as ISO 14001 Certification. An essential aspect includes informing stakeholders about an organization's efforts to reduce the environmental impacts of its productive activities (Hunt and Auster 1990; Azzone and Bertele 1994). While organizations can be classified into three stages or categories (i.e. reactive; preventive; and proactive) da Silva and Jabour (Chapter 3) argued that classifying organizations based on their environmental management stage can, in practice, be more complex than this.

Communicating environmental risks and the decision making process is brought into focus by Laboy-Nieves and Oms (see Chapter 4) who assessed tsunami hazards in the Caribbean Region. Due to a dynamic tectonic environment, high population density, and extensive development in coastal municipalities, the area of Puerto Rico is at a significant risk for serious earthquakes and tsunamis (von Hillebrandt and Huérfano

2006) as shown by the recent catastrophe in Haiti. The tsunamigenic potential of a massive submarine landslide in two carbonate amphitheater south of the Puerto Rico Trench was evaluated. The majority of the islanders of the Caribbean lives and works in the coastal zone as a result of an increasing sea level-based tourism industry. The structures that support residential settlements and the economic connection in these coastal areas reflect a high degree of vulnerability to a strike by a tsunami. Laboy-Nieves and Oms (Chapter 4) examined the urgent need to increase public awareness on the possible impact that a major tsunami could have in the Caribbean, placing emphasis on Puerto Rico. The question was raised about who is responsible for the high risk developments along the coastlines of the islands in the Caribbean Region, such as Puerto Rico and Hispaniola (Beatley et al. 2002). By allowing urban construction to occur in flooding areas, islanders also become willing participants and must bear some responsibility of the possible consequences of the tsunami danger. Morales (2004) noted that this lack of public awareness is in part due to poor and errant environmental administration.

A legitimate role of the government is to take action as effectively as possible to reduce risks. To quantify health risk reduction, Bosque and Schaffner (Chapter 5) argued that we must determine the value that society is willing to pay in order to save a human life. Cost-benefit analysis (CBA), for instance, has been employed to enhance the risk management decision making process (Tietenberg and Lewis 2009). The "Value of Statistical Life" has also been used to assess the mortality benefits (i.e., mortality reduction) of environmental and safety regulations. Governments have used the CBA in the evaluation of regulations and actions that will affect the environment and the general public throughout risk reduction.

Furthermore, Schlör et al. (Chapter 6) noted that to determine whether or not we are living in a sustainable way, information is needed about the ecological, economic and social conditions of the environment in which we live. They report that genuine savings (GS), an indicator developed by the World Bank, measures the rate of savings of an economy taking into account investment in human capital, depletion of natural capital and damage caused by carbon dioxide emissions. However, the present GS does not consider water as a renewable resource. To solve this dilemma Schlör et al. combined the modified Adjusted Genuine Savings (AGS) with the Human Development Index (HDI). This type of combined analysis is particularly significant for developing countries (i.e. they get information on their real welfare, which enables them to protect natural resources more effectively and thereby to improve their socioeconomic development perspective). The AGS enables an institution to determine whether the growth is caused by the consumption of fixed capital, or resource depletion such as water, or the damage caused by CO_2 emissions. Hence, AGS/HDI provides a critical new dimension for the relative interpretation of economic development (Pillarisetti and van den Bergh 2010).

1.3 MANAGEMENT OF SURFACE AND GROUNDWATER RESOURCES

Numerous emerging states are characterized by unstable weather situations such as extreme variations in seasonal and annual rainfall. This can be compounded by poor water management practices and an inadequate infrastructure (Silveira 2002). Recurrent catastrophic floods, for instance, have huge social and economic effects.

The improvement of management practices in emerging states must be based on experiences in the collection of long term data series, the use of extended numerical tools, and the elaboration of realistic methods. Latin America, the most water resource rich area of the world, paradoxically experiences local shortages in supply and quality of water, due to negative impacts from pollutants, deforestation, damming, mining and the pressure from an increasing human population (Hillstrom and Hillstrom 2004). Phillips (Chapter 7) argued that while countries such as Costa Rica have an admirable record of good governance and management of forest resources, they have not sufficiently addressed how to manage their water resources. The same fresh water management challenges even confound countries with the most developed economies in the world (Jackson et al. 2001). Phillips concluded that balancing population and economic growth against the availability of economic resources will continue to challenge the ability of developing states to manage their surface and groundwater resources. To be successful, the solution will require a broad-based comprehensive adaptive management plan involving all stakeholders (Gregory et al. 2006). A good example is the large-scale environmental restoration project occurring in the Everglades in Florida (Redfield 2000).

Dorval et al. (see Chapter 8) have focused on Port-au-Prince, the capital of Haiti, to illustrate hydrological processes on urban catchment based on measurements and hydrological models. Their discussions highlight the importance of hydrological data collection activities which enable a further understanding of the catchment behavior. Emphasis is given on the need of integration of actors involved in storm water drainage. One good example from a facility in a developed country is the Experimental Observatory for Urban Hydrology Program in Lyon (France). It was built by a partnership of a multidisciplinary team representing universities, research laboratories, engineers, and the community (OTHU 2009). The Program is based on long term approaches to understand and interconnect natural sciences with hydraulics, social science and economics.

Fifi et al. in Chapter 9 discussed the impact of surface water runoff on the aquifers. The infiltration of urban storm water runoff is considered as one of the main factors in the deterioration of soils and groundwater quality. This situation is very critical in developing countries where urban surface runoff is seriously contaminated by organic and inorganic pollutants. In addition to metals or microbiological pollutants, groundwater from coastal aquifers of developing nations is also exposed to seawater contamination. This situation not only leads to a scarcity of groundwater resources, but also to human health risks for consumers. To understand and to manage these risks, Fifi et al. reviewed the impacts of surface runoff on groundwater by analyzing the plain of Cul-de-sac aquifer which contributes approximately 50% of the water supply of the urban community of Port-au-Prince.

In the USA, Europe, Japan and other developed countries, environmental risk analysis (ERA) has rapidly become not only a scientific framework for examining problems of environmental protection and remediation but also a tool for setting standards and formulating guidelines in modern policies (Ganoulis and Simpson 2006). This environmental tool may also be applied in developing countries for sustainable groundwater management. Then, the risk management of groundwater resources is defined through assessments of groundwater vulnerability, aquifer potential and source protection areas. In this context, five key stages can be identified for a decision-making framework for groundwater management in cities: investigation of groundwater management issue's;

identification and quantification of water users and uses; knowledge of aquifers behavior; planning and implementation; and monitoring and evaluation.

1.4 CONSERVATION OF LAND, AIR, WATER AND CULTURAL ASSETS

True sustainability development includes not only ecological elements of biodiversity conservation and ecological restoration, but also the social and economic aspects. In Chapter 10, del Amo-Rodríguez et al. presented a framework called "biocultural resource management" based on many years of academic and practical experiences in the field of ethnoecological restoration and resource management. They considered actual land use, local socio-economic problems and the expectations of the people involved, emphasizing cultural factors. These authors reported on a case study from the Totonacapan area in Veracruz State in tropical Mexico, the assessment of ethnoecological risks. This culture, as well as the ecosystems of the region is at high risk due to poverty and environmental degradation (Medellín 1988). A participative land use plan was the starting point for achieving better management. At the municipal level, categorization of the different types of land use allowed them to determine the causes of degradation. To conduct the preliminary diagnosis they employed satellite images and reviewed relevant literature to generate different thematic maps. Once the main types of land use were determined, they established the degree of conservation and degradation and the level of impact on natural resources. For each category of land use, planning and management activities were proposed considering quality and ecological fragility, based on the degree of conservation and degradation of the different categories of land use. Finally, ethnoecological strategies were designed and suggested in collaboration with community members (del Amo-Rodríguez et al. 2008a, 2008b). Let us next consider the air that we breathe.

On a global scale, respiratory conditions result in a large amount of disability and premature death. Carro-Anzalotta and San Miguel-Rivera (Chapter 11) reviewed the use of aerosols and their impacts on climate and human health. Air pollution is a major risk factor especially in crowded cities in developing states. It is directly associated with millions of deaths per year (Buckeridge et al. 1998). Pandey and Nathwani (2003) indicated that the mortality risk in most polluted area is 17% higher than in the least polluted scenario. Estimates of exposure and risk for acute respiratory infections, chronic obstructive pulmonary disease, and lung cancer, in several epidemiological studies, show that over 1.6 million premature deaths, and nearly 3% of the global burden disease, were attributable to indoor air pollution from solid fuels in 2000 (Ezzati 2005). Epidemiologic evidence suggests that air contamination, especially fine combustion source pollution, common to many urban and industrial environments, is an important risk factor for cardiopulmonary disease and mortality. Reviewers often note that evidence of health effects due to acute or short-term exposures is stronger that the evidence for chronic effects due to longer-term exposure (Pope 2000). People with heart or lung diseases, children and older adults are the most likely to be affected by air pollution exposure. Carro-Anzalotta and San Miguel-Rivera (Chapter 11) recommended that governments need to make better use of the scientific community as well as in raising awareness among citizens about the effects of aerosols on human health and the environment.

Modern air pollutants associated with road traffic and the use of chemicals for domestic, food and water treatment and pest control, are rarely present in excessive large concentrations, so effects on health are usually far from immediate or obvious (Briggs 2003). Ramos-Jusino and Laboy-Nieves in Chapter 12 discussed the need for poly aromatic hydrocarbons (PAHs) monitoring and how the bromeliad *Tillandsia* can be employed to assist in this effort. They demonstrated the use of *Tillandsia* to reduce pollutants, monitor toxics and to improve indoor air quality in high traffic and dense residential areas. Ramos-Jusino and Laboy-Nieves (Chapter 12) noted that it is important to educate people about how roadways and vehicles may impact human health and natural resources.

Oil pollution of water resources (i.e. ecosystems) is another area of major concern. In some environments, petroleum hydrocarbons persist indefinitely whereas under another set of conditions the same hydrocarbons may be completely bio-degraded within a few hours or days (Kaufman and Robertson 2010). Remediation technologies have the greatest potential for risk reduction if they are effective in removing the more carcinogenic, high molecular weight compounds. Bioremediation techniques can be considered a promising alternative to clean oil spills by using microbial processes (Rosa and Triguis 2007). Villafañe-Deyack and Laboy-Nieves (Chapter 13) argued that bioremediation works well for restoring soils and aqueous phases contaminated with petroleum hydrocarbons and polychlorinated biphenyl's (PCBs), but that its implications for risk assessment are uncertain. For petroleum hydrocarbons in soil and PCBs, international regulatory guidelines on the manage-ment of risks from contaminated sites are now emerging (Sutter II 2007).

Balthazard-Accou et al. (Chapter 14) and Emmanuel et al. (2004) have inves-tigated the contamination of drinking water supplies, as well as the socioeconomic conditions of Haiti. The aquifer of the Plaine du Cul-de-Sac at Port-au-Prince, and that of the Plaine des Cayes provide a large proportion of the drinking water supplies for the city. However, the water reserves have been found to be contaminated by *Cryptosporidium* oocysts (i.e. intestinal parasites found in humans and domestic animals). These opportunistic agents are the major cause of digestive pathology in HIV-infected individuals in developing countries and cause acute diarrhea in children under five. In Haiti, *Cryptosporidium* is responsible for 17% of acute diarrheas observed in infants under 2 years of age and 30% of chronic diarrheas in patients infected with HIV. The transmission of *Cryptosporidium* oocysts in young children, HIV-infected individuals, and people living in poor socioeconomic conditions is probably due to the consumption of water or contaminated food. In order to try and alleviate this problem, Balthazard-Accou et al. (Chapter 14) made three recom-mendations: the adoption of national drinking water guidelines; the protection of groundwater resources through sanitary inspection for control of *Cryptosporidium* contamination; and treatment of raw water using adsorption techniques.

1.5 RISK MANAGEMENT AND WASTEWATER REUSE

Urban market gardening using wastewater is an area for concern (Kouam-Kenmogme et al. Chapter 15; Jimenez et al. 2010). Many people in developing nations have none or incomplete schooling, and lack knowledge about how to use wastewater. There is a need for education as a fundamental step in the strategy for reduction of the sanitary impacts specifically within the framework of the reuse of wastewater in urban farming (UNHSP 2008). Nongovernmental organizations (NGOs), for example, can accompany market gardeners on site in the various phases of production in order to weed out poor practices and behaviours that are conducive to the spread of waterborne disease. Moreover, Kouam-Kenmogme and collaborators (Chapter 15) argued that it is important that people follow elementary rules of hygiene and wash the market garden produce thoroughly with drinking water before eating them. The gardeners must work with appropriate safety wear like boots and gloves, and avoid direct contact with wastewaters as much as possible. Using well water could reduce microbial contamination of crops; hence gardeners and vendors would benefit by setting up fountains in the bottomlands where they wash the products before taking them to market. This simple wash with good quality water leads to a logarithmic reduction of pathogens (Xanthoulis 2008).

In a related study, Cosme-Colón et al. (Chapter 16) reported on the application of low-cost sorbents to remove chromium from industrial wastewater discharges. Removal of toxic heavy metal contaminants from industrial wastewater is one of the most important environmental issues facing society today. Such pollutants do not degrade biologically, and as a result they accumulate in the food chain (Laboy-Nieves 2009). Chromium contamination of water, especially Cr(VI), its most toxic, mutagenic and carcinogenic form, is a persistent and serious global threat to human health and natural water resources. There is an urgent need for effective low-cost techniques for chromium removal from industrial wastewater, in particular since operating costs are increasing, and as environmental laws become more stringent. Sorption technology appears to be the most promising, eco-friendly and economically suitable alternative to control this problem. Cosme-Colón et al. (Chapter 16) highlighted environmental issues of chromium in industrial wastewater discharges, assessing the use of low-cost sorbents for removal of this pollutant, and have analyzed risk management with regards to pollutant removal and the protection of human and environmental health. They sustained that it is of utmost importance to consider holistic approaches including stakeholders participation and using iterations, if new information is developed that could change the needs, or the nature of one part or of the whole process being assessed. Any meaningful effort should recognize that humans are an integral part of the landscape, particularly in urban settings (Laboy-Nieves 2009).

In many countries, urban effluents are subjected to physiochemical and biological treatments. Anaerobic digestion continues to be the most widely used wastewater management process because it represents a sustainable system and a suitable method for developing countries. Cáceres-Acosta et al. in Chapter 17 reported on the viability of an upflow anaerobic sludge process for wastewater treatment. The technique combines physical and biological processes with anaerobic degradation such as hydrolysis, fermentation, acetogenesis and methanogenesis. The advantages of this reactor process include conservation of the environment, low energy consumption, and biogas and low sludge production.

1.6 CONCLUDING REMARKS

There is a need to improve the communication of environmental risks to the general public as well as to gain a better understanding of the decision making process as it relates to sustainable development and risk management. Both of these will play major roles in achieving changes in attitudes that contribute to better environmental awareness in our society. We can argue that the average person often does not fully understand the basis of environmental risks or how to deal with them. Ecological schooling in particular is imperative for the youthful populace who will be potential decision makers. Compounding the issue of water resources availability is the contamination of water supplies and a lack of effective wastewater treatment facilities in many developing states. Effective management of surface and groundwater resources is thus a key feature for the sustainable development of emerging nations, as well as being crucial for individual wellbeing. We therefore need to work together in seeking long-term solutions to environmental problems. It is a major challenge facing our society.

REFERENCES

Azzone G, Bertele U. 1994. Exploiting green strategies for competitive advantage. Long Range Planning 27(6): 69–81.

Beatley T, Brower DJ, Schwab AK. 2002. An introduction to coastal zone management. Devon, UK: Island Press. 329 p.

Briggs D. 2003. Environmental pollution and global burden of disease. British Medical Bulletin [Internet].[cited 2010 Jan 13]; 68: 1–24. Available from: http://bmb.oxfordjournals.org/cgi/reprint/68/1/1.pdf

Buckeridge D, Gozdyra P, Ferguson K, Schrenk M, Skinner J, Tam T, Amrhein C. 1998. A study of the relationship between vehicle emissions and respiratory health in an urban area. Geographical & Environmental Modelling 2(1): 23–42.

del Amo-Rodríguez S, Vergara-Tenorio MC, Ramos-Prado JM, Jiménez-Valdés L, Ellis EA. 2008a. Plan de ordenamiento ecológico de participación comunitaria del Municipio Zozocolco, Veracruz. Editorial de la Universidad Veracruzana. 131 p.

del Amo-Rodríguez S, Vergara-Tenorio MC, Ramos-Prado JM, Jiménez-Valdés L, Ellis EA. 2008b. Plan de ordenamiento ecológico de participación comunitaria del Municipio Espinal, Veracruz. Editorial de la Universidad Veracruzana. 114 p.

Emmanuel E. 2004. Evaluation Des Risques Sanitaires Et Ecotoxicologiques Liés Aux Effluents Hospitaliers. Thèse, Institut National Des Sciences Appliquées De Lyon. 257 p.

Ezzati M. 2005. Indoor air pollution and health in development countries. Lancet 366: 104–6.

Ganoulis J, Simpson L. 2006. Environmental risk assessment and management: promoting security in the Middle East and the Mediterranean region. In: Morel B, Linkov I, editors. Environmental Security and Environmental Management: The Role of Risk Assessment, Springer, p. 245–253.

Gregory R, Ohlson D, Arvai J. 2006. Deconstructing adaptive management: Criteria for applications to environmental management. Ecological Applications 16(6): 2411–2425.

Hillstrom K, Hillstrom LC. 2004. Latin America and the Caribbean: A continental overview of environmental issues. Santa Barbara (CA): ABC-CLIO's The World's Environment Series. p 131–153.

Hunt CB, Auster ER. 1990. Proactive environmental management: avoiding the toxic trap. MIT Sloan Management Review 31(2): 7–18.

Jabbour CJC, Santos FCA, Nagano MS. 2009. Análise do relacionamento entre estágios evolutivos da gestão ambiental e dimensões de recursos humanos: survey e estado-da-arte. Revista de Administração da Universidade de São Paulo - RAUSP [Internet]. [cited

2010 Mar 14];44(4) 342–364. Available from: http://www.rausp.usp.br/download. asp?file=v4404342.pdf

Jackson RB, Carpenter SR, Dahm, CN, McKnight DM, Naiman RJ, Postel, SL, Running SW. 2001. Water in a changing world. Ecological Applications 11(4): 1027–1045.

Jimenez B, Mara D, Carr R and Brissaud F. 2010. Wastewater treatment for pathogen removal and nutrient conservation: suitable systems for use in developing countries. In: Dreschsel P, Scott CA, Raschid-Sally L, Redwood M, Bahri A, editors. Earthscan/IWMI/IRDC. Wastewater irrigation and health. 432 p.

Kaufman L, Robertson C. 2010 May. In Gulf Oil Spill, Fragile Marshes Face New Threat. New York Times; 1–2. <http://www.nytimes.com/2010/05/02/us/02spill. html?partner=rss&emc=rss>

Laboy-Nieves EN. 2009. Environmental management issues in Jobos Bay, Puerto Rico. In: Laboy-Nieves EN, Schaffner F, Abdelhadi AH, Goosen MFA, editors. Environmental Management, Sustainable Development and Human Health. London: Taylor and Francis. p 361–398.

Medellín SM. 1988. Arboricultura y silvicultura tradicional en una comunidad totonaca de la costa [dissertation]. [Veracruz (Mexico) Instituto Nacional de Investigaciones Sobre Recursos Bióticos. 347 p.

OTHU. Research Field Observatory in Urban drainage. [Internet]. [updated 12 03 2009]. France; [cited 2010 04 27]. Available from: http://www.graie.org/othu/.

Pandey MD, Nathwani JS. 2003. Canada wide standard for particulate matter and ozone: cost-benefit analysis using a life quality index. Society for Risk Analysis 23(1): 55–67.

Pillarisetti JR, van den Bergh JCM. 2010. Sustainable nations: what do aggregate indexes tell us? Environmental Development and Sustainability 12: 49–62.

Pope CA. 2000. Review: epidemiological basis for particulate air pollution health standards. Aerosol Science and Technology 32: 4–14.

Redfield, GW. 2000. Ecological research for aquatic science and environmental restoration in south Florida. Ecological Applications 10(4): 990–1005.

Rosa AP, Triguis JA. 2007. Bioremediation process on Brazil shoreline. Environmental Science and Pollution Research 14(7): 470–476.

Silveira ALL. 2002. Problems of modern urban drainage developing countries. Water Science and Technology 45(7): 31–40.

Sutter II GW. 2007. Ecological Risk Assessment. Boca Raton, Florida (USA): CRC Press. 538 p.

Tietenberg T, Lewis L. 2009. Environmental and natural resource economics.. Boston (MA). Addison Wesley.

[UNHSP] United Nations Human Settlements Programme UN-HABITAT and Greater Monetor Sewerage Commission. 2008. Atlas of excreta, wastewater sludge, and biosolids management: Moving forward the sustainable and welcome uses of a global resource. In: LeBlanc RJ, Matthews P, Richard RP, editors. Nairobi (Kenya): UN-HABITAT. 632 p.

[USDHHS] Department of Human Health and Services (US); Environmental Health Policy Comity. 1997. Fundamentos de Evaluación para los Programas de Comunicación de Riesgos a la Salud y sus Resultados. [Principles of health risk and their results in the risk communication programs assessment] Washington (DC).

von Hillebrandt C, Huérfano V. 2006. Emergent tsunami warning system for Puerto Rico and the Virgin Islands. In: Mercado-Irizarry A, Liu P, editors. Caribbean Tsunami Hazard. New York. World Scientific. p 231–243.

Xanthoulis D. 2008. Low-cost wastewater treatment. Development of Teaching and Training Modules for Higher Education. ASIALINK. EuropeAid. Contract VN/Asia-Link/012(113128) 2005–2008. 443 p.

Coastal erosion in Cataño, Puerto Rico (@Eddie N. Laboy-Nieves)

Chapter 2

Communicating environmental risks in developing countries

Karlo Malavé-Llamas and María del C. Cotto-Maldonado

SUMMARY

Communicating risk information can be a daunting challenge and when environmental factors are involved the challenge becomes a real trial. A reason could be the perception of the environment by the majority of the people as intangible and pristine; so how this benign being can pose a risk or threat to the population? The increased visibility of global change, terrorism, industrialization and catastrophic emergencies underscores the potential for environmental related human tragedy along with economic, social, and political consequences. Therefore, the environmental risk communication must be targeted, understandable, and effective without inadvertently provoking hostility, mistrust or panic. Because risk, is far from being a purely scientific issue, other important factors must be addressed before delivering the information. An important component of the risk is perception. Risk is perceived according to the characteristic of the hazard, personal and social context. Many low-income communities as Hispanic, Afro-Americans and other minorities have faced and continue to face disproportionate environmental exposures and disease. Sometimes the industries, news and governmental agencies fail in the effort to overcome the prevalent public distrust. Environmental risk communication is a powerful and necessary tool giving the information to the public in a clear and objective way.

2.1 INTRODUCTION TO RISK COMMUNICATION

Hazard communication and the involved risks for humans and their assets is a commonplace activity which occurs in a multitude of "arenas". However, when environmental risks plus developing countries are incorporated into its equation, the task in hand becomes monumental. The origin of risk communication was centered on debates about risks associated with waste disposal, toxic chemicals and heavy metals, air and water pollution, nuclear power, electric and magnetic fields, oil spills, food additives, radon in homes, and biotechnology (Chess 1995; McBeth and Oakes 1996; Burger et al. 1999; Elliot et al. 1999; Grobe et al. 1999; McDaniels et al. 1999; Baron et al. 2000; Longo and Alberini 2006).

The USA National Academy of Sciences defines risk communication as, "an interactive process of exchange of information and opinion among individuals, groups, and institutions". It involves multiple messages about the nature of risk and other messages, but not strictly about risks (Covello 1998). The World Health Organization refers to risk communication as "an interactive process of exchange of information and opinion on risk among risk assessors, risk managers and other interested parties" (WHO 2010). Those definitions suggest a discussion between all the interest parties or "two-way communication". Other definitions are based in the exchange of information and opinions. This means, that, *risk communication* refers to a social process by which people become informed and empowered about hazards and risks and subsequently are influenced towards behavioral changes, therefore, they can participate in the decision-making process about risk issues. Risk communication is a community-based adaptation that addresses the societal, cultural, environmental, political, and economic characteristics of a population (Basu and Dutta 2008; Ebi 2009; Ford et al. 2009). It can be understood that the complexity of the risk communication trend strives in the nature, intricacy and variability of actors, which may be audience or authorities (Quick et al. 2009a). Taking this into account it is important, that the "experts" in our respective fields, remember that informing and communicating about risks is more likely to succeed when treated as a two-way process. When participants are seen as legitimate partners, and when people's attitudes and worldview's regarding environment and technology are respected, the message gets to the audience with less subjectivity (Quick et al. 2009b). The integration of the theory of planned behavior and attitude functions can provide a more detailed explanation and more precise practical guidance regarding behavioral prediction (Wang 2009). This is particularly true in the case of risk controversies. Acceptance of risks is not an information/education issue; it results from a societal discourse. This concept, has a holistic grasp, is culturally important to a particular group, and what that group or individuals identify as needs.

2.2 RISK PERCEPTION

The public perception of risk is crucial for its communication (Rimal et al. 2005; Wong 2009). According to Sandman (1987) risk perception is subjective and culturally based, he developed the equation Risks = Hazard + Outrage, and stated that "the risk that kill you are not necessary the risk that angers and frighten you". This equation was further clarified, defining the outrage as the level of public anger and fear about an environmental risk issue (Covello and Sandman 2001; Leshner et al. 2009; Quick et al. 2009a). Examples of some outrages are listed in Table 2.1.

Although perception is of great importance for risk communication, it is influenced by individual needs and wants. The hierarchical theory of needs of Abraham Maslow presents the importance of the satisfaction of the basic needs to reach self actualization (Jackson et al. 2004). When needs are identified and satisfied the communication channel to discuss risk opens up. Maslow (1970) though that all needs have the same importance but the lower ones are the bases for the development of the upper ones. Rowan (1998) presents some changes to the theory but basically agree with the concepts of the satisfactions of the basic needs. Figure 2.1 presents Maslow's pyramid of hierarchic needs. According to Maslow people have needs and

Table 2.1 Factors influencing risk perception (adapted from DPHHS 1994).

Acceptable Risk	Unacceptable Risk
• Voluntary	• Imposed
• Under personal control	• Controlled by other people
• Having a clear benefits	• Having little or no benefits
• Distributed fairly	• Unfairly distributed
• Come from natural sources	• Man made
• Statistically measured	• Catastrophic
• Coming from a trusted source	• Coming by an untrusted source
• Familiar source	• Exotic source
• Mayor population affected are adults	• Mayor population affected are children
• Understandable	• Confusing
• Immediate	• Delayed (future)
• Reversible	• Permanent
• Do not evoke fear	• Evoke fear or anxiety
• Personal stake	• Impersonal
• Moral	• Immoral

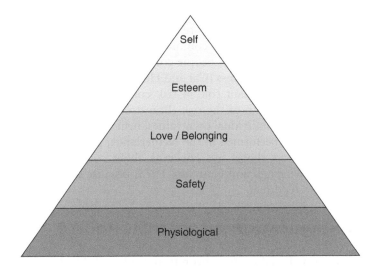

Figure 2.1 Maslow pyramid of needs (adapted from Jackson et al. 2004).

perform specific actions to satisfy it (Sunrow 2003). The needs are expressed in five levels and organized as a pyramid. Each level in the pyramid represents a need to be satisfied and its necessary to satisfy the needs in the lower level to rise to the upper ones. The five levels that all people experiment are: the physiological needs (as food, shelter and sex), safety and security, belonging (as friendship and love), self-esteem, and self-actualization (Sunrow 2003; Yang 2003; Zalenski and Raspa 2006). Some investigators suggest that the satisfaction of the needs is indispensable not only to reproduce or genetic survival but also to transmission of the culture (Yang 2003) and have implications in the humanitarian sector (Bierneld et al. 2006). When people

Table 2.2 Risk communication models (Adapted from Glik 2007).

Risk Communication Model	Characteristics
Perception	Determines the role of risk to influence the levels of concern, anger, anxiety, fear, hostility, and outrage, which in turn can significantly change attitudes and behavior positively or negatively.
Mental Noise	Refers to how stressed people process information and how changes in information are assimilated affect their communication.
Negative Dominance	Deals with the processing of negative and positive information in high-concern situations.
Trust Determination	Refers to the importance of trust in order to achieve the goals of education and consensus building imperative for an effective risk communication plan.
Socio-Psychological	Identifies a set of messages, characteristics, social influences and factors which determine whether the process is effective.

perceive their safety is compromised by the risk at hand, risk communications will not be possible until this need is satisfied (Bränström and Brandberg 2010).

If the benefit of a risk is out of context, the parties are deprived of the opportunity to make a decision having a real view of the risk. To be effective in the risk communication and to involve the community is necessary refashioned the tools of the assessment. This indicates that people could accept involuntary exposure and risk if they are involved in the process (Cai and Hung 2005).

Mass media play an important role in risk communication because they have great influence in the perception of the risk (Bränström and Brandberg 2010). This awareness develop from the individual own values, believes and experiences (Beacher et al. 2005). In a research performed in the Netherlands, people living in areas near hazardous industrial plants report less concerns about contamination that people living in other areas (Wiegman and Gutteling 1995) because they tend to reject and ignore the mass media reporters.

2.3 RISK COMMUNICATION THEORETICAL MODELS

Risk communication is based on five theoretical models (Table 2.2) which describe how risk information is processed, perceive, and how decisions are made (Covello 1998, 2001). It is crucial to acknowledge these models, because they provide a foundation for coordinating effective communication in high-concern situations.

2.3.1 The perception model

Many factors affect the perception of risk, and can alter the magnitude of the observed risk (Wildavsky and Dake 1990; Renn et al. 1992; Rogers 1997; Covello 1998; Turner et al. 2006; Berkley et al. 2009). As inferred from Table 2.1, there are at least 15 risk perception factors that have direct relevance to risk communication (Rimal and Morison 2006). Levels of concern tend to be most intense when the risk is perceived to be involuntary, inequitable, detrimental, or incontrollable, associated with untrustworthy individuals and organizations, or dreaded adverse irreversible outcomes. When the previous risk perception is involved or thrown into the mix, it is referred as the "outrage" factors. Therefore it widely accepted that an individual's perception of risk is based on a combination of hazard (e.g., mortality and morbidity statistics) and outrage (Sandman 1987). Outrage factors take on strong moral and emotional overtones, predisposing an individual to react emotionally, increasing the levels of perceived risk. Research on risk perception suggests that specific activities should be undertaken as part of a risk communication effort (Johnson 1993; Glik 2007). The successful collection and evaluation of empirical data (surveys, focus groups, or interviews about stakeholder judgments) is crucial for the interpretation of risk perception factors. Effective risk communication approaches requires mutual understanding of the interested or affected parties regarding stakeholder perceptions and the expected levels of concern, worry, fear, hostility, stress, and outrage.

2.3.2 The mental noise model

This model strives on how people under pressure process information. This is important because it is well established that when people are in a state of trepidation, their ability to process information effectively and efficiently is severely impaired (Baron et al. 2000). When people consider that what they value is being endangered, they undergo an ample range of emotions, ranging from anxiety to anger (Liu 2009). The emotional and mental agitation generated by these strong feelings produce what is denominated as mental noise. Risk exposures, associated with negative psychological attributes are also accompanied by severe mental noise (Maslow 1970; Neuwirth et al. 2000; Glik 2007), interfering with a person's ability to engage in rational or logical discourse.

2.3.3 The negative dominance model

The processing of information in high-concern situations is the basis for this model. The relationship between negative and positive information is asymmetrical, with negative information weighting more. The negative dominance model is coherent with a central theorem of modern psychology that states that people put additional value on negative outcomes (Maslow 1970). That is why negative messages should be counterbalanced by a larger number of positive or solution-oriented messages (Covello 1998; Glik 2007). Communications that contain negative words (*no, not, never, nothing, none*) and words with negative connotations, have a propensity to receive closer attention, are remembered longer, and have greater impact than positive messages (Covello 1998). The use of unnecessary negative wording with stakeholders in high-concern situations can be detrimental due that they could hold positive or solution-oriented information. Hence, risk communication is most effective when it focus on what is being done not on what cannot or have not been done.

2.3.4 The trust determination model

The need to establish trust is a common thread in all risk communication strategies (Renn and Levine 1991; Peters 1997; Slovic 1999). Education and consensus-building is an achievable goal, only when trust emerges as a byproduct of ongoing actions, listening, and communication skill (Peters et al. 1997; Linebarger and Piotrowski 2008). To establish or maintain trust, third-party endorsements from responsible sources should ideally be undertaken, as well as the use of four trust determination factors: caring and empathy; dedication and commitment; competence and expertise; and honesty and openness (Slovic 1999). Assessment studies show that small group settings, such as information exchanges and public workshops, are the most effective scene for communicating trust factors and divulging authenticity as an education tool (Covello 1998; Eisenberg et al. 2006; Nan 2008; Petraglia 2009).

The trust transference principle states that a lower trusted source more often than not takes on the credibility of the highest trusted source that takes the same position on the issue (Covello 1998; Nan 2009). Citizen advisory groups, health and safety professionals, scientists, and educators, are perceived to have high to medium trust on health, safety, and environmental issues (USEPA 1990; Aldoory 2006). This fact enables them to communicate effectively, even when communication barriers exist. However, individual trust overrides organizational trust. Trust in individuals from reliable organizations may significantly increase or decrease depending on how they present themselves (verbally and non-verbally) and how they interact with others (Cai and Hung 2005). Perceptions of trust decreases by actions or communications that emphasize disagreements, lack of coordination, insensitivity, unwillingness to acknowledge risks or to share information, and negligence in fulfilling responsibilities between information specialists or managers (Chess et al. 1995).

2.3.5 The socio-psychological model

This model aims to recognize content clarity and acceptance of information as well as the personal attributes of people that could influence the flow of risk communication (Glik 2007; Dutta and Basu 2008; Dutta and d Souza 2008; Han et al. 2009). Three overlapping processes need to be considered and linked: (1) how people deal with hazards, (2) how risk information is processed and evaluated and (3) how accepted information affects risk perception, evaluation and behavior. Interactive risk communication is far more likely to be effective, as a two-way communication pathways were information is exchange between the affected group and the "expert" group (Rohrmann and Chen 1999).

2.4 RISK COMMUNICATION INTO ACTION

Difficulties arisen for communicating effectively in high concern situations and are apparent during the exchange of information about the nature, type, magnitude, significance, control, and management of risks (Covello and Sandman 2001). The process needs to address the strengths and weaknesses of the various channels through which risk information is communicated, given or presented to the target audience: *press releases, public meetings, hot lines, web sites, small group discussions, information exchanges, public exhibits and availability sessions, public service announcements, and other print and electronic materials* (Alberini et al. 2007; Liu 2007).

Fukamizu (2007) questioned risk communication from the perspective of the political philosophy of democracy, stating that in the majority of situations, the definition of risk communication isn't satisfied. Because of its complexity risk communication is divided two types (Wrench 2004). The first one is called "technical model", which emerges for the natural science so that risk messages could be expressed by experts. The second, "the democratic one", emerges from social sciences as a criticism of the technical model (Wrench 2004).

Effective communication is critical for the successful resolution of health, safety, or environmental controversies (Glik 2007). Fear, anxiety, distrust, anger, helplessness, or frustrations are expressed in high-concern situations, as substantial barriers to effective communication (Fishhoff 1995; Covello 1998; Covello et al. 2001). These emotions charge and change the communication environment and rules complicating the communication scenario (Bränström and Brandberg 2010).

Covello (2006) expressed seven cardinal rules to be used during the risk communication process. For an effective process the public should be involved and treated as a partner. It's necessary not to only listen but to appreciate and respect public knowledge. All parties ought to be open, honest and work with other credible sources. The media has an important role for sharing information between all parts; it should consider the "3Cs": context, complexity and consequences (Glik 2007; Stryker et al. 2008; Ford et al. 2009). Therefore, it is important to meet the needs of the media, observe verbal and body language, and to speak clearly and with compassion (not with pity) to plan and assess the risk communication process.

2.5 ENVIRONMENTAL PROBLEMS AND RISK COMMUNICATION

One of the relevant issues during the process of risk communication is the knowledge of cultural behavior (Mortenson et al. 2006). Researchers like, el Katsha and Watts (1997) studied the prevalence and incidence of schistosomiasis in two Nile delta villages in Egypt as an environmental problems associated with the growth of the population, the absence of drainage provisions, and a high water table. The study revealed that poor communication among government agencies, community practices (like washing dishes in irrigation water channels) with unknowledgeable exposition to the disease's parasite, and the expensive bill of potable water, are crucial issues to reach effective risk communication goals. Another example related to the effectiveness of risk communication was reported by Taylor et al. (2009) during the massive bird mortality in the coastal town of Esperance in Australia. Native birds began to die and the community was very concern about the possibility of environmental pollution in the area. The Esperance Emergency Response Team collected environmental data for a scientific risk assessment. Although the community received nearly 80 reports, the above authors concluded that data collected was insufficient or suitable to determine if an environmental contamination occurred in the area (Taylor et al. 2009). This case exemplifies that quantity of data do not determine if the information is accepted and has a meaning for the community. This is consistent with Grabill and Simmons (1998) who reported that people separate risk assessment from risk communication and "this separation can lead to unethical and oppressive risk communication practices because the public is separated from fundamental risk decision making process."

The increase in international exchanges and the advances in communications are attributes of the contemporary globalization of the economy. Political instability, environmental degradation, poverty, and problems in the labors market are factor that increase the emigration of millions of people to other countries (Howson and Fineberg 1998). These exchanges increase the probability of new infectious diseases, exposure to unknown toxic substances, outbreaks of food poisons, and violent attacks as the chemical and biological terrorism (Howson and Fineberg 1998; Covello and Sandman 2001; Aldoory 2006; Aldoory and van Dyke 2006). The Global Thinking Project could provide a pathway to apply communication technology to discuss local environmental issues and concerns with people in different part of the world (Hassard 1997). A study conducted in China and Australia (Rohrmann and Chen 1999) revealed that real difference in risk perception could be observed between different cultures and between people of the same culture but different professions. Informed decision-making about risks is proportional to the effectiveness of the communication strategy and relies on building or re-building trust and dialogue among stakeholders, and reaching consensus (Morgan et al. 1992; Covello 1998). It has been demonstrated that unresolved conflicts, inadequate planning, preparation, resources, skill, and practice are major barriers to successful risk communication (Fischhoff 1995; Chess et al. 1995). Government officials, industry representatives, and scientists, often complain that risk information response by non-experts and lay people is irrationally and they do not accurately perceive and evaluate risk information (Jasanoff 1993). Conflicts are exacerbated by complex, confusing, inconsistent, or incomplete risk messages, distrust in the information sources, selective and biased reporting by the media and psychological heuristic factors that affect how risk information is processed (USEPA 1990; Renn and Levine 1991; Sjoberg 2000; Moeller 2006). Experts and risk management authorities acquire information about the interests, concerns and perspectives of stakeholders by inquiring as depicted in Figure 2.2.

2.6 RISK ASSESSMENT

Knowing how to communicate information is essential in the risk communication process. The complexity of this process requires a sound structure and organization for the design of an integral work plan, focusing on social, epidemiological, environmental, behavioral, organizational and administrative elements (USDHHS 1997). A message isn't effective if the person receiving it cannot understand what is being communicating. Different audiences perceive differently the communicated information, thus the risk assessment process should ponder cultural diversity, values, language and communication skills of those receptors, to prepare a community profile and determine the target audience (USDHHS 1997; NACCHO-ATSDR 2002). It is necessary to present information in verbal forms or fact sheets, but face to face communication is the most effective form to divulgate the message, especially to small audiences. The feedback of information between the audience and the communicator is the stone of the risk communication (NACCHO-ATSDR 2002).

Developing countries ought to promote the use of the risk communication process depicted in Figure 2.3, especially if it is done by people of the same country.

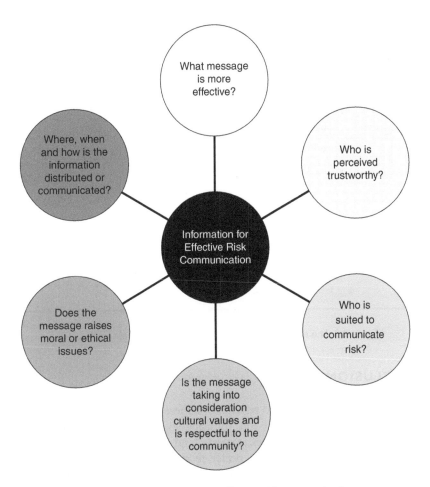

Figure 2.2 Information required for an effective risk communication process.
(adapted from Covello et al. 2001; Fischhoff 2006)

Many models used for risk communication come from developed countries, needing a modification and adaptation according the reality of the dwellers from developing countries (NACCHO-ATSDR 2002). For instance, Latin America is a broadly Spanish-speaking region, but in Spanish, like in many other languages people use idioms to describe the same thing. The term "habichuela" (bean in English) is used in Puerto Rico but "frijol" is used in Cuba and Mexico. Cultural behaviors like the "machismo" are necessary to know because it allow an efficient way to reach some specific sector (male or female) of the community that could be expose to a hazard (USDHHS 1997). Developed countries are facing a trans-cultural emigration and the natives of these countries increase the challenge to be effective in the risk communication, as reported by NACCHO-ATSDR (2002) in the United States, where cultural diversity is always present.

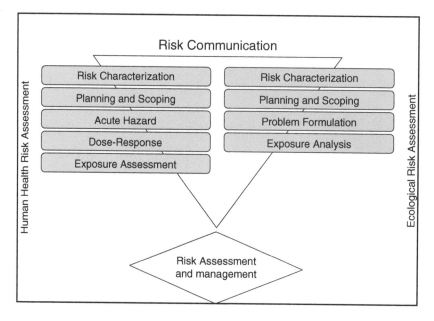

Figure 2.3 The risk assessment and management process (EPA 2010).

2.7 CONCLUSION

In this chapter we discuss risks and risk factors; these can be defined more narrowly by using technical means or more broadly by using socio-psychological parameters. The definitions also have their audiences; experts tend to prefer a focused and narrower approach, while public groups often prefer more comprehensive and holistic definitions. This dichotomy of views brings difficulty and concerns to all interested in giving or getting the actual facts so that decision making could be more accurate. This caught particular importance when it was realized that the risk management policies proposed by experts and specialized agencies were not necessarily acceptable to the wider public.

It is commonly accepted that if risk communication is to be more successful there has to be better dialogue and trust among all parties, predominantly from government officials, recognized experts and other legitimate groups in society and of course the general public. This change in perspective has meant that risk communication has had to become more integrated into the democratic and political processes, which in turn has forced decision-making on risks, particularly by governments, to become more open, transparent and democratic and based on the population needs. This in turn has raised such important issues as public trust in governments and expert agencies, freedom and availability of information in the public domain, mechanisms for public consultation, and roles of scientific experts and advisory committees. Therefore risk communication has come to mean much more than the mere passing on of information; it is a two way processes of information exchange between all involve parties, including the promotion of public dialogue between different stakeholders, conflict resolution, and risk prevention.

How risks and risk factors are defined and perceive is how they need to be assessed, that way the risk communication process will be effective. Taking that into thought it is also important to remember that risk communication has six main components: the objectives; the messages; target audiences; information sources and presentation; communications and conflict resolution. The complexity of the process is increased by the adjuvant effect that personal characteristic, perception of risk and type of risk is involved. This is evidenced by the five theoretical models which describe how risk information is processed, perceive, and how decisions are made. The perception model determines the role of risk to influence the levels of concern, anger, anxiety, fear, hostility, and outrage, which in turn can significantly change attitudes and behavior positively or negatively. The mental noise model refers to how stressed people process information and how changes in information are assimilated affect their communication. The negative dominance model deals with the processing of negative and positive information in high-concern situations. Trust determination assesses the importance of trust in order to achieve the goals of education and consensus building imperative for an effective risk communication plan and the socio-psychological model identifies a set of messages, characteristics, social influences and factors which determine whether the process is effective. It is crucial to acknowledge these models, because they provide a foundation for coordinating effective communication in high-concern situations.

Then, satiations of risks involving environmental justice, health and pollution, because of the broadness and imperceptive nature (air pollution is measure but not always seen) of the problem, bring uncertainly and fear to the community. Educational level and the link it has with economical level sometimes make the problem and difficulty of the task to developing countries bigger. This present an extraordinary communication challenges. However it is possible to develop an effective risk communication strategy for such events. It would be a disastrous error to underestimate the importance and necessity of developing, by consensus among organizations, communities and individuals a risk communication plan to address environmental health problems in developing countries.

REFERENCES

Alberini A, Tonin S, Turvani M, Chiabai A. 2007. Paying for Permanence: Public Preferences for Contaminated Site Cleanup. Journal of Risk and Uncertainty 34(2): 155–178.

Aldoory L, van Dyke M. 2006. The roles of perceived "shared" involvement and information overload in understanding how audiences make meaning of news about bioterrorism. Journalism and Mass Communication Quarterly 83(2): 346–361.

Barkley-Patton J, Goggin K, Liston R, Bradley-Ewing A, Neville S. 2009. Adapting Effective Narrative-Based HIV-Prevention Interventions to Increase Minorities' Engagement in HIV/AIDS Services. Health Communication 24(3): 199–209.

Baron J, Hershey JC, Kunreuther H. 2000. Determinants of priority for risk reduction: The role of worry. Risk Analysis 20(4): 413–428.

Basu A, Dutta MJ. 2008. The Relationship Between Health Information Seeking and Community Participation: The Roles of Health Information Orientation and Efficacy. Health Communication 23(1): 70–79.

Beecher N, Harrison E, Goldstein N, McDaniel M, Field P, and Susskind L. 2005. Risk Perception, Risk Communication, and Stakeholder Involvement for Biosolids Management and Research. Journal of Environmental Quality 34: 122–128.

Bierneld M, Lindmark G, McSpadden LA, Garrett MJ. 2006. Motivations, concerns, and expectations of Scandinavian health professionals volunteering for humanitarian assignments. Journal of Disaster Management and Response 4(2): 49–58.

Bränström R, Brandberg Y. 2010. Health Risk Perception, Optimistic Bias, and Personal Satisfaction. American Journal of Health Behavior 34(2):197–205.

Burger J, Pflugh KK, Lurig L, Von Hagen LA, Von Hagen S. 1999. Fishing in urban New Jersey: Ethnicity affects information sources, perception, and compliance. Risk Analysis 19(2): 217–229.

Cai DA, Hung CJF. 2005. Whom do you trust? A cross-cultural comparison. In: Cheney G, Barnett GA, editors. International & multicultural organizational communication. Cresskill (NJ): Hampton Press. p.73–104.

Chess C, Salomone KL, Hance BJ, Saville A. 1995. Results of a national symposium on risk communication: Next steps for government agencies. Risk Analysis 15(2): 115–125.

Covello VT. 2006. Risk Communication and Message Mapping: A New Tool for Communicating Effectively in Public Health Emergencies and Disasters. Journal of Emergency Management 4(3): 25–40.

Covello VT, Peters RG, Wojtecki JG, Hyde RC. 2001. Risk Communication, the West Nile Virus Epidemic, and Bioterrorism: Responding to the Communication Challenges Posed by the Intentional or Unintentional Release of a Pathogen in an Urban Setting. Journal of Urban Health: Bulletin of the New York Academy of Medicine 78(2): 382–391.

Covello VT, Sandman PM. 2001. Risk communication: Evolution and revolution. In: Wolbarst A, editors. Solutions to an Environment in Peril. Baltimore (MD): John Hopkins University Press. p.164–178.

Covello VT. 1998. Risk perception, risk communication, and EMF exposure: Tools and techniques for communicating risk information. In: Matthes R, Bernhardt JH, Repacholi MH, editors. Risk Perception, Risk Communication, and Its Application to EMF Exposure: Proceedings of the World Health Organization/ICNRP International Conference; 1998 May. Vienna (AT): International Commission on Non-Ionizing Radiation Protection. p. 179–214.

Dutta MJ, Basu A. 2008. Meanings of Health: Interrogating Structure and Culture. Health Communication 23(6): 560–572.

Dutta MJ, de Souza R. 2008. The Past, Present, and Future of Health Development Campaigns: Reflexivity and the Critical-Cultural Approach. Health Communication 23(4): 326–339.

Ebi KL. 2009. Facilitating Climate Justice through Community-Based Adaptation in the Health Sector. Environmental Justice 2(4): 191–195.

Eisenberg EM, Baglia J, Pynes JE. 2006. Transforming Emergency Medicine Through Narrative: Qualitative Action Research at a Community Hospital. Health Communication 19(3): 197–208.

el Katsha S. Watts S. 1997. Schistosomiasis in two Nile delta villages: and anthropological perspective. Tropical Medicine and International Health 2(9): 846–54.

Elliot SJ, Cole DC, Krueger P, Voorberg N, Wakefield S. 1999. The power of perception: Health risk attributed to air pollution in an urban industrial neighborhood. Risk Analysis 19(4):621–633.

[EPA] Environmental Protection Agency (USA). 1990. Public Knowledge and Perceptions of Chemical Risks in Six Communities: Analysis of a Baseline Survey. Washington (DC).

[EPA] Environmental Protection Agency (USA). 2010. Superfund Risk Assessment [Internet]. Available from: http://www.epa.gov/oswer/riskassessment/risk_superfund.htm.

Fischhoff B. 1995. Risk perception and communication unplugged: Twenty years of progress. Risk Analysis 15(2): 137–145.

Fischhoff B. The McGraw-Hill Homeland Security Handbook [New York (NY)]: McGraw-Hill; 2006. Chapter 30, The Psychological Perception of Risk; p.463–492.

Ford LA, Crabtree R, Hubbell A. 2009. Crossing Borders in Health Communication Research: Toward an Ecological Understanding of Context, Complexity, and Consequences in Community-Based Health Education in the U.S.-Mexico Borderlands. Health Communication 24(7): 608–618.

Fukamizu M. 2007. Risk Communication and Deliberative Democracy: How Democratic is Risk Communication? Journal of the Graduate School of Letters. [Internet]. [cited 2010 Feb28] 2:65–67. Available from: http://ci.nii.ac.jp/naid/110006624336/en.

Glik DC. 2007. Risk Communication for Public Health Emergencies. Annual Review of Public Health 28:33–54.

Grabill JT, Simmons WM. 1998. Toward a Critical Rhetoric of Risk Communication: Producing Citizens and Role of Technical Communicators. Technical Communication Quarterly 7(4): 415–441.

Grobe D, Douthitt R, Zepeda L. 1999. A model of consumers' risk perceptions toward recombinant bovine growth hormone (rbGH): The impact of risk characteristics. Risk Analysis 19(4): 661–673.

Han PKJ, Moser RP, Klein WMP, Beckjord EB, Dunlavy AC, Hess BW. 2009. Predictors of Perceived Ambiguity About Cancer Prevention Recommendations: Sociodemographic Factors and Mass Media Exposures. Health Communication 24(8): 764–772.

Hassard J. 1997.Teaching students to think globally. Journal of Humanistic Psychology 37(1): 24–63.

Howson CP, Fineberg HV. 1998. The pursuit of global health: the relevance of engagement for developed countries. Lancet 351(9102): 586–90.

Jackson T, Jager W, Stalg S. 2004. Beyond Insatiability: Needs Theory, Consumption and Sustainability [Internet]. Guildford (UK): Centre for Environmental Strategy, University of Surrey (UK); [cited 2010 Feb28]. Available from: http://portal.surrey.ac.uk/pls/portal/docs/PAGE/ENG/RESEARCH/CES/CESRESEARCH/ECOLOGICAL-ECONOMICS/PROJECTS/FBN/BEYONDINSATIABILITY.PDF

Jasanoff S. 1993. Bridging the two cultures of risk analysis. Risk Analysis 13(2): 123–129.

Johnson BB. 1993. 'The mental model' meets 'the planning process': wrestling with risk communication research and practice. Risk Analysis 13(1): 5–8.

Leshner G, Bolls P, Thomas E. 2009. Scare' Em or Disgust 'Em: The Effects of Graphic Health Promotion Messages. Health Communication 24(5): 447–458.

Linebarger DL, Piotrwski JT. 2008. Evaluating the Educational Potential of Health PSAs with Preschoolers. Health Communication 23(6): 516–525.

Liu BF. 2007. Communicating with Hispanics about crises: How counties produce and provide Spanish-language disaster information. Public Relations Review 33(3): 330–333.

Liu M. 2009. The intrapersonal and interpersonal effects of anger on negotiation performance: A cross-cultural investigation. Human Communication Research 35: 148–169.

Longo A, Alberini A. 2006. What Are The Effects of Contamination Risks on Commercial and Industrial Properties? Evidence from Baltimore, Maryland. Journal of Environmental Planning and Management 49(5): 713–737.

Maslow AH. 1970. Motivation and Personality. New York (NY): Harper and Row.

McBeth MK, Oakes AS. 1996. Citizen perception of risks associated with moving radiological waste. Risk Analysis 16(3): 421–427.

McDaniels TL, Gregory RS, Fields D. 1999. Democratizing risk management: Successful public involvement in local water management decisions. Risk Analysis 19(3): 497–509.

Moeller SD. 2006. Regarding the pain of others: Media, bias and the coverage of international disasters. Journal of International Affairs 59: 173–196.

Morgan G, Fischhoff B, Bostrom A, Lave L, Atman CJ. 1992. Communicating risk to the public. Environmental Science and Technology 26(11): 2048–2056.

Mortenson S, Liu M, Burleson BR, Liu Y. 2006. Exploring cultural and individual differences (and similarities) related to skilled emotional support. Journal of Cross Cultural Psychology 3: 366–385.

[NACCHO-ATSDR] National Association of County and City Health Official (USA); Agency for Toxic Substances Disease Registry (US) 2002 Assessment to Action: An tool for improving the health of communities affected by hazardous waste. Washington (DC).

Nan X. 2008. The influence of liking for a public service announcement on issue attitude. Communication Research 35(4): 503–528.

Nan X. 2009. The influence of source credibility on attitude certainty: Exploring the moderating effects of timing of source identification and individual need for cognition. Psychology and Marketing 26(4); 321–332.

[NRC] National Research Council (USA). 1996. Understanding Risk: Informing Decisions in a Democratic Society. Washington (DC): National Academy Press (US).

Neuwirth K, Dunwoody S, Griffin RJ. 2000. Protection motivation and risk communication. Risk Analysis 20(5): 721–733.

Peters RG, Covello VT, McCallum DB. 1997. The determinants of trust and credibility in environmental risk communication: An empirical study. Risk Analysis 17(1): 43–54.

Petraglia J. 2009. The Importance of being authentic: persuasion, narration, and dialogue in health communication and education. Health Communication 24(2): 176–185.

Quick BL, Bates BR, Quinlan MR. 2009a. The Utility of Anger in Promoting Clean Indoor Air Policies. Health Communication 24(6): 548–561.

Quick BL, Bates RB, Romina S. 2009b. Examining Antecedents of Clean Indoor Air Policy Support: Implications for Campaigns Promoting Clean Indoor Air. Health Communication 24(1): 50–59.

Renn O, Bums WJ, Kasperson JX, Kasperson RE, Slovic P. 1992. The social amplification of risk: Theoretical foundations and empirical applications. Journal of Social Science Issues 48; 137–6.

Renn O, Levine D. 1991. Credibility and trust in risk communication. In: Kasperson and Stallen, editors. Communicating Risks to the Public. Dordrecht (NL): Kluwer Academic Publishers.

Rimal RN, Lapinski M, Cook R, Real K. 2005. Moving toward a theory of normative influences: How perceived benefits and similarity moderate the impact of descriptive norms on behaviors. Journal of Health Communication 10: 433–450.

Rimal RN, Morrison D. 2006. A uniqueness to personal threat (UPT) hypothesis: How similarity affects perceptions of susceptibility and severity in risk assessment. Health Communication 20: 209–219.

Rogers GO. 1997. The dynamics of risk perception: How does perceived risk respond to risk events? Risk Analysis 17(6): 745–757.

Rohrmann B, Chen H. 1999. Risk perception in China and Australia: an exploratory cross-cultural study. Journal of Risk Research 2(3): 219–241.

Rowan J. 1998. Maslow amended. Journal of Humanistic Psychology 38(1): 81–112.

Sandman PM. 1987. Risk Communication: Facing Public Outrage. U.S. Environmental Protection Agency Journal. [Internet]. [cited 2010 Feb 28]. p. 21–22. Available from: http://www.wpro.who.int/internet/files/eha/toolkit/web/Technical%20References/Risk%20Communication%20and%20Public%20Information/Introduction%20to%20risk%20perception.doc.

Sjoberg L. 2000. Factors in risk perception. Risk Analysis 20(1): 1–11.

Slovic P. 1999. Trust, emotion, sex, politics, and science: Surveying the risk-assessment battlefield. Risk Analysis 19(4): 689–701.

Stryker JE, Moriarty CM, Jensen JD. 2008. Effects of Newspaper Coverage on Public Knowledge About Modifiable Cancer Risks. Health Communication 23(4): 380–390.

Sumrow A. 2003. Motivation: a new look at an age-old topic. Journal of Radiology Management 25(5): 44–47.

Taylor S, Papadopoulos C, Vieillet S, Di Marco P. 2009. Assessing Community Health Risks: Proactive Vs Reactive Sampling. American Journal of Environmental Science (6): 695–696.

Turner MM, Rimal RN, Morrison D, Kim H. 2006. The role of anxiety in seeking and retaining risk information: Testing the risk perception attitude framework in two studies. Human Communication Research 32: 1–27.

[USDHHS] Department of Human Health and Services (USA); Environmental Health Policy Comity. 1997. Fundamentos de Evaluación para los Programas de Comunicación de Riesgos a la Salud y sus Resultados. [Principles of health risk and their results in the risk communication programs assessment] Washington (DC).

Wang X. 2009. Integrating the Theory of Planned Behavior and Attitude Functions: Implications for Health Campaign Design. Health Communication 24(5): 426–434.

Weigman O, Gutteling JM. 1995. Risk Appraisal and Risk Communication: Some Empirical Data from the Netherlands Reviewed. Basic & Applied Social Psychology 16(1/2): 227–249.

Wildavsky A, Dake K. 1990. Theories of risk perception: Who fears what and why. Daedalus 112: 41–60.

[WHO] World Health Organization. 2010. Risk communication. [Internet] World Health Organization. Available from: http://www.who.int/foodsafety/micro/riskcommunication/en/index.html.http://www.wvdhhr.org/bphtraining/courses/cdcynergy/content/activeinformation/resources/epa_seven_cardinal_rules.pdf.

Wrench J. Development and Validity Testing of the Risk Communication Style Scale [abstract]. In: [unknown editor]. Annual Meeting of the International Communication Association; 2004 May 27, New Orleans. New Orleans (LA).

Wong N. 2009. Investigating the Effects of Cancer Risk and Efficacy Perceptions on Cancer Prevention Adherence and Intentions. Health Communication 24(2): 95–105.

Yang KS. 2003. Beyond Maslow's culture-bound linear theory: a preliminary statement of the double-Y model of basic human needs. Journal of Nebraska Symposion on Motivation 49: 175–255.

Zalenski RJ, Raspa R. 2006. Maslow's hierarchy of needs: a framework for achieving human potential in hospice. Journal of Palliative Medicine 9(5): 1120–1127.

Tourist feeding a monkey at the Loja Zoo (Ecuador), even though the warning for not to do it
(@Eddie N. Laboy-Nieves)

Chapter 3

Environmental management in Brazil: Reducing organizational risks

Eliciane Maria da Silva and Charbel José Chiappetta Jabbour

SUMMARY

The pollution and the environmental impact created by the current productive process, the premature obsolescence and the unnecessary disposal of products are relevant issues for Brazilian companies. This chapter presents background information about environmental management in Brazil bearing in mind the environmental risks produced by the productive process of organizations. Basic concepts of environmental management and their stages in organizations, such as reactive, preventive and proactive, are described. Also several levels of industrial environmental consciousness are presented. Next, according to the proactive stage, this chapter describes the concepts of the traditional functional areas related to safety and health, human resources, production and operations, marketing, finance and their influence on environmental management. These concepts can help industries reduce the environmental impacts they generate. We conclude by presenting some studies that show how Brazilian organizations have been managing the environmental risk. These studies reveal that Brazilian companies, to become environmentally proactive, still have a lot to accomplish as far as environmental risk management is concerned. This statement has implications for environmental education field of local universities, responsible for training the future risk management professionals. Conclusions and reflections on this issue are also provided.

3.1 INTRODUCTION TO ENVIRONMENTAL MANAGEMENT IN BRAZIL

Brazil is known for its natural resources abundance, its rich diversity of fauna and flora species and for holding the largest portion of the Amazon rainforest in its territory. Nevertheless, like in other parts of the world and in developed countries, the majority of this natural richness can be reduced as a direct or indirect consequence of industrial activities.

The productive processes consume natural resources such as raw material and basic inputs. The assets and products generated by industrial activities produce environmental impacts even after leaving the factories and reaching the market, e.g. cars and electronic equipment, which are constantly produced. The current consumption and production system can be considered unsustainable: products are manufactured and discarded as soon as possible, leading to more intense and faster cycles of pollution and environmental impacts. These evidences together with more recent issues such as global warming, have brought great uncertainty about the future of humanity and the quality of the environment in which we live resulting in growing risks to social actors in general (Hoffman 2005; Boiral 2006).

Environmental risk awareness has also been verified among Brazilians. In a survey about the greatest threats facing humanity, Brazilians considered "pollution and environmental problems" the most urgent issues, ahead of more relevant domestic social problems such as "income inequality" (Cenci 2009). Although the findings of that survey are inconclusive, the industries in the country should be prepared to withstand the "green movement" in the internal and external markets. Brazilian companies have to take part in activities aimed at reducing environmental risks and the risks associated to such uncertainties.

Brazil has nearly 1669 companies with ISO 14001 Certification, but that number represents a declining tendency, given that in December 2005 there were 2500 certified companies. ISO 14001 is a norm published by ISO that certifies the greening of companies world-widely (Massoud et al. 2010). This norm establishes requirements and a framework to guide companies on how to become environmentally adequate. When compared to other countries (Brazil, Russia, India and China), Brazil had more certifications in December 2008 than Russia (720 certifications), but trailed India (3281 certifications) and China (39,195 certifications). The industrial sectors with the most certifications were the automotive, petrochemical and chemical, all with international presence and generators of considerable environmental impact (RMAI 2006). The company with the most certifications is Petrobrás, with 41 business units certified in ISO 14001. Nearly 50% of the Brazilian certifications are in the state of São Paulo, which has the most industrial activity in Brazil. Except for China, Brazil stands out in terms of ISO 14001 certifications when compared to countries with similar per capita income (refer to per capita income of countries in Constanza 2007).

To countenance the environmental dilemma in Brazil, this Chapter presents theoretical bases that can help industries to reduce the environmental impacts generated by their operations. Several levels of industrial environmental consciousness are presented. The inclusion of environmental issues in the organizational structure is also investigated to demonstrate how each organizational area can contribute to reducing environmental risks related to industrial activities. Lastly, we present empirical evidences reported by the best environmental management researchers in Brazil in order to provide readers with a broad spectrum of how Brazilian organizations are dealing with environmental problems and inherent risks to their productive processes.

3.2 STAGES OF ENVIRONMENTAL AND STRATEGIC MANAGEMENT

The development of environmental consciousness in companies has brought changes in their organizational routine and is called environmental management (Jabbour et al. 2009). According to Haden et al. (2009), environmental management can be understood as "the organization-wide process of applying innovation to achieve sustainability, waste reduction, social responsibility, and a competitive advantage via continuous learning and development and by embracing environmental goals and strategies that are fully integrated with the goals and strategies of the organization". In practical terms, when an organization decides to manage its environmental aspects, it should choose from a series of environmental management practices for planning and organization, operations, and communication (González-Benito and González-Benito 2006).

The planning and organizational practices include environmental considerations in the decisions-making process, the establishment of environmental policies, and the adoption of environmental management tools such as the ISO 14001 Certification, known worldwide. Operational practices include the need of rethinking and continuous improvement of transformation activities in organizations, especially in manufacturing industries. Some examples of these practices are related to the analysis, change, or maintenance of machinery and equipment, decisions about product development and manufacturing processes, and the logistic system adopted by the company. With respect to communication, the practice focus implies informing the stakeholders about the organization efforts to reduce the environmental impacts of its productive activities.

Not all the organizations that claim to be engaged with the environment have an environmental management system that can support such engagement (Hunt and Auster 1990). In other words, the environmental management can have different stages of development and maturity. A number of studies has identified this diversity of environmental development stages in different organizations (Hunt and Auster 1990; for example, Azzone and Bertele 1994; Hart 1995; Berry and Rondinelli 1998). Each level of environmental management is called "environmental management stage". In general, organizations can be classified into three stages: (1) reactive; (2) preventive; and (3) proactive. However, organizations can also occupy intermediate positions among these stages. Hence, classifying organizations based on their environmental management stage can be, in practice, more complex than it was presented here.

In the reactive stage the organization tends only to respond to legal impositions that arise from environmental problems generated by its productive process. This is what Hunt and Auster (1990) call "fire fighter". The organization keeps causing pollution due to its productive process, but it tends to employ new equipment, such as smokestack and filtering membranes, to reduce the negative far reaching externalities of the pollution. In this context, the environmental issue can be considered a non-returnable extra expense, and it tends to influence the organization structure.

The preventive stage major objective implies to reduce or avoid pollution during the productive stage. An example of this decision is the usage of machinery and equipment in the productive process that continuously collects residues generated by previous stages, to further reuse them in other productive processes. Another example is to use water-based paints during the manufacturing painting process instead of mineral-based paints that cause more pollution.

At last, organizations in the proactive stage of environmental management can understand better the environmental improvements presented so far, and, in addition, they make intensive use of the practices described by González-Benito and

González-Benito (2006). These organizations continuously evaluate the environmental risks of their activities in terms of market and strategy in order to develop cleaner productive processes and reducing environmental impact-oriented products. They are focused on the continuous improvement of their environmental management activities. This approach tends to generate long-term benefits, such as the stakeholders' respect and loyalty, operational cost reduction by the substitution of manufacturing materials, and the creation and use of competitive advantages (Berry and Rondinelli 1998).

Proactive environmental management tends to create opportunities of environmentally appropriate innovations (Porter and Linde 1995), which are offered in green consumer markets more aware of environmental causes (Layrargues 2000), improving the company's institutional image and, consequently, its financial results (Miles and Covin 2000). Moreno et al. (2003) reinforced that environmental management in companies may stimulate the development of either cost reduction or differentiation strategies.

Environmental issues in manufacturing should be thought about carefully in terms of equality when they are dealt together with the consolidated performance objectives or competitive priorities of production (Angell and Klassen 1999). It is worth considering that environmental aspects are not incompatible with other competitive priorities of manufacturing. In fact, it is an objective that can stimulate the performance of other competitive dimensions (Jiménez and Lorente 2001). Thus, environmental considerations create potential to increase consolidated performance objectives of production function. Cost is favoured by reducing the consumption of raw materials, attempting to find eco-efficient process technology and substituting raw materials for reusable and recyclable materials (Seiffert and Loch 2005). Quality is enlarged and becomes more robust due to incorporating environmental dimension into the principles of quality management (Aboulnaga 1998). Including this dimension in the innovation strategy of companies leads to the creation of a wider range of products as well as the fact that it contributes to increasing the flexibility of lines and mixes of products (Porter and Linde 1995). Environmental accidents are prevented, ensuring that the ordered production is not delayed, which ensures speed of delivery (Hunt and Auster 1990), 1990). Hence, an organization can continuously improve its environmental practices as well as decreasing the respective uncertainties and reducing the risk of operating in a world with fast-growing environmental problems (Gupta and Piero 2003).

To increase the environmental management level to the proactive stage, a company that top management decision-makers provide effective support to those practices that best commit to the environment (Zutshi and Sohal 2004). This support should be made easier by a continuous spread and assessment of the environmental policy. Upper lever managers tend to encourage every area of the organization to rethink their activities considering the environment and reducing associated risks. The organizational areas that provide support to environmental management are the following:

3.3 ENVIRONMENTAL MANAGEMENT AND FUNCTIONAL AREAS

The proactive stage represents the environmental management scenario that best reduces the uncertainties of the relationship between the organization and environment (Boiral 2006). To maximize effectiveness, proactive environmental management requires structural changes in the organization, as depicted in Figure 3.1. The organizational structure can be understood as the distribution of activities, different authority levels, and communication means of the company.

- Safety and health: In general, it consists of members of the Internal Committee for Accident Prevention (CIPA) and health professionals, such as physicians and phonoaudiologists. CIPA develop risk maps, and programs for the fire safety, environmental risk prevention, fundamental medical and occupational health control, hearing conservation and, guidance on the appropriate use of collective and individual protection equipments. It is an area of great interface with the environmental management since the events that cause accidents in the workplace can reach the environment creating significant impacts.
- Human resources: this area is responsible for attracting workers, developing performance policies (Hayes and Wheelwright 1984), training, assessment, promotion, transfer, dismissing and remunerating, personal motivation, and

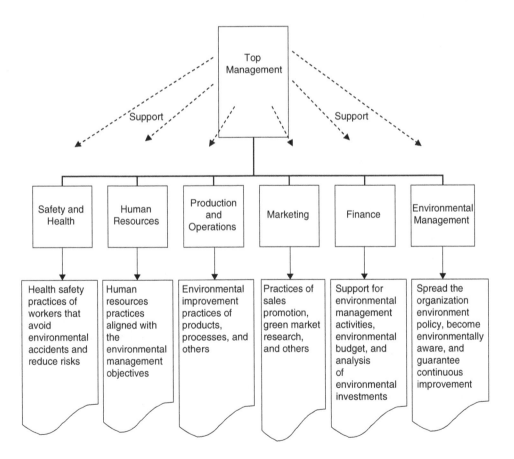

Figure 3.1 Inclusion of the environmental dimension in the organizational structure.

selecting and employing procedures (Fine and Hax 1985). It is considered the key to the success of the environmental practices. The alignment of the human resources practices, such as training, performance assessment and rewards, can motivate workers to meet the objectives of continuous improvement of the environmental performance of the organization (Jabbour and Santos 2008).

- Production and operations: this area is responsible for supplying or manufacturing products and services for the business. Managers make decisions regarding operations and their connection with other functions. They plan and control the production process an its interface with the organization and the external environment (Schroeder 2007). Due to its transforming capacity related to the selection, processing, and manufacturing of products, the production area can be considered the one that causes most environmental impacts. In order to mitigate such impacts, the production area can adopt preventive environmental technologies and techniques to develop and manufacture environmentally enhanced products. Some examples of these techniques are: Design for Environment, Life Cycle Analysis and Design for Disassembly (Angell and Klassen 1999). Barbieri (2004) stated that environmental management is developed in production in its initial phase. Hunt and Auster (1990) reported that production function was the first to incorporate the environmental dimension. Hart (1995) noted that environmental management in companies progresses along a continuous development of competences in environmental and production management, such as pollution control and developing sustainable products. Thus, the pioneering experience of production function is important for the development of environmental competences of all functional areas of a company.
- Marketing: it is related to the "process of planning and executing the conception, pricing, and promotion and distribution of ideas, goods, and services to create exchanges that meet individual and organizational objectives" (Ferrell and Lucas 1987). It should operate close to the environmental and production areas in order to have an environmentally adequate organization offering environmentally enhanced products to the consumers (Polonsky and Rosenberger III 2001; Rex and Baumann 2007; Sharma et al. 2010).
- Finance: the funding of management. The financial area is responsible for registering, monitoring and controlling the past financial transactions and current operations. It also deals with the search for funds to meet current and future needs. This area provides budget to make feasible the performance of the environmental management and the analysis of investments related to the environment in several different areas of the company. A recent study demonstrated that the adoption of environmental management practices is related to the highest financial indexes of companies (Molina-Azorin et al. 2009).
- Environmental management: it is responsible for establishing a policy framework to encourage incorporation of the environmental dimension products, processes, and organizational strategies. It supports upper management and makes decisions about adopting environmental technologies that can bring benefits to the whole company, such as effluent treatment stations. It should keep records of the company's efforts in terms of environmental management and compare its activities with the competitors.

3.4 RISK MANAGEMENT

By managing the environment, companies are more likely to reduce the risks related with it. However, are there evidences that Brazilian firms are proactively managing environmental risks? Studies on environmental management in Brazil are incipient. Jabbour et al. (2008) reported that among all articles published between 1996 and 2005 in mainstream administration and management journals in the Country, only 2.5% dealt with environmental management in organizations. It was until 2005 that a journal focused on social-environmental management. The pioneer study of Maimon (1994) concluded that the businesses should have their environmental management activities inspected by the public sector, and that the environmental statements of organizations were not reliable at the time the study was carried out. Many of them neglected environmental issues but claimed to have adopted proactive approaches. The author adds that the organizations with better environmental performance were subsidiaries of multinational or domestic companies focused on exporting. In another pioneer study, Donaire (1994) stated that the environmental issue was incorporated in companies changing their organizational structure. Among the companies investigated in the case studies, those with the best environmental performance operated in the industrial sectors that tend to pollute more, for instance chemical industries. The environmental management area of the companies in this sector tends to be stronger and exert more authority in the organizational structure.

Polizelli et al. (2003) investigated the environmental management of 5 major companies in the telecommunication sector in Brazil. The conclusions indicate that those companies were improving their environmental management practices focusing on proactivity, but this process presented some gaps such as the lack of environmental management practices related to the marketing management. Jabbour et al. (2009) also found continuity and discontinuities in the evolution of the environmental management of 94 companies with ISO 14001 certification. According to the authors, the environmental management characteristics vary according to the intensity of the human resources support to the environmental management objectives.

Palma and Nascimento (2005) analyzed 315 questionnaires answered by students of an undergraduate administration course in Brazil to verify whether they accept or reject products made from recycled materials. Around 80% of the respondents affirmed having bought such products. Nevertheless, Stirbolov and Rossi (2003) carried out an exploratory study to verify whether Brazilian consumers incorporated the environmental dimension in their purchase decisions. The result obtained was negative indicating that environmental aspects are not a decisive factor in the behavior of the Brazilian consumer.

Silva et al. (2008) analyzed whether the set of environmental management practices can be considered a new competitive priority for manufacturing in addition to manufacturing's traditional competitive priorities. The authors used a survey with sixty-five Brazilian companies with ISO 14001 Certification. The results revealed that organizations have been incorporating environmental management practices in manufacturing. These companies follow a production prevention perspective, focused on the reduction, reuse and recycling of materials as well as on the replacement of raw materials, components and suppliers. Thus, a trend can be perceived towards incorporating environmental management practices in typical logistics operations. Nevertheless, it was possible to observe that these environmental management practices could influence classic competitive priorities of manufacturing. However, for environmental management to be considered a complete and new competitive priority, companies need to consider the incorporation of other environmental management practices, with a more strategic orientation and that meet the expectations of Brazilian society.

Masullo and Lemme (2009) examined the relationship between the internationalization of the Brazilian companies and the adoption of environmental management practices and certifications. The results indicate that the stronger the internationalization process, the deeper is the adoption of environmental practices. Campos et al. (2009) investigated 109 companies to verify whether the achievement of ISO 14001 Certification would bring better financial results. Among all companies, 45 stated that they did not have an environmental management system; 40 indicated having it but not the certification; and 24 had the ISO 14001 environmental management certification. Companies with ISO 14001 exhibit superior financial results.

In another study, 315 administration students, who had undertaken environment management courses, were interviewed (Silva et al. 2009). The authors found a low environmental awareness among those who will be the future administrators in Brazilian companies. The outcome makes us rethink the environmental education in Brazil.

This set of empirical evidences demonstrates that for Brazilian companies been become environmentally proactive, they still have a lot to accomplish as far as environmental risk management is concerned. This statement has implications for environmental education field of local universities, responsible for training the future risk management professionals.

3.5 CONCLUSION

In the near future, the implementation of environmental management practices will be a qualifying criterion for an organization due to its needs and the pressure of environmental impacts caused by manufacturing products and providing services. Customers will demand that companies take responsibility for the management of residues of the consumed products, such as packages and components with a high risk of environmental degradation. The risk and uncertainties related to the environment will become higher in industrial sectors that explore natural resources, such as oil (Kolk and Pinkse 2005; Sullivan 2010).

The increase of the worldwide uncertainty about the environmental future of the planet also runs through the Brazilian reality. Brazil, known for its rich biodiversity, has a consolidated economic growth and is one of the most politically stable countries in Latin America. The review of one of the most important studies carried out in the environmental management area revealed that, Brazilian companies still need awareness about the benefits of environmental practices and about the organizational risks and uncertainties that an inadequate management can cause. The adoption of such practices influences the classical organizational areas resulting in changes in processes, products, and routines. At first, they can create resistance to the changes since initial investments are substantial, but the recovery on these investments becomes competitive advantages which can last long.

If the environmental dimension is not incorporated in the current organizational context, it can generate potential environmental risks increasing the uncertainties in a planet that is becoming more and more polluted due to industrial activities. On the other hand, the reduction of risks and uncertainties is related to a more proactive environmental management despite the challenges that this approach poses to organizational managers. Environmental risk tends to increase and its management ought to be a central issue among scholars and professionals.

REFERENCES

Aboulnaga IA. 1998. Integrating quality and environmental management as competitive business strategy for 21st century. Environmental Management and Health 9(2): 65–71.

Angell LC, Klassen RD. 1999. Integrating environmental issues into the mainstream: an agenda for research in operations management. Journal of Operations Management 17(5): 575–598.

Azzone G, Bertele U. 1994. Exploiting green strategies for competitive advantage. Long Range Planning 27(6): 69–81.

Barbieri JC. 2004. Gestão ambiental empresarial. São Paulo: Saraiva. p. 178–190.

Berry MA, Rondinelli DA. 1998. Proactive corporate environmental management: A new industrial revolution. Academy of Management Executive 12(2): 38–50.

Boiral O. 2006. Global warming: Should companies adopt a proactive strategy? Long Range Planning 39(3): 315–330.

Campos LMS, Grzebieluckas C, Selig P. 2009. As empresas com certificação ISO 14001 são mais rentáveis? Uma abordagem em companhias abertas no Brasil. Revista Eletrônica de Administração - Read [Internet]. [cited 2010 Mar 14];13(1) 1–24. Available from: http://read.adm.ufrgs.br/edicoes/pdf/artigo_581.pdf

Cenci DR. 2009. Conflitos socioambientais urbano-metropolitanos: cidadania, sustentabilidade e gestão no contexto da RMC – Região Metropolitana de Curitiba [Thesis]. [Curitiba (PR), Brazil]: Universidade Federal do Paraná. p. 130–142.

Constanza R. 2007. Toward ecological economy. Chinese Journal of Population, Resources and Environment 5(4): 20–25.

Ferrell OC, Lucas GH. 1987. An evaluation of progress in the development of a definition of marketing. Journal of the Academy of Marketing Science 15(3): 12–23.

Fine CH, Hax AC. 1985. Manufacturing strategy: a methodology and an illustration. Interfaces 15(6): 28–46.

González-Benito J, González-Benito O. 2006. A review of determinant factors of environmental proactivity. Business Strategy and the Environment 15(2): 87–102.

Gupta M, Piero T. 2003. Environmental management is good business. Industrial Management 45(5): 14–19.

Haden SSP, Oyler JD, Humphreys JH. 2009. Historical, practical, and theoretical perspectives on green management: An exploratory analysis. Management Decision 47(7): 1041–1055.

Hart SL. 1995. A natural-resource-based view of the firm. Academy of Management Review 20(4): 986–1014.

Hayes RH, Wheelwright SC. 1984. Restoring our competitive edge: competing through manufacturing. New York: John e Wiley. p. 316–332.

Hoffman AJ. 2005. Climate change strategy: The business logic behind voluntary greenhouse gas reductions. California Management Review 47(3): 21–46.

Hunt CB, Auster ER. 1990. Proactive environmental management: avoiding the toxic trap. MIT Sloan Management Review 31(2): 7–18.

Jabbour CJC, Santos FCA. 2008. Relationships between human resource dimensions and environmental management in companies: proposal of a model. Journal of Cleaner Production 16(1): 51–58.

Jabbour CJC, Santos FCA, Nagano MS. 2009. Análise do relacionamento entre estágios evolutivos da gestão ambiental e dimensões de recursos humanos: survey e estado-da-arte. Revista de Administração da Universidade de São Paulo - RAUSP [Internet]. [cited 2010 Mar 14];44(4) 342–364. Available from: http://www.rausp.usp.br/download.asp?file=v4404342.pdf

Jiménez JB, Lorente JJC. 2001. Environmental performance as an operations objective. International Journal & Production Management 21(12): 1553–1572.

Kolk A, Pinkse J. 2005. Business responses to climate change: Identifying emergent strategies. California Management Review 47(3): 6–20.

Layrargues PP. 2000. Sistemas de gerenciamento ambiental, tecnologia limpa e consumidor verde: a delicada relação empresa-meio ambiente no ecocapitalismo. Revista de Administração de Empresas - RAE [Internet]. [cited 2010 14 Marc];40(2) 80–88. Available from: http://www16.fgv.br/rae/redirect.cfm?ID=990

Massoud MA, Fayad R, Kamleh R, El-Fadel M. 2010. Environmental Management System (ISO 14001) Certification in Developing Countries: Challenges and Implementation Strategies. Environmental Science & Technology 44(6): 1884–1887.

Masullo DG, Lemme C. 2009. Um exame da relação entre o nível de internacionalização e a comunicação ambiental nas grandes empresas brasileiras de capital aberto. Revista Eletrônica de Administração - READ [Internet]. [cited 2010 Mar 14];15(3) 1–23. Available from: http://www.read.ea.ufrgs.br/edicoes/pdf/artigo_598.pdf

Miles MP, Covin JG. 2000. Environmental marketing: A source of reputational, competitive, and financial advantage. Journal of Business Ethics 23(3): 299–311.

Molina-Azorin JF, Claver-Cortes E, Lopez-Gamero MD, Tari JJ. 2009. Green management and financial performance: a literature review. Management Decision 47(7): 1080–1100.

Moreno EC, Lorente JC, Jiménez JB. 2003. Gestión ambiental y ventaja competitiva: el papel de las capacidades de prevención de lacontaminación y lagestión de recursos humanos. In: Iberoamerican Academy of Management, editor, Management in Iberoanerican Coutries: Current Trends and Future Propects. Proceedings of the 3rd annual conference Iberoamerican Academy of Management; 2003 Dec 7-10; São Paulo: FGV-SP. p. 1–16.

Polizelli DL, Petroni LM, Kruglianskas I. 2003. Gestão ambiental nas empresas líderes do setor de telecomunicações no Brasil. Revista de Administração da Universidade de São Paulo - RAUSP 38(1): 46–57.

Polonsky MJ, Rosenberger III PJ. 2001. Reevaluating green marketing: A strategic approach. Business Horizons 44(5): 21–30.

Porter ME, Linde CVD. 1995. Green and competitive: ending the stalemate. Harvard Business Review 73(5): 120–134.

Rex E, Baumann H. 2007. Beyond ecolabels: what green marketing can learn from conventional marketing. Journal of Cleaner Production 15(6): 567–576.

RMAI. 2006. ISO 14001. [Internet]. [updated 2006 May 25]. São Paulo (SP), Brazil: Revista Meio Ambiente Industrial; [cited 2006 Dec 12]. Available from: http://www.rmai.com.br/v3/Editions.aspx?idEdition=21

Schroeder RG. 2007. Operations management: contemporary concepts and cases. New York: McGraw-Hill/Irwin. p. 46–52.

Seiffert MEB, Loch C. 2005. Systemic thinking in environmental management: support for sustainable development. Journal of Cleaner Production 13(12): 1197–1202.

Sharma A, Iyer GR, Mehrotra A, Krishnan R. 2010. Sustainability and business-to-business marketing: A framework and implications. Industrial Marketing Management 39(2): 330–341.

Silva EM, Jabbour CJC, Santos FCA, De Castro M. 2008. Análise da relação entre a dimensão ambiental e as prioridades competitivas tradicionais de produção: um estudo em empresas com certificação ISO 14001. In: ANPAD Associação Nacional dos Programas de Pós Graduação, editor. Proceedings of the 32nd annual meeting da ANPAD - ENANPAD; 2008 Sept 6-10; Rio de Janeiro (RJ), Brazil: ANPAD. p. 1–16.

Silva HMR, Carvalho S, Gonçalves-Dias SLF, Teodósio ASS. 2009. Consciência ambiental: um estudo exploratório sobre suas implicações para o ensino de administração. Revista de Administração de Empresas Eletrônica - RAE-eletrônica [Internet]. [cited 2010 Mar 14];8(1) 1–23. Available from: http://www.scielo.br/pdf/raeel/v8n1/a04v8n1.pdf

Sullivan R. 2010. An assessment of the climate change policies and performance of large European companies. Climate Policy 10(1): 38–50.

Zutshi A, Sohal AS. 2004. Adoption and maintenance of environmental management systems: Critical success factors. Management of Environmental Quality 15(4): 399–419.

Hacienda Buena Vista coffee plantation and environmental management project in Ponce, Puerto Rico (@Eddie N. Laboy-Nieves)

Chapter 4

Tsunami hazards: A case study of risk management in the Caribbean Region

Eddie N. Laboy-Nieves and Julio A. Oms

SUMMARY

The islands of the Caribbean lie in an active plate boundary zone with a historical record of devastating earthquakes and tsunamis. These phenomena are mainly triggered by the interaction of the North American Plate and the Caribbean Plate at their boundaries, which can produce major seismic events with results such as the recent catastrophe in Haiti. The tsunamigenic potential of a massive submarine landslide in two carbonate amphitheater south of the Puerto Rico Trench was evaluated. Concerns about the interpretation of hazards and risks are presented. The majority of the islanders of the Caribbean lives and works in the coastal zone as a result of an increasing sea level-based tourism. The structures that support residential settlements and the economic connection in these coastal areas reflect a high degree of vulnerability to a strike by a tsunami. This chapter examines the urgent need to increase public awareness on the possible impact that a major tsunami could have in the Caribbean, placing emphasis on Puerto Rico. The authors presents challenges for establishing a tsunami warning system, to educate coastal communities about their vulnerability, and to orient them on ways for mitigating the risks.

4.1 INTRODUCTION

For millions of years, Puerto Rico (66°N, 18°W, 9497 km²) and many other islands located in the Caribbean Region, has shown a wide variety of events that characterized the evolution of its geology and climatology (Picó 1969; Pindell and Barrett 1990; Applegate and Ransom 2010). Hurricanes, earthquakes and tsunamis have deeply affected the islands' social and natural history (Reid and Taber 1920; Meyerhoff 1933; Picó 1969; Calais 2004) by claiming thousands of lives and causing billions of dollars in structural and infrastructural damages over the past four centuries. Major earthquakes (i.e. magnitude greater than 7.0 in the Richter scale) and tsunamis are the least known by local residents, mainly because the last major earthquake occurred 92 years ago, and because contrary to hurricanes, there is not a certain way for predicting where or when they may occur. Historical records (Table 4.1) show that major earthquakes have struck the Caribbean several times, although the

Table 4.1 Historical occurrence of major earthquakes in the greater Caribbean Region (magnitude in the Richter scale),* tsunami-generating tremors (adapted from Mueller et al. 2004).

Earthquake location	Date	Magnitude
Haiti	2010	7.0
Martinique	2007	7.3
Dominican Republic	2003	6.5
Dominican Republic	1953	6.9
Mona Canyon, Puerto Rico	1946*	7.5
Dominican Republic	1946*	8.1
Mona Canyon, Puerto Rico	1918*	7.5
Anegada Trough, Puerto Rico	1867*	7.5
Puerto Rico Trench	1787	8.1
Mona Canyon, Puerto Rico	1670	~7.0

locations and sizes of events that have occurred more than a few decades ago are not well documented (Mueller et al. 2004). Both, anecdotic chronicles and scientific data recall the aftershock sequence of those earthquakes, which usually continues for months, if not years (Picó 1969; Applegate and Ransom 2010). The frequency of aftershocks will diminish with time, but damaging earthquakes and the probability of tsunamis will remain a threat to the Caribbean Region and its inhabitants.

After the December 2004 tsunami that devastated the Indonesia region, residents of Puerto Rico became more worried about seismic and/or tsunami hazards in the island. This collective concern was very recently enlarged, following the catastrophes associated to the January and February 2010 Haitian and Chilean earthquakes, respectively, and the moderate (5.8 magnitude) earthquake on May 16, 2010 which became the most recent strong seismic strike felt in nearly nine decades (Alvarado-Vega 2010a). The Enriquillo Fault that caused the Haitian earthquake is part of a seismically active zone between the North American and Caribbean tectonic plates (Figure 4.1); the plates are sliding past each other at a speed of 20 mm/yr (Calais 2004; Mann 2006). The earthquake undoubtedly relieved some stress on the fault segment that ruptured during that event, but the extent of rupture along the fault is unclear at this time. Fault slip models, preliminary radar surface deformation measurements, and examination of satellite and airborne imagery for surface rupture suggest that the segment of the fault east of the Haitian January 12, 2010 epicenter and directly adjacent to Port-au-Prince did not slip appreciably in this event. This implies that the fault zone near Port-au-Prince still stores sufficient strain to be released as a large, damaging earthquake and a potential tsunami within the lifetime of structures built during the reconstruction effort.

The experience of the recent Port-au-Prince, Haiti earthquake reveals a need for a better understanding of the nature and extent of earthquake and tsunami hazards in the Caribbean Region. The arc of islands that forms the Greater and Lesser Antilles is seismically active due to the relative motion between the plates, like that encircling the Pacific Ocean (Applegate and Ransom 2010). As can be inferred from López-Marrero and Villanueva-Colón (2006), the risk at which the inhabitants of the Caribbean Region are subjected to calamities and devastation from earthquakes and tsunamis, is a function of the occurrence probability, the magnification of the effects

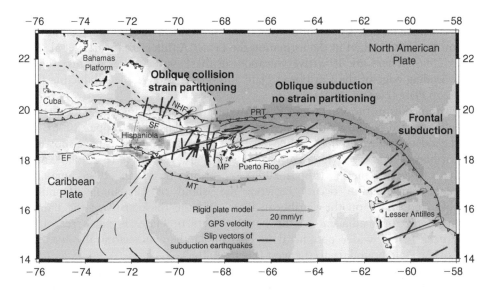

Figure 4.1 Tectonic framework of the Caribbean defined by the subduction of the North American Plate beneath the Caribbean Plate. EF: Enriquillo Fault; SF: Septentrional Fault; NHF: North Hispaniola Fault; Puerto Rico Trench; MP: Mona Pasaje; MT: Muertos Trench; LAT: Lesser Antilles Trench (Calais 2004).

due to anthropogenic alterations, and the level of social readiness for the appropriate contingencies.

4.2 CARIBBEAN TECTONICS AND EARTHQUAKES

Caribbean geology is complicated by wide dispersal over many geographic elements, some with poor accessibility, tropical weathering, young volcanic cover, and some factors just poorly studied (James et al. 2009). It is widely accepted that the Caribbean Plate formed in the Pacific and migrated between the Americas (Pindell and Barrett 1990). But alternative interpretations consider that the Plate formed in place and includes extended continental crust; hybrids of these ideas also exist all compiled in James et al. (2009). The Plate appeared in the mid-Cretaceous and has moved 1100 km eastward about 20 mm/yr relative to the North American and South American plates (Pindell and Barrett 1990, Hedges 2001, López 2006). The formation of an arc at the trailing edge of the Plate, coupled with the west-dipping 'Great Arc' subduction zone in the east, anchored the Caribbean in the mantle reference frame (Pindell et al. 2006).

The West Indian Archipelago extending from the Greater Antilles (Cuba, Jamaica, Hispaniola and Puerto Rico) to the Lesser Antilles were formed by andesitic volcanism resulting from the subduction of the North American Plate beneath the Caribbean Plate (Calais 2004, Figure 4.1). Underwater volcanoes (seamounts) that enlarged with time to emerged eventually above the water level as islands (Meyerhoff 1933). In the early Cenozoic (~60 million year ago) the proto-Antilles began to collide with the Bahamas platform (part of the North American Plate) and fused.

This initiated a transform fault south of Cuba and northern Hispaniola, adding to the geological complexity of the Region. The eastern boundary of the Caribbean Plate is characterized by a subduction zone of 850 km length and curvature of 450 km that has resulted in the formation the Lesser Antilles arc (Martin et al. 2005; López 2006). There are 19 active volcanoes in the Lesser Antilles because they are at the leading edge of the Caribbean Plate and directly above the subducting North American Plate (Hedges 2001).

Earthquakes in the Caribbean Region typically result from the sudden rupture of faults stressed by the motion of tectonic plates (Calais 2004; López 2006). GPS-based geodetic studies on the Caribbean islands and mainland areas have shown that the Caribbean Plate is moving east-northeastward at a rate of 18 to 20 +/−3 mm/yr relative to North America (López 2006). According to Mann (2006), maximum interplate transpression occurs between the island of Hispaniola (Dominican Republic and Haiti) on the Caribbean Plate, and the thick crust of the Bahamas carbonate platform adjacent to the North American Plate. For this reason, the earthquake belt is wider peripheral to Hispaniola than in tectonically less active parts of the Plate boundary to the west (Figure 4.2).

The displacement between the Caribbean and the North American plates occurs in a 3,200 km boundary and creates high levels of stresses in the Earth's crust. These

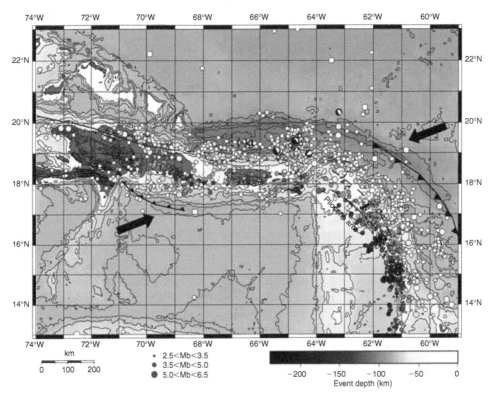

Figure 4.2 Location of earthquakes as a function of depth and size in the northeastern Caribbean. Barbed lines represent subduction zones; arrows represent the direction of relative plate motion (ten Brink 2003).

stresses build up with time, causing the Earth's crust in the vicinity of seismic faults to deform like a rubber band, without breaking (Calais 2004). When these stresses are too high (exceeding the strength of faults), they are released by a sudden slip on a fault: the "rubber band" breaks, an earthquake happens. Calais (2004) described the relative motion between the Caribbean and North American plates by three major active tectonic structures (Figure 4.1). This strike-slip plate boundary ranks with the great seismogenic strike-slip plate boundaries of the world including the San Andreas Fault of California (1,500 km), the Alpine Fault of New Zealand (600 km in length), and the 1,000 km North Anatolian Fault of Turkey (Mann 2006). The plates grind past one another at rates that in human terms would be considered extremely slow (20 mm/yr), but are significant in geologic terms (Mann 2006).

The Septentrional Fault, a major left-lateral strike-slip fault that follows the northern coast of Haiti (offshore) and continues eastward into the Dominican Republic, is responsible for the uplift of the Cordillera Septentrional and for active folding and faulting at its contact with late Neogene to Holocene units of the Cibao Valley (López 2006). The Enriquillo Fault, the second major left-lateral strike-slip fault in Hispaniola, is well-exposed in Haiti, where it is marked by a 200 km long narrow valley striking east-west through the southern peninsula. The Enriquillo Fault ends abruptly in south-central Hispaniola and connects southeastward to low angle thrust motion at the western termination of the Muertos Fault. A number of historical earthquakes, like the January 2010, have occurred on the Enriquillo Fault. The North Hispaniola Fault extends offshore the Hispaniola in a roughly east-west direction. This fault is continuous with the Puerto Rico Trench (the deepest part of the Atlantic Ocean: 800 kilometers long, maximum depth = 8,605 meters, Figure 4.3) to the east

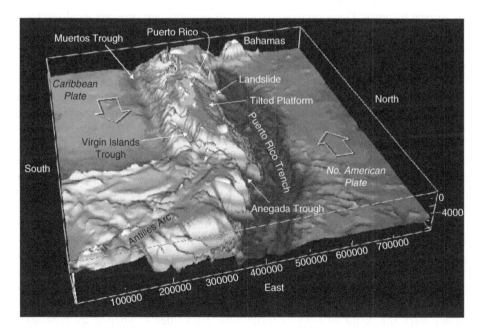

Figure 4.3 Location of the Puerto Rico Trench and relative movement of the Caribbean Plate (Dillon et al. 1999).

which is the site of recent strike-slip and low-angle thrust faulting and a very strong (−400 mGals) negative gravity anomaly (Calais 2004). Both the North Hispaniola and Puerto Rico Trench faults mark the site of subduction for slabs of the Atlantic lithosphere beneath Hispaniola and Puerto Rico, respectively. GPS velocities in Puerto Rico, St. Croix, and the Lesser Antilles show that these areas move largely as a single block part of the Caribbean Plate (Calais 2004; López 2006).

Earthquakes generated along transcurrent faults contribute to seismic hazards in Central America and the Caribbean islands, more than subduction earthquakes, and have the potential to generate tsunamis (Burbach et al. 1984; Rodríguez 2007). With respect to tsunamis, it is crucial to acknowledge that these events are not limited to earthquakes, but they can be originated by submarine landslides and volcanic eruptions, both hazards present in the Caribbean. Some of the active faults are capable of producing earthquakes magnitude 7.0 which could trigger tsunamis (McCann 1998), either by the tremors or by the potential to generate landslides. Scientists are still unable to assess marine underwater landslide hazards, nor able to predict their occurrence following a nearby earthquake to evaluate their tsunamigenic potential (Ward 2001). A thorough scientific examination of these events is required to establish a reliable warning system to alert coastal communities of any and all imminent dangers associated with locally generated tsunamis (Mercado et al. 2002).

4.3 SUBMARINE LANDSLIDES

Submarine landslides are of great concern to marine scientists, geologists, oceanographers, hazards mitigation officials, and state and federal agencies responsible to deal with natural hazards. These underwater land abatements can generate localized tsunamis of catastrophic proportions that could be more damaging than the tsunamis generated by earthquakes (Fritz et al. 2009). For instance, on 8 July 1958, an 8.3 magnitude earthquake along the Fairweather Fault triggered a major subaerial rockslide into Gilbert Inlet at the head of Lituya Bay on the south coast of Alaska. A mudflow landslide was generated by these earthquakes that impacted the water at high speed creating the highest wave run-up in recorded history (i.e. 524 m), which caused total forest destruction and erosion down to bedrock (Fritz et al. 2009). On 26 December 2004 a 9.0 magnitude earthquake ruptured the oceanic basin of Indonesia for more than 1,300 kilometers long, and displaced the seafloor by about 10 meters horizontally and five meters vertically. The Fault rupture and the trillions of tons of rock that collapsed along hundreds of miles, contributed to generate the worst tsunami in recorded history, with a toll around 200,000 human casualties (Athukorala and Resosudarmo 2006).

Deposits of submarine landslides have been mapped offshore in the Puerto Rico Trench and the Mona Canyon (Scalon and Mason 1996; ten Brink et al. 2006). Figure 4.4 shows the carbonate platform of two large amphitheater-shaped scars tilted 45° at the south slope of the Puerto Rico Trench: the Arecibo and the Loíza amphitheaters, 50 and 30 km across, respectively. Several smaller scarps and slope failures are identified along the edge of this platform. Data suggest that the amphitheaters were the result of single catastrophic landslides or shaped by continuous retrograde slumping of smaller segments (Bunce and Fahlquist 1962; Scalon and Mason 1996; Schwab et al. 1991; Mercado et al. 2002; ten Brink et al. 2006). It is postulated that the amphitheater-shaped slide scars could be the result of giant

Figure 4.4 Landslides amphitheaters south of the Puerto Rico Trench.
Source: http://woodshole.er.usgs.gov/project-pages/caribbean/tsunami.html.

submarine slope failures with a displaced volume of over 910–1500 km^3 (Schwab et al. 1991; Mercado et al. 2002; ten Brink et al. 2006).

Mercado et al. (2002) modeled the flooding that could have been caused by a landslide which was mapped in the Puerto Rico Trench. Although the lateral extension of these slides is less than those associated with earthquakes, the height of these waves can be substantially larger. Computer simulations suggest that if a significant landslide occurs in the Arecibo Amphitheater, a 20 meters high tsunami wave could arrive to the coastal Arecibo area in 3.5 minutes. After reporting that waves from the catastrophic December 2004 tsunami that affected the Indonesian region were recorded in tide gauges from Puerto Rico (Thomson et al. 2007), the media and the scientific community has expressed concerns of what would be the potential implications if a greater single landslide event were to occur in the Caribbean Region nearby the Puerto Rico Trench.

4.4 TSUNAMIS

Undersea earthquakes, volcanic eruptions, and landslides generate tsunamis when the water column is disturbed and attempts to find a stable position. Tsunamis can be caused either by nearby earthquakes or by seismic activity with epicenters thousands of kilometers distant from the land areas they finally affect (Scheffers et al. 2005; López 2006). They can travel at great velocities (v $= \sqrt{gD}$, where g is the acceleration due to gravity and D is the water depth) of the order of 700 km/h and exhibit large wavelength ($\lambda = vT$, where T is the period) of hundreds of kilometers (Hunt 2005).

There are three types of earthquake tsunami scenarios described by Curtis and Pelinovsky (1999): local (have up to 24 minutes of travel time), regional (take between 24 minutes and two hours), and distant or teletsunamis (i.e. have travel times of over two hours). The magnitude of a tsunami at its source is directly related to the earthquake's depth and intensity (Hunt 2005). When it arrives at a coastline, the effect is influenced by offshore seafloor conditions, wave direction, and coastline configuration. Wavelengths are accentuated in bays, particularly where they have relatively shallow depths and topographic restrictions.

Historical documents describing fatalities or economic loss to societies due to tsunami impacts in the Caribbean are numerous; 56 events have been reported over the past 500 years (Lander et al. 2002). The most recent event in 1946 following a magnitude−8.1 earthquake off the northeast coast of the Dominican Republic killed more than 1,800 people (López 2006). However, earthquakes originating in the Puerto Rico and Hispaniola trenches are not the only tsunami threats on the Atlantic coast of the Americas. In 1755, an earthquake in an undersea fracture zone off Portugal generated a tsunami that reached the Caribbean Region, resulting in run-ups of more than 7–10 m in the northern part of several of the Lesser Antilles (Zambo and Pelinovsky 2001).

Large submarine landslides in the Caribbean Region are known to produce deadly local tsunamis (ten Brink et al. 2006). The collision of the North American and the Caribbean Plate has caused the tilting of the limestone platform deposits of northern Puerto Rico during the last 3.3 million years (refer to Figure 4.4). These formerly horizontal layers, deposited near the water surface, now are inclined to the north and have subsided to depths of more than 4.5 kilometers (2.8 miles). Great slabs of limestone, as much as 70 kilometers (43 miles) wide, have broken off and slid into the Puerto Rico Trench (Dillon et al. 1999). There are 19 active and dormant (likely to erupt again) volcanoes in the Eastern Caribbean (Figure 4.5), some like the Kick'em Jenny, an active underwater volcano near Grenada (Martin et al. 2005), posses a great potential to generate landslides that can be generated by underwater eruptions. Islands such as Grenada, St. Vincent, St. Lucia, Martinique, Dominica, Guadeloupe, Montserrat, Nevis, St. Kitts, St. Eustatius and Saba have 'live' volcanic centers, while other islands such as Anguilla, Antigua, Barbuda, Barbados, British Virgin Islands, most of the Grenadines and Trinidad & Tobago (which are not volcanic) are close to volcanic islands and are, therefore, subject to volcanic hazards such as severe ash fall and volcanically-generated tsunamis. Historical records alarmingly show that the Caribbean Region has suffered four times more casualties from tsunamis than Hawaii, Alaska, and the entire west coast of the United States combined (Lander et al. 2002), thus it is imperative to institutionalize a tsunami warning system for the Region.

4.5 TSUNAMI RISKS IN PUERTO RICO

Risks such as storm surges or rising sea levels that result in coastal inundation threaten nearly every coastline around the world. In the Caribbean Region, coastal hazard investigations have mainly emphasized on hurricane and earthquake hazards, with relative little emphasis to tsunamis, perhaps because the time scales are different

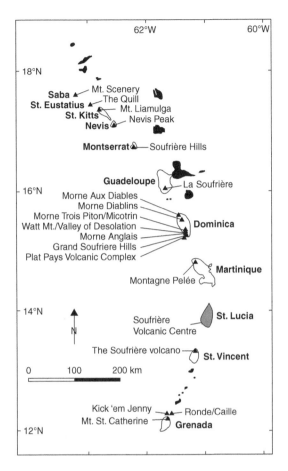

Figure 4.5 Location of volcanoes in the Lesser Antilles.
Source: http://www.uwiseismic.com

for these phenomena. However, all of the known causes for the formation of local and regional tsunamis are present in the Caribbean: earthquakes, submarine landslides, submarine volcanic eruptions, subaerial pyroclastic flows into the ocean, and distant major tsunamis (ten Brink et al. 2006).

If the risk of storm surge is added to coastal hazards, then local islanders are highly vulnerable to be affected by coastal inundation. More than half of the 56 earthquakes and tsunamis listed by Lander et al. (2002) have occurred between May and November, the hurricane season of the Caribbean. It is of particular interest that on November 18, 1867, 20 days after Puerto Rico was devastated by Hurricane San Narciso, a strong regional earthquake occurred with an approximate magnitude of 7.5 on the Richter scale. This produced a tsunami that ran inland almost 150 meters in some low lying areas of the island (Picó 1969). Bender et al. (2010) predicts nearly a doubling of the frequency of category 4 and 5 storms by the end of the 21st century (Figure 4.6). Although there is no relation between hurricanes and tsunamis, the largest increase of intense hurricanes is projected by Bender et al. (2010) for

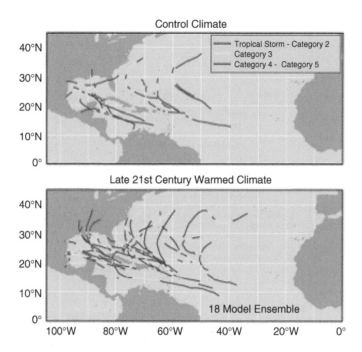

Figure 4.6 Tracks for all storms reaching category 4 or 5 intensity, for the control and the warmed 18-model ensemble conditions (Bender et al. 2010).

the northern Caribbean, coincidentally, the area more prone to tsunamigenic events in Puerto Rico. In order to strive with the forces driving coastal inundation hazards, nations with coastlines ought to deal with preparedness for storm surge and tsunamis risk reduction. Contrary to storm surges, tsunamis are certainly the most challenging event for risk management, because it demands a warning lead time measured usually in minutes (von Hillebrandt and Huérfano 2004).

Puerto Rico established in 1972 its Coastal Zone Management Program (CZMP) to provide guidance for public and private sustainable developments in the coastal zone. The CZMP defined the Island coastal zone as the strip of land extended one kilometer (or more if necessary) inland to protect key natural littoral systems. The partnership among governmental agencies, universities, and the Puerto Rico Seismic Network developed tsunami flood limits maps for the coastal zone (Figure 4.7). According to Díaz (2008), this zone comprises 1278 km of coastline, where ~64% of the island population (~4 million residents) lives. This zone is characterized by having 40% urban land area, eight major ports, eight airports, 1700 km of sanitary infrastructure, 81 industrial parks, and produces around $30 billion in the Island's economy (~52% of the Gross Internal Product). Given that destructive tsunamis have been relatively rare in Puerto Rico, it is easy to overlook the hazard for the present highly populated coastal zone of the island, until a tsunami related disaster occurs.

Since 2000 the Puerto Rico Seismic Network has been developing a Tsunami Warning System (TWS) for the island as well as the US Virgin Islands. Identifying and mitigating the hazards posed by local and distant tsunamis involves the full

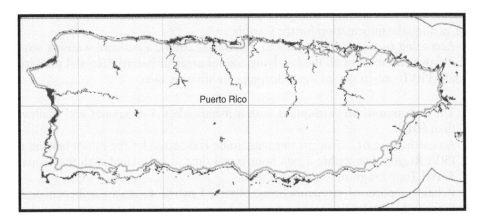

Figure 4.7 The coastal zone of Puerto Rico (gray countour lines) and tsunami flood areas (dark patches)
Source: http://poseidon.uprm.edu/

integration of three key components: hazard detection and forecasting, threat evaluation and alert dissemination, and community preparedness and response. In the case of Puerto Rico, the TWS focuses on tsunamis originating from earthquakes. However, von Hillebrandt and Huérfano (2006) argued that the TWS cannot detect other tsunamigenic sources, such as submarine landslides. The priorities of the System include: establishing a continuous operation to monitor earthquakes, placing a network of tsunami-ready tide gauges throughout Puerto Rico, strengthening the communication systems between the lead emergency management agencies, and educating the public about this phenomenon. The possibility of extending these operations to other parts of the Caribbean has also been contemplated.

The detection and location of tsunamigenic earthquakes is carryout by the Puerto Rico Seismic Network (PRSN) established in 1974. It presently consists of almost 30 seismic stations installed from Mona Island to the west of Puerto Rico thorough Anegada Island in the British Virgin Islands to the east. In January 2003 the automatic earthquake processing system (known as Early Bird) developed by the West Coast and Alaska Tsunami Warning Center, was installed in the PRSN, tailored to detect and locate earthquakes in Puerto Rico, the Caribbean Sea, and the Atlantic Ocean. Once an earthquake is detected, its information is distributed automatically over the Internet. The PRSN personnel receive the information on pagers and cell phones and disseminate that information to emergency management agencies and weather offices (von Hillebrandt and Huérfano 2006).

IOC (2009) reported that the Pacific Tsunami Warning Center provides an interim tsunami warning service for the Caribbean Region. This duty should now be assumed by the Caribbean Tsunami Center (CTC), established in February 2010 at the University of Puerto Rico, Mayagüez Campus. In the Caribbean and adjacent regions there are over 110 seismic stations with real-time seismic data exchange capabilities. Significant progress has also been made towards the establishment of the Caribbean Tsunami Information Centre in Barbados by 2009 with funding provided through the United Nations Development Program. The Puerto Rico

Seismic Network (PRSN) is operating in 24/7 basis, providing earthquake and tsunami information and warning for Puerto Rico and the Virgin Islands as well as post-earthquake information for the Caribbean.

According to von Hillebrandt and Huérfano (2006), a tsunami warning would call for an evacuation of all the low lying coastal areas of Puerto Rico and the Virgin Islands (PRVI) only if one of the following conditions is met:

- The detection of an earthquake with a magnitude 6.5 or greater and shallower than 60 km
- An earthquake of 7.5 or greater magnitude is detected by the PRSN beyond the PRVI Region, but within a two hour travel time as calculated with the Tsunami Travel Time Program (Gusiakov 2000)
- Reliable reports of tsunamis being observed in the Eastern Caribbean or the Western Mid Atlantic Ocean.

4.6 TSUNAMI RISK MANAGEMENT CONCERNS FOR PUERTO RICO

Due to a dynamic tectonic environment, high population density, and extensive development in the 44 coastal municipalities, the case study area of Puerto Rico is at a significant risk for serious earthquakes and tsunamis (von Hillebrandt and Huérfano 2006). The 60 km wide surface between the Puerto Rico Trench and the northern coastline of the island is deep (~8 km), with a 4,000-meter-high slope separating the deepest fore arc region from a gently-tilted carbonate platform shore (ten Brick et al. 2006). Computer wave model simulations have shown that a major submarine landslide in the Arecibo Amphitheater (refer to Figure 4.4) could result in a local catastrophic tsunami capable of impacting the Arecibo shoreline within 3.5 minutes, and producing run-up waves of up to 20 meters (Mercado et al. 2002). Although Puerto Rico has an Early Tsunami Warning System (von Hillebrandt and Huérfano 2006), it has not received the deserved attention by the emergency management agencies of the Commonwealth government, nor by the high risk communities from the coastline. Another concern is that as of February 2010, only 20% of the coastal municipalities have been certified with a Tsunami Ready Program (Alvarado-Vega 2010a).

Mitigation of coastal hazards from locally generated tsunamis in Puerto Rico and neighboring countries in the Caribbean Region will be difficult, because of the relative short travel time of waves generated in the Puerto Rico Trench or surrounding volcanic areas that could impact nearby densely populated coastal communities. Due to such extremely short response times, it is crucial that communities react quickly and move to higher grounds during strong earthquakes or during sudden recession of the sea. Huérfano (2003) and Mercado and Justiniano (2003) modeled 269 earthquake scenarios associated with faults in Puerto Rico, and reported that the tsunami hazard is greatest in the western part of the Island and then, in order of decreasing exposure in the north, south and east coasts. The models predict that these local tsunamis will exhibit short travel times (i.e. can affect areas within minutes after the earthquake), and have potentially high waves (6 to 9 meters) close to the epicenter which will decrease quickly with distance. These tsunami scenarios were used to

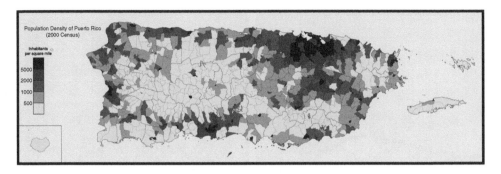

Figure 4.8 Population density of Puerto Rico. *Source: US 2000 Census.*

update the tsunami inundation maps of Puerto Rico which can be accessed at http://poseidon.uprm.edu, and represent the maximum inundation from local faults.

Natural disasters caused by hurricanes and tsunamis are dramatic evidence that the design codes of coastal structures, including those in the inundation zone, need to take these extreme events into account. History has shown that the Caribbean has the potential for disastrous tsunamis. Computer models demonstrate that submarine landslides, which have no detectable precursors (like tsunamis triggered by earthquakes), will be generated locally, producing tsunamis that could reach dense populated areas within minutes (Mercado et al. 2002). The comparison of Figures 4.7 and 4.8 reveals that most of the population of Puerto Rico is concentrated along the north coast, which together with the west coast, is highly vulnerable to tsunamis (Huérfano 2003). Tourist facilities are also concentrated mainly in this coastline. Therefore tourists, visitors and residents need to be advised that they are staying in a tectonically active area with a long historic record of infrequent large earthquakes and accompanying tsunamis (Mann 2006). While it is easier to establish a mitigation program after a major disaster, the Caribbean has the opportunity to prepare for and lessen the effects of any impending disasters (Lander et al. 2002).

The environmental loads on a structure include hydrostatic pressure, fluid impingement, form and viscous drag, and impact due to waterborne debris. Yim and Cheung (2006) modeled the interaction of storm waves and tsunamis with structures. They concluded that Lagrangian models may provide the best solution for the development of a sophisticated and robust code for simulation of tsunami wave basin experiments and prototype events. More importantly, it allows for an exact means of tracking the fluid-structure interfaces, which determines: the energy input to the wave field by the wave generator; the wave forces on the coastal structures and floating debris; and energy dissipation at the bottom boundary and the beach which may contain porous media and/or movable sediment.

Figure 4.9 shows two aerial photos of the Coco Beach Peninsula in the northern coastline of Puerto Rico. Both pictures depict the limit of the tsunami flood as it appears in the map developed by the Federal Emergency Management Administration and the Puerto Rico Seismic Network. The picture on the right taken in 2010 shows significantly more residential and commercial built up areas in the coastal zone compared to that from 1999. These two pictures also demonstrate that government

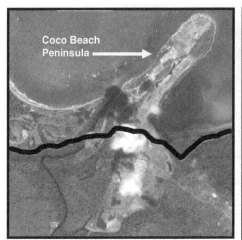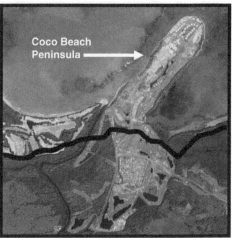

Figure 4.9 Comparison of the inland limits (black bold line) of a tsunami flood (maximum run up = 5.6 m) in the northern shore of the Municipality of Río Grande with built-up residential and commercial areas in 1999 (left) and 2010 (right).

agencies apparently approved the construction of residences and hotels within an inundation zone. Even though the technical literature on disaster prevention and response compiled by Lashley and Houple (2000); the effect of tsunami hydrostatic forces upon structures (Yim and Cheung 2006); the records of the penetration of storm surges and tsunamis (Mann 2006); the existence of a four decades Coastal Zone Management Program (Díaz 2008); and the institutionalization of a Tsunami Warning System (von Hillebrandt and Huerfano 2006; Alvarado-Vega 2010b), it can be inferred from the difference in structural developments shown in Figure 4.9, that Puerto Rico's government authorities and residents have assumed an attitude of complacency about tsunami hazards.

The construction of permanent structures in coastal flood hazard areas along the coastline of Puerto Rico is very common (Mann 2006; Beatley et al. 2002; Alvarado-Vega 2010b). Like Coco Beach, there are also many areas in the Dominican Republic, Jamaica and other islands of the Caribbean with tourist destinations at sea level. These areas have a high threat of been engulfed by a tsunami (Mann 2006). This risky behavior could be an indicator of reprehensible and irresponsible actions of the permitting authorities, contractors, and insurance entities who, for the sake of economic interest, endanger people, coastal ecosystems, and structures. Beatley et al. (2002) concluded that without the intention to inflict harm, state regulatory and political institutions tend to encourage coastal development for economic gain of specific stakeholders, despite it increases the danger to the environment, and places people and property at risk of coastal hazards. Beatley et al. (2002) also argued that hazard insurance encourages coastal growth because these developments are subsidized. For instance, owners of property damaged by coastal storms and flooding often are allowed to rebuild in the same or equally hazardous locations.

The damage-rebuild-damage cycle accounts for many damage claims and with apparently few incentives for avoiding them. Laboy-Nieves (2009) has noted that there are still many obstacles to overcome environmental ignominies, including ignorance of the decision-making process, weak law enforcement, challenges of building laws by developers, poor planning by emergency management agencies, and public indifference. He also argues that the flora and the fauna, as well as the abiotic component of littoral ecosystem, are the silent victims of anthropogenic activities. These natural components need to be set aside, because they form the "green belt" that softens the momentum of tidal waves, reduces beach erosion, and diminishes the tsunami flow pressure, preventing the washing out of structures (Hirashi 2006).

Thirty years ago, the Department of Natural Resources and a private consultant firm published a book about coastal flood hazards in Puerto Rico (DNR and Field 1980). Only one paragraph within the entire volume addressed the danger of tsunamis. However, at the date of that publication, the knowledge and experiences about tsunamis in the Caribbean and abroad was relatively ample. This fact raises again the question about who is responsible for the high risk developments along the coastlines of the islands in the Caribbean Region, such as Puerto Rico and Hispaniola (Beatley et al. 2002). By allowing urban construction to occur in flooding areas, islanders also become willing participants and must bear some responsibility of the possible consequences of the tsunami danger. Morales (2004) noted that this lack of public awareness is in part due to poor and errant environmental administration.

There are serious social, economic, political and environmental consequences if a tsunami impacts the highly urbanized coastline of Puerto Rico (López-Marrero and Villanueva-Colón 2006) and the neighboring islands. The 2004 Indonesian and the 2010 Haitian and Chilean calamities are clear examples that the construction of urban, industrial, touristic, and commercial facilities in coastal plains aggravates the risk to human life and property (Laboy-Nieves 2008). Seismic waves can submit soils to sudden compaction or a complete loss of strength by liquefaction of soil grains. Surface effects after liquefaction have been observed including buildings settling and tilting, submerging islands, dry land becoming large lakes, roads and other filled areas settling, differential movement occurring between bridges and their approach fills, and trucks and other vehicles sinking into the ground (Hunt 2005). Tsunamis may also cause saline intrusion, such as in the northern coastal aquifer of Puerto Rico (Laboy-Nieves 2008), and consequently affect the general health of people, the quality of water for agricultural and industrial activities, and ecological degradation, as reported by Ravisankar and Poongothai (2008). Hence, there is a need to increase environmental awareness, and to aim for social, economic and environmental harmony.

It has been recommended by von Hillebrandt and Huérfano (2006) that, in order to continue supporting the tsunami warning system in Puerto Rico and Caribbean islands, the following factors need to be considered for future legislation:

- Changes and modifications to the existing Puerto Rico Land Use Plan so that it incorporates tsunami vulnerability and hazards.
- Adoption and implementation of the federal provisions and regulations delineated in the National Tsunami Hazard Mitigation Program (NTHMP) and the Puerto Rico Tsunami Warning and Mitigation Program.

- Amendment of existing laws and statutes in Puerto Rico to adopt and validate the above two Programs to delineate the public policy to regulate all projects proposed for development in high-risk coastal areas.
- Develop a tsunami guideline for the Caribbean Region to educate and aware coastal communities about their vulnerability, and to orient them to mitigate risks.

Disasters often reflect a failure in communicating science to governments and applying scientific knowledge on everyday life, or a gap in the coordination of emergency management institutions (NTHMP 2001). Since 2009, the US National Oceanic and Atmospheric Administration, is providing budgetary support to Puerto Rico to implement the National Tsunami Ready Program. This program has proven to be an excellent venue to promote and validate tsunami readiness and has been widely accepted by local public officials and the media (PRSN 2010).

According to Anwar et al. (2008), vulnerability and natural hazards determine the degree of risk. While the magnitude and intensity of natural hazards are beyond human control, vulnerability is a factor influenced by anthropogenic processes. Laboy-Nieves (2009) concluded that new schemes evolving out of coastal governance have not only incorporated trans-disciplinary science, but they are beginning to address the commitments and sacrifices necessary to achieve a balance between human uses of coastal resources and ecological integrity. The challenge will be to preserve ecosystem functions and use natural capital while simultaneously sustaining local communities, social and formal institutions, as well as economies and markets. Therefore, it is imperative that disaster management agencies in the Caribbean Region establish partnerships with academic and scientific institutions to help increase educational and awareness level of decision-makers and the general public, about the vulnerability of islanders to tsunamigenic and earthquake events, but also on the role of different social groups and institutions to collaborate to reduce the risk of devastation.

4.7 CONCLUDING REMARKS

The Caribbean Region shelters the three main elements that can provoke a tsunami: earthquakes, submarine landslides, and marine volcanic eruptions. The tectonic dynamics of the subduction of the North American Plate underneath the Caribbean plate have been extensively studied and identified as the main tsunamigenic factor of the Region, which has experienced more than 50 tsunamis in nearly 500 years of historical records. The probability of massive landslides in two carbonate amphitheater south of the Puerto Rico Trench, require the urgent need for applied research for tsunami risk management. Given the sparse temporal and spatial distribution of large underwater landslides, prediction is a crucial aspect of hazard assessment and hazard mitigation. The linkage between relatively new marine geology tools and mathematical models, and the adoption of principles recommended for hazards mitigation, will enable a broader assessment of ocean floor stability, the data collection for reducing uncertainties, and a much accurate contingency to tsunamigenic events. Consequently, the dangers for at-risk human, structural and infrastructural groups can be reduced by implementing the appropriate methods.

The rapid increase in population in the northern Caribbean Plate boundary zone to the present level of 54 million people could imply that future tsunamis and earthquakes may be more devastating than the ones historically recorded. In the case of Puerto Rico, it has been 92 years since the last major tsunamigenic earthquake. Almost four generations have passed and not until very recently, due primarily to the Indonesian, Haitian and Chilean catastrophes, have the Caribbean island's inhabitants realized their vulnerability to tsunamigenic events. A wake up call is raising the awareness of all stakeholders in the Caribbean Region about the real threat of tsunamis, specially, when such a large percentage of the population lives near the coast and within two kilometers inland. The rarity of tsunamis should not make people complacent about the need to develop effective warning systems. Tsunamis cannot be prevented, but their potential for disaster can be mitigated. Doing so requires foresight and advanced planning, not just emergency relief. Suitable and consistent legislation, appropriate building designs and construction codes, holistic land-use planning, public education, reliable community preparedness, and effective early warning systems can all reduce the impact of disasters.

ACKNOWLEDGMENTS

Contributor: Mr. Arturo Torres, US Geological Survey; Mr. Ricardo Ríos-Menéndez, Universidad del Turabo
Reviewer: Dr. Daniel Laó Dávila, Puerto Rico Seismic Network

REFERENCES

Alvarado-Vega J. 2010a. Puerto Rico unprepared for tsunamis. Puerto Rico Daily Sun [Internet]. Available from: http://www.prdailysun.com/news/PR-unprepared-for-tsunamis-expert-says.

Alvarado-Vega J. 2010b. Early morning quake startles Puerto Rico residents. Puerto Rico Daily Sun [Internet]. Available from: http://www.prdailysun.com/index.php?page=news.article&id=1274065080.

Anwar HZ, Post J, Strunza G, Brikmann J, Gebert N. 2008. Role of the community's vulnerability at local level and its contribution to tsunami risk in Indonesia: study case at Padang Municipality. In: Proceedings of the International Conference on Tsunami Warning (ICTW). Bali, Indonesia. November 12–14, 2008.

Athukorala P, Resosudarmo BP. 2006. The Indian Ocean tsunami: Economic impact, disaster management, and lessons Asian Economic Papers 4(1): 1–39.

Applegate D, Nassif RC. 2010 [cited 2010 Jan 31]. USGS issues assessment of aftershock hazards in Haiti. [Internet]. Reston, Virgina: U.S.A. Geological Survey (USGS). Available from: http://www.usgs.gov/newsroom/article.asp?ID=2385&from=rss_home

Beatley T, Brower DJ, Schwab AK. 2002. An introduction to coastal zone management. Devon, UK: Island Press. 329 p.

Bender MA, Knutson TR, Tuleya RE, Sirutis JJ, Vecchi GA, Garner ST, Held IM. 2010. Modeled impact of anthropogenic warming on the frequency of intense Atlantic hurricanes. Science 327: 454–458.

Bunce ET, Fahlquist DA. 1962. Geophysical investigation of the Puerto Rico Trench and outer ridge. Journal of Geophysical Research 67: 3955–72.

Burbach GV, Forhlich C, Pennington WD, Matumoto T. 1984. Seismicity and tectonics of the subducted Cocos Plate. Journal of Geohysical Research 89(B9): 7719–35.

Calais E. 2004 [cited 2010 Jan 31]. GPS Campaign in the Dominican Republic, October 12–18, 2003, data analysis and preliminary results. [Internet]. Indiana, USA: Purdue University. Available from: http://web.ics.purdue.edu/~ecalais/projects/caribbean/dr2003/report_DR_2003_WWW.html

Curtis GD, Pelinovsky EN. 1999. Evaluation of tsunami risk for mitigation and warning. Science of Tsunami Hazards 17(3): 187–192.

Díaz E. 2008 [cited 2010 Feb 19]. Coastal Zone Management Program. [Internet] Commonwealth of Puerto Rico: Department of Natural and Environmental Resources. Available from: http://drna.gobierno.pr

Dillon W, ten Brink U, Frankel A, Rodríguez R, Mueller C. 1999. Seismic and Tsunami Hazards in Northeast Caribbean. American Geophysical Union 80(26): 309–310.

[DNR] Department of Natural Resources, Field RM. 1980. Coastal flood hazards and responses in Puerto Rico: an overview. Department of Natural Resources, Coastal Zone Management Program. 116 p.

Fritz HM, Mohammed F, Yoo J. 2009. Lituya Bay landslide impact generated mega-tsunami 50th anniversary. Pure and Applied Geophysics 166: 153–175.

Gusiakov V. 2000. Tsunami Travel Time Calculation Program for the Caribbean Region, Version 2.4. Tsunami Laboratory, Institute of Computational Mathematics and Mathematical Geophysics, Siberian Division of the Russian Academy of Sciences. Novosibirsk, Russia. CD.

Hedges SB. 2001. Biogeography of the West Indies: An overview. In: Woods CA, Sergile FE, editors. Biogeography of the West Indies: Patterns and perspectives. London: CRC Press. p 15–33.

Hiraishi T. 2006. Coastal damage due to the Indian Ocean tsunami and its defense by greenbelt. Book of Abstracts of the Second International Workshop in Coastal Disaster Prevention: 27–28.

Huérfano V. 2003 [cited 2010 Jan 30]. Modes of faulting in the local zone of Puerto Rico. [Internet]. Mayagüez, Puerto Rico. University of Puerto Rico. Available from: http://poseidon.uprm.edu/

Hunt RE. 2005. Earthquakes. In: Hunt RE, editor. Geotechnical engineering investigation handbook. London: Taylor and Francis. p 893–996.

[IOC] Intergovernmental Oceanographic Commission. 2009. Five years after the tsunami in the Indian Ocean: from strategy to implementation. UNESCO. 24 p.

James HH, Lorente MA, Pindell JL. 2009. The Origin and Evolution of the Caribbean Plate. England. The Geological Society of London Special Publication 328. 868 p.

Laboy-Nieves EN. 2008. Ética y Sustentabilidad Ambiental en Puerto Rico. Actas del Foro Internacional de Recursos Hídricos. INDRHI-República Dominicana. p 63–74.

Laboy-Nieves EN. 2009. Environmental management issues in Jobos Bay, Puerto Rico In: Laboy-Nieves EN, Schaffner F, Abdelhadi AH, Goosen MFA, editors. Environmental management, sustainable development and human health. London: Taylor and Francis. p 361–398.

Lander JF, Whiteside LS, Lockridge PA. 2002. A brief history of tsunamis in the Caribbean. Science of Tsunami Hazards 20(2): 57–94.

Lashley B, Houple H. 2000. Caribbean disaster information: a bibliography. Caribbean Disaster Information Network (CARDIN). University of the West Indies. Mona, Jamaica. 52 p.

López AM. 2006. Tectonic studies of the Caribbean: Pure GPS Euler vectors to test for rigidity and for the existence of a northern Lesser Antilles forearc block; constraints for tsunami risk from reassessment of the April 1, 1946 Alaska-Aleutians and August 4, 1946 Hispaniola events.[dissertation]. [Indiana (USA)]: Northwest University. 250 p.

López-Marrero TM, Villanueva-Colón N. 2006. Atlas Ambiental de Puerto Rico. San Juan. Editorial Universitaria. p 133–134.

Mann P. 2006. The risk of tsunamis in the northern Caribbean. PHI KAPPA PHI FORUM 86(1): 1–25.

Martin JW, Wishner K, Graff JR. 2005. Caridean and sergestid shrimp from the Kink'em Jenny submarine volcano, southeastern Caribbean Sea. Crustaceana 78(2): 215–21.

McCann WR. 1998 [cited 2010 Jan 31]. Tsunami hazard of Western Puerto Rico from local sources: characteristics of tsunamigenic faults. . [Internet]. Mayagüez, Puerto Rico. University of Puerto Rico. Available from: http://poseidon.uprm.edu/

Mercado A, Justiniano H. 2003 [cited 2010 Jan 31]. Tsunami coastal flood mapping for Puerto Rico and adjacent islands: Final Report for Task 1, Puerto Rico Tsunami Warning and Mitigation Program 077. [Internet]. Mayagüez, Puerto Rico. University of Puerto Rico. Available from: http://poseidon.uprm.edu/

Mercado A, Grindlay NR, Lynett P, Liu PL. 2002 [cited 2010 Jan 31]. Investigation of the potential tsunami hazard on the North Coast of Puerto Rico due to submarine landslides along the Puerto Rico Trench. [Internet]. Mayagüez, Puerto Rico. University of Puerto Rico. Available from: http://poseidon.uprm.edu/

Meyerhoff HA. 1933. Geology of Puerto Rico. San Juan. The University of Puerto Rico. p 19–90.

Morales B. 2004. Hacia un antropocentrismo menos radical: el caso de Puerto Rico. In: Galanes-Valdejuli L, Aledo-Tur A, Domínguez-Gómez JA, editors. Ética y ecología. San Juan: Editorial Tal Cual. p 147–152.

Mueller CS, Frankel AD, Petersen MD, Leyendecker EV. 2004 [cited 2010 Jan 31]. Seismic Hazard Maps for Puerto Rico and the U. S. Virgin Islands. [Internet]. Golden, Colorado: U.S. Geological Survey (USGS). Available from: http://earthquake.usgs.gov/hazards/products/prvi/documentation

NTHMP. 2001. Designing for tsunamis: Seven principles for planning and designing for tsunami hazards. [Internet]. Washington, DC. USA: NOAA. Available from: http://nthmp-history.pmel.noaa.gov/Designing_for_Tsunamis.pdf

Picó R. 1969. Nueva Geografía de Puerto Rico. San Juan. Editorial Universitaria. p 79–82.

Pindell JL, Barrett SF. 1990. Geological evolution of the Caribbean region: a Plate tectonic perspective. In: Dengo G, Case JE, editors. The Caribbean Region. The Geology of North America. USA. Geological Society of America. 405–432.

Pindell JL, Kennan L, Stanek KP, Maresch WV, Draper G. 2006. Foundations of Gulf of Mexico and Caribbean evolution: eight controversies resolved. Geologica Acta 4: 89–128.

[PRSN] Puerto Rico Seismic Network. 2010 [cited 2010 Mar 10]. Caribbean Tsunami Warning System. [Internet]. Puerto Rico: Puerto Rico Seismic Network. Available from: http://www.prsn.uprm.edu/English/tsunami/index.php.

Ravisankar N, Poongothai S. 2008. A study of groundwater quality in tsunami affected areas of Sirkashi Taluk, Nagapattunam District, Tamilnadu, India. Science of Tsunami Hazards 27(1): 47–55.

Reid HF, Taber S. 1920. The Virgin Islands earthquake of 1867–1868. Bulletin of the Seismological Society of America 10: 20–25.

Rodríguez CE. 2007. Earthquake-induced landslides. In: Bundschuh J, Alvarado GE, editors. Central America: Geology, resources, and hazards. London: Taylor and Francis. p 1217–55.

Scalon KM, Masson DG. 1996. Sedimentary Processes in a Tectonically Active Region: Puerto Rico North Insular Slope. In: Gardner JV, Field M, Twichell DC, editors. Geology of the U.S. Seafloor: the View from GLORIA. Boston: Cambridge University Press. p 123–134.

Schefferst A, Scheffers S, Kelletat D. 2005. Paleo-Tsunami Relics on the Southern and Central Antillean Island Arc. Journal of Coastal Research 21(2): 263–273.

Schwab WC, Danforth WW, Scanlon K, Masson D. 1991. A giant submarine slope failure on the northern insular slope of Puerto Rico. Marine Geology 96: 237–246.

ten Brink US. Ocean Explorer [Internet]. 2003. [cited 2010 Feb17]; Figure 2. Location of earthquakes as a function of depth and size in the northeastern Caribbean. Available from: http://oceanexplorer.noaa.gov/explorations/03trench/trench/trench.html

ten Brink US, Geist E, Andrews BD. 2006. Size distribution of submarine landslides and its implication to tsunami hazard in Puerto Rico. Geophysical Research Letters 33(4)L11307: 1–4.

Thomson RE, Rabinovich AB, Krassovski MV. 2007. Double jeopardy: Concurrent arrival of the 2004 Sumatra tsunami and storm-generated waves on the Atlantic coast of the United States and Canada. Geophysical Research Letters 34 (L15607): 1–6.

von Hillebrandt C, Huérfano V. 2006. Emergent tsunami warning system for Puerto Rico and the Virgin Islands. In: Mercado-Irizarry A, Liu P, editors. Caribbean Tsunami Hazard. New York. World Scientific. p 231–243.

Ward SN. 2001. Landslide Tsunami. Journal of Geophysical Research 106(11): 201–215.

Yim SC, Cheung KF. 2006. Storm wave and tsunami interaction with structures. Book of Abstracts of the Second International Workshop in Coastal Disaster Prevention: 10–13.

Zambo N, Pelinovsky EN, 2001. Evaluation of tsunami risks in the Lesser Antilles. Natural Hazards and Earth System Sciences 1: 221–231.

Chapter 5

Cost-benefit analysis to enhance the risk management decision making process

Adalberto Bosque and Fred C. Schaffner

SUMMARY

Society faces a number of risks in such areas as health, transportation, and the work environment. Individuals and the general public must make choices on how to manage risk and use their limited resources. The utilization of the Cost-Benefit Analysis (CBA) process can improved risk management decisions while providing a mechanism to measure non-market goods and services that otherwise would be ignored. A legitimate role of the government is to take action as effectively as possible to reduce these risks. In order to quantify health risk reduction, we must determine the value that society is willing to pay in order to save a human life. Several methods have been developed to quantify risk reductions and many projects and regulations might impose costs to society in exchange for a risk reduction. The "Value of Statistical Life" for instance has been used to assess the mortality benefits (i.e., mortality reduction) of environmental and safety regulations. When gains or losses from an action are accruing over time, discounting methods are used to determine the present value of those gain or losses. Governments have used and will continue to use the CBA in the evaluation of regulations and actions that will affect the environment and the general public throughout risk reduction. Society needs to understand the benefits of the CBA so as to optimize its risk management initiative.

5.1 INTRODUCTION

Industrial, commercial, agricultural and domestic activities and their consequential production of solid wastes, chemical pollutants and cultural eutrophication have increased the adverse risk to people's health and the environment. Like many other localities, Puerto Rico faces serious challenges from diverse environmental stressors including solid waste contamination of unprotected aquifers and coastlines (Figure 5.1A), industrial solvent contamination of groundwater (Figure 5.1B), cultural eutrophication-generated impacts on wildlife resources and agricultural resources (Figure 5.2), and a myriad of other causes. For example, Laguna Cartagena, located at Laguna Cartagena National Wildlife Refuge in southwestern Puerto Rico is a dystrophic wetland system that receives highly eutrophic agricultural drain water from

Figure 5.1 (A) Municipal solid waste landfill in Culebra Island, Puerto Rico; (B) Groundwater treatment system at the Fiber's public supply well superfund site in Guayama, Puerto Rico.

Figure 5.2 Mortality of horses after the emanation of H_2S gas in Laguna Cartagena, Puerto Rico.

the Lajas Valley Irrigation System. The US Fish and Wildlife Service drained the lagoon in January of 2005 to repair the outlet water level control structure. This resulted in a lush growth of terrestrial vegetation on the lagoon bottom. As reported by Schaffner (2006), a 13 cm rainfall event in May 2005 completely refilled the lagoon, drowned the terrestrial vegetation, and triggered massive bacterial decomposition that generated large amounts of toxic H_2S gas, killing horses trapped in a 4 m wide strip along the lagoon's shoreline (Figure 5.2).

The use of information on human health risk in the urban planning process would assist in the improvement of a healthy and sustainable urban environment (Poggio et al. 2008). Full accounting of the environmental costs and benefits of human activities is important towards obtaining the greatest possible benefits of the developing knowledge based economy (Sabau 2010; Kubiszewsk et al. 2010; Vatn 2010). Projects or regulations might impose cost on society in exchange for a risk reduction. Millions of dollars are invested to reduce the risks that threaten human health and the environment. As resources become more limited, informed decisions will have to be undertaken so as to maximize net social benefits. To determine net social benefits, one must compare the risk reduction costs associated with a project or regulation versus the risk reduction benefits obtained by society (Shanmugam 2000). Reduction in risk of death, for example, is arguably the most important benefit underlying many health, safety, and environmental legislative mandates (Alberini 2001).

Several methods have been proposed for estimating the implicit price for life and health (Alberini 2005; Shanmugam 2000). These include the cost of illness, human capital, insurance, court award and compensation, and portfolio approaches. The two principal methodologies to calculate the prices are the contingent valuation which is a stated preference approach, and the revealed preference method (Shanmugam 2000). Furthermore, if society is to receive a risk reduction benefit, "Willingness to Pay" will be used to assess the benefit for such a reduction. If society is willing to accept an increase in risk, then "Willingness to Accept" should be used. Brady (2007a) discussed three commonly used methods for determining willingness to pay for risk reduction: wages differential estimation, willingness to pay for risk reduction products, and contingency valuation surveys.

"Risk Treatment" is the method of selecting and implementing measures to change risk. Risk treatment measures can include avoiding, optimizing, transferring or retaining risk. Having identified and evaluated the risks, the next step involves the identification of suitable actions for managing these risks, the evaluation and assessment of their results or impact and the specification and implementation of treatment plans. The treatment of risk in the policy process involves the identification and quantification of the risks; and deciding how much risk is acceptable (Tietenberg and Lewis 2009). Cost Benefit Analysis will assist us to evaluate the alternative being considered to reduce risk and will thus enhance risk management decisions.

This chapter presents an overview of the Cost-Benefit Analysis (CBA) process. It describes various methods used to conduct a CBA for risk reduction. Information about the Value of a Statistical Life (VSL) to quantify risk, the use of discount rates, and the risk management process is analysed.

5.2 COST-BENEFIT ANALYSIS

Since the 1980s, USA agencies have been subjected to executive orders that emphasize cost benefit analysis (CBA) as a principal tool for assessing major proposed regulations (Jenkins et al. 2001). Measures of public preferences toward risk are critical to evaluate civic policies on many safety, environmental, and health issues (Ashenfelter and Greenstone 2004). The proper value of the risk reduction benefits for government policy is society's willingness to pay for the benefit (Viscusi and Aldy 2003). In addition, CBA is a tool that is frequently used by economists who analyze regulations (Hahn and Tetlock 2007). For instance, federal agencies in the USA, are required to prepare Regulatory Impact Analyses (RIAs) for every major regulatory action they undertake. RIAs have come to mean the use of economic scrutiny in particular, benefit-cost or cost-effectiveness analysis to examine the implication of government regulations (Harrington and Morgenstern 2004).

One example of CBA is the US Congress's requirement that the US Environmental Protection Agency (USEPA) examines the costs and benefits of the Clean Air Act retrospectively from 1970 to 1990 and then prospectively from 1990 to 2010. Congress has also required USEPA to assess the social benefits and costs of every major new regulation (Matthews and Lave 2000). In 2006 USEPA stated that for some of the most recent Clean Air Act decisions such as the particulate matter decisions that have led to rules recently, more than 90% of the monetized benefits come out of the mortality valuations. Another example is the use of the VSL by USEPA in the justification of the 2000 Diesel Sulphur Rule, where the VSL accounted for nearly 90% of the estimated annual total benefits from improved air quality (Shogren and Stamland 2002).

The first step in evaluating the benefit of pollution abatement is to estimate the effect of pollution emissions on things we care about. The next step is monetizing estimated damages (Matthews and Lave 2000). USEPA reports of the effectiveness of the Clean Air Act show that the benefits of cleaner air have far exceeded the cost (Matthews and Lave 2000). It also looks for opportunities to deal with the issue of updates on cost-effectiveness analysis, mortality valuation, and efforts to include economic questions in their large-scale health survey (USEPA 2006). The report "Regulatory Impact Analysis for NAAQS for Particulate Matter" provided information on the economic value due to reducing premature fatalities based on exposure to particulate matter (USEPA 2001).

It can be argued that governments should protect their citizens, and to do so it must act effectively and as thoroughly as possible so as to reduce the risk that it citizens might experience. The use of the CBA will allow us to determine if the benefits received by society exceed the costs. CBA will also allow for an assessment of the value of the benefits received thought improvement in environmental quality and health as well as a reduction in mortality. Balmford et al. (2002) noted that it is difficult to quantify some benefits provided by the environment because they are not captured by conventional, market-based economic activity and analysis. Hence it is important to understand the advantages of the CBA.

5.3 COST-BENEFIT ANALYSIS METHODS

When estimating the social benefits of non-market goods, economists use two main valuation methods: revealed preferences (cost of illness, human capital, hedonic property and wages, averting behaviour), and stated preferences (contingent valuation and behaviour, conjoint analysis, stated choices). The Contingent Valuation Method (CVM) is the most commonly used stated preferences (Haab and McConnell 2003). The stated preference approach, which largely relies on primary data collected with surveys (Boyle 2003b), has as its most commonly used form the CVM for measuring the value of non-market goods (Nocera et al. 2002). In contingency valuation, respondents are asked to report their willingness to pay (WTP) for a specified-and hypothetical risk reduction (Alberini 2005; Brady 2007a; Farber and Griner 2000) such as for an improvement in environmental quality, health or safety, or in the provision of a public good (Alberini 2001). It is a survey-based method frequently used for placing monetary values on environmental goods not bought and sold in the marketplace (Carson 2000; Whitehead 2000). Carson et al. (2003) stated that CVM is a survey approach designed to create the missing market for public goods by determining what people's WTP would be for specified changes in the quantity or quality of such goods or, more rarely, what they would be willing to accept in compensation for well-specified degradations in the provision of these goods.

CV can be used to construct economic values for a wide array of tangible and intangible objects (Boyle 2003b). This method could also be used to assess risk reduction of illness and death. The WTP approach to valuing health is a measure of the amount that individuals are willing to pay for various perceived gains such as improvements in health, or the prevention of an impaired health state, or a reduction in risk of an adverse event (Abelson 2007). Nijkamp et al. (2008) and Tietenberg and Lewis (2009) showed that revealed preference techniques seek to obtain preferences from actual observed market-based information. These methods are based on actual observable choices and from which actual resource values can be directly inferred (Tietenberg and Lewis 2009). These methods also are known as indirect methods and use observed behaviour to measure or infer economic values (Adamowicz et al. 2008).

The hedonic method, like any other nonmarket valuation methods, depends on observable data resulting from the actual behaviour of individuals. The method examines the market prices for consumer goods, real estate, and/or wage rate for jobs (Adamowicz et al. 2008). These prices depend in part of the environmental and safety attributes of alternative goods, houses or jobs. The hedonic method for non-market valuation relies on market transactions for the products whose characteristics vary to determine the value of key underlying characteristics (Taylor 2003). The hedonic property value and wage approach share the characteristic that they use a multiple regression analysis statistical technique (Tietenberg and Lewis 2009). The most developed indirect approach to measure WTP for risk reduction to life is through wage differential, which represent an implicit valuation of a statistical life (Field and Field 2006).

Another revealed preference approach employed to quantify risk reduction value is the averting method. The cost of illness (COI) approach estimates values for morbidity based on the concept that an individual would be willing to pay at least as much as the cost of treating an illness in order to avoid getting it. COI is the *ex post* sum of various identifiable costs, such as loss of work income and medical expenses, but usually does not account for pain and suffering. The value of health is the increase in the earnings and avoidance of medical expenses of individuals as a result of improved health (Abelson 2007).

Defensive or averting behaviour includes actions that reduce exposure to contamination and measures that will mitigate adverse effects of exposure. Actions taken to avoid or reduce damages from exposure to contamination are another category of economic losses. Choices of defensive behaviours reveal something about the value of avoiding damages (Dickie 2003; Tietenberg and Lewis 2009) indicate that averting expenditures can provide a lower-bound estimate of the damage cause by contamination because people would not normally spend more to prevent a problem that caused by the problem itself. Models of defensive behaviour focus on expenditures that people would take in order to offset adverse effects caused by, for example, contamination (Boyle 2003a).

In the standard human capital approach, it is assumed that the value to society of an individual's life is measured by future production potential, usually calculated as the present discounted value of expected labor earnings. Calculations are made to determine the present value of the total monies that an individual would have earned over the rest of his working days. The human capital approach can be used to value mortality by estimating the value of premature death in terms of forgone earnings. When gains or losses from an action are accruing to individuals over time, discounting methods are typically used. Discounting is a procedure that deducts future values of a particular good in order to determine the present value of the stream of benefits or costs in relation to the benefits or costs at different times in the future. The values of future effects are thereby adjusted to render them comparable to the values placed on current consumption, costs, and benefits, reflecting the fact that a given amount of future consumption is worthless than the same amount of consumption today (USEPA 2001).

The revealed preference approach infers the hedonic value of an environmental good affecting the value of a market such as air quality (Adamowicz et al. 2008). This method relies on property values and wage data (Shanmugam 2000). Examples of revealed preference approaches are the wage-risk and the consumer-market studies. It has been documented thought labour market studies that the VSL could also be derived from the wage compensating differentia (Field and Field 2006). Gathering primary, site specific data is costly and as such, a popular alternate method is to conduct a "benefit transfer" (Plummer 2009). Benefit transfer is the practice of using benefit estimates from one or more existing studies to value changes in a similar good or service in a different time or place in lieu of directly estimating the benefit (Delavan and Epp 2001). The benefit transfer relies on information from previous studies. Monetary value transfers require original studies from which values are based and justified (USEPA 2000; Spash and Vatn 2006).

5.4 VALUE OF STATISTICAL LIFE

As a means of determining risk, the Value Statistical Life (VSL) has been used to assess the value that people are willing to pay for risk reduction, and is the value of mortality risk (Eeckloudt and Hammitt 2001). VSL is a statement of how much a given population is prepared to give to decrease the total amount of expected premature death by one person (=one life) (Brady 2007b). Using evidence on market choices that involve implicit tradeoffs between risk and money, economists have developed estimates of the VSL (Viscusi and Aldy 2003) based on the valuation of changes in the level of risk exposure, rather than the valuation of the life of a specific individual. It is an estimate of the monetary benefit of avoiding an anonymous death (Brady 2008; Mrozek and Taylor 2002). The VSL should approximately correspond to the value that people place on their lives in their private decisions (Brannon 2005; Viscuzi and Aldy 2003, 2007; Viscuzi 2008).

The VSL is a key input for computing the mortality benefits of environmental and safety policies that save lives (Alberini 2005). An appropriate value of risk reduction benefits for government policy is society's willingness to pay. In the case of mortality risk reduction, the benefit is the value of the reduced probability of death that is experienced by the affected population (Viscusi and Aldy 2003). The VSL is by far the main class of benefits used to defend new federal environmental, health, and safety rule (Shogren and Stamland 2002).

Willingness to Accept calculates the VSL by using a hedonic wage equation intended to reveal the added compensation required to attract individuals interested in riskier work. Workers are willing to accept additional risks of death in the workplace if they can obtain a monetary compensation affording them the same utility level (Dionne and Michaud 2002). In wage-risk studies, workers are assumed to be willing to give up income for enhanced workplace safety, or to require (i.e. accept) income for taking on more risk (Abelson 2007). Workers in high-risk occupations demand higher wages in order to be persuaded to assume the risk (Tietenberg and Lewis 2009). Recent estimates of the VSL-age relationship show an inverted-U shaped pattern (Viscusi 2008; Viscusi and Aldy 2007). The VSL rises with age, reaches a peak, and then eventually tapers off. Although valuation of risk should increase with income, wage studies have not been appropriate for assessing this relationship (Shanmugam 2000; Greenstone and Gallagher 2005).

Stated preference methods have been used frequently for the evaluation of risk reductions related to mortality in order to obtain estimates of the VSL, and increasingly also to value morbidity endpoints (USEPA 2006). While imperfect, these methods provide policy makers with information on how the general public might trade off income against reduction in the risk of specified health effects. While implementing risk reducing regulation, society could receive benefits through reduced illness and mortality. In undertaking cost-benefit analysis, the US government has monetized the risk of death through the idea of VSL at about $6.1 million in 2003 (Sunstein 2003). The tradeoff between money and small risk of death is VSL, which has become the standard for assessing the benefit of risk and environmental regulations (Viscusi 2008).

5.5 USE OF DISCOUNT RATES

Discounting in public policy evaluation is normally referred to as social discounting or discounting using the social rate of interest (USEPA 2001). The health effects of environmental policies often have quite different time patterns of incidence. The health benefits of environmental policies with long time horizons do not differ from other benefits in terms of the principles that should be applied in valuing them. Society's willingness to pay for these benefits is the appropriate benefit concept. A fundamental difference arises, however, because this willingness to pay amount should reflect society's rate of time preference with respect to health risk. Benefit cost analysis of government projects that reduce health risks over an extended period of time requires an estimate of the value of future life. This in turn requires a discount.

When gains or losses from an action are accruing to individuals over time, discounting methods are typically used. Discounting is a procedure that deducts future values of a particular good in order to determine the present value of the stream of benefits or costs in relation to the benefit or costs at different time in the future. It is a process whereby the values of future effects are adjusted to render them comparable to the values placed on current consumption, cost, and benefits, reflecting the fact that a given amount of future consumption is worth less than the same amount of consumption today (USEPA 2001).

Empirical estimates of the discount rate for risk reductions, like empirical estimates of values of other non-marketed goods, can be obtained either by inferring a discount rate from market transactions or by eliciting it directly through the use of surveys. During the early 1970s the USA Office of Management and Budget published a circular that required, with some exceptions, all government agencies to use a discount rate of 10% in their benefit/cost analysis. A revision issued in 1992 reduced the required discount rate to 7% (Tietenberg and Lewis 2009).

5.6 RISK MANAGEMENT

Contamination of our environment, including the groundwater and soil, has been of concern due to its adverse impact to people health and the environment. Great amount of resources which in most cases are limited need to be allocated to deal with the risk posed by the contaminants or actions. While there are many definitions of the word risk, the United States Environmental Protection Agency (USEPA) considers risk to be "the chance of harmful effects to human health or to ecological systems resulting from exposure to an environmental stressor". A stressor is any physical, chemical, or biological entity that can induce an adverse response. Stressors may adversely affect specific natural resources or entire ecosystems, including plants and animals, as well as the environment with which they interact (USEPA 2010).

Khadam and Kaluarachchi (2003) note that the health risk assessment process will assist in the allocation of limited resources. Poggio et al. (2008) indicated that risk assessment characterized the potential adverse effect of human exposure to environmental hazards. The assessment identifies the hazard posed by a particular action, quantify its probability and determine its likely consequences. Once the appraisal is completed the risk management part of the process will used the data to reduce the hazard.

The estimates obtained in the threat evaluation are used as a basis for deciding on alternatives to decrease, eliminate or manage the peril under consideration (Eduljee 2000). Once the potential public health consequences of an exposure to a particular substance or action are characterized it is time to implement the risk management process. Risk characterization integrates the information from the hazard identification, dose-response, and exposure assessments, using a combination of qualitative information, quantitative information, and information regarding uncertainties. At least seven factors affect and inform risk management decisions (Figure 5.3). Each factor passes through four analytical steps to integrate the information for a risk management decision (USEPA 2000).

Ecosystems include three types of services: provisioning, regulating, and cultural (Richmond et al. 2007), and in the face of increasing human pressure on the environment, these benefits should act as powerful incentives to conserve nature, yet evaluating them has proved difficult because they are not captured by conventional, market-based economic activity and analysis (Balmford et al. 2002).

Individuals and society must make choices on how to manage risk and use their limited resources. In order to make risk-informed choices we need to comprehend the implication of a risk, make a decision if it requires management and its cost, and then implement the decision to reduce risk to an acceptable level (Pollard et al. 2008). Most environmental decisions are about comparison of risk and reward, loss versus gain. For risk inform decisions try to understand the meaning of the risk, decide whether it requires management and what it might cost, and the implement the decision (Pollard et al. 2008).

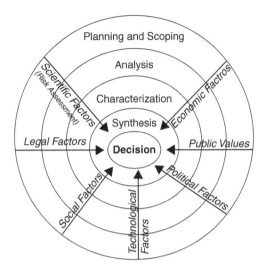

Figure 5.3 Risk management decision framework (USEPA 2000).

The use of the CBA will improve risk management decision making while providing a mechanism to measure non-market goods and services that otherwise would be ignored. Additionally, helping individuals recognize the value of nature's services will increase their investments in conservation (Daily et al. 2009). With quantitative assessment of ecosystem services, and incentives for landowners to provide them, these services will be less likely to be ignored by those making land use and land management decisions (Nelson et al. 2009). Choices are based on complex tradeoffs and thus, the value to society of a change in the quantity or quality of an environmental good or service can be better appreciated when it is accessible to measurement (Muradian et al. 2010; Pascual et al. 2010).

5.7 CONCLUSIONS

Cost-Benefit Analysis (CBA) determines whether society will be better off if a policy or action is implemented. Several methods have been developed in order to conduct CBA. CBA compares costs and benefits to society of policies, programs, or actions to protect or restore the environment. CBA measures the net gain or loss to society from a policy or action. CBA compares the expected costs and benefits of a program by converting both into dollars, a common unit of measurement. Governments should protect their citizens, and therefore act effectively to reduce the risk that its citizens might experience. The use of the CBA will allow us to determine if the benefit received by society exceeds the costs thus whether the action is to be supported. CBA will also allow an assessment of the value of the benefits received thought improvement in environmental quality, health improvement or mortality reduction and will allow us to improve our risk management decisions.

REFERENCES

Abelson P. 2007. Establishing a monetary value for lives saved: issues and controversies. In: Perkins F, Dobes L, Harrison M, Abelson P, editors. Proceedings of the Conference in Delivering Better Quality Regulatory Proposals Through Better Cost Benefit Analysis 2007 No 21. Canberra Au. Sydney University.

Adamowicz WL. Chapman D, Mancini G, Munns W, Stirling A, Tomasi T. 2008. Valuation methods. In: Stahl R, Kapustka L, Munns W, Bruins R, editors. Valuation of ecological resources. Pensacola (FL): CRC Press, Taylor and Francis Group. p. 59–96.

Alberini A. 2001. Willingness to pay for mortality risk reduction: the robustness of vsl figures from contingency valuation studies. Risk Analysis 32(1): 242–245.

Alberini A. 2005. What is a life worth? robustness of vsl values from contingent valuation surveys. Risk Analysis 4(1):783–800.

Ashenfelter O, Greenstone M. 2004. Estimating the value of a statistical life: the importance of omitted variables and publication bias. Princeton University. Working Paper 479.

Balmford A, Bruner A, Cooper P, Costanza R, Farber S, Green R, Jenkins M, Jefferiss V, Madden J, Madden J. 2002. Economic reasons for conserving wild nature. Sciences Compass: 950–953.

Bellavance F, Dionne G, Lebeau M. 2007. The value of a statistical life: a meta-analysis with a mixed effects regression model. Canada research chair in risk management. Working Paper 06–12.

Boyle K. 2003a. Introduction to revealed preference methods. In: Boyle K, Brown T, editors. 2003. A primer on nonmarket valuation. Dordrecht (The Netherlands): Kluwer Academic Publishers. p. 258–268.

Boyle K. 2003b. Contingency valuation in practice. In: Champ P, Boyle K, Brown T, editors. 2003. A primer on nonmarket valuation. Dordrecht (The Netherlands): Kluwer Academic Publishers. p. 111–170.

Brady K. 2007a. An expressed preference determination of college students. valuation of statistical lives: methods and implication. Logan UT: Utah State University.

Brady K. 2007b. Mortality risk reduction theory and the implied value of life. Logan UT:. Utah State University.

Brady K. 2008. The value of human life: a case for altruism. department of economics. Logan UT: Utah State University.

Brannon I. 2005. What is a Life Worth ? Regulation (Winter 2005): 60–63.

Bromquist G. 2004. Self protection and averting behavior, values of statistical lives, and benefit cost analysis of environmental policy. Review of Economics of the Household 2: 89–110.

Carson R, Mitchell R, Hanemann M, Kopp Raymond, Presser S, Ruud P. 2003. Contingency valuation and lost passive use: Damages from the exxon valdez oil spill. Environmental and Resource Economics 25: 257–286.

Champ P, Boyle K, Brown T. 2003. A primer on nonmarket valuation, Kluwer Academic Publishers. Dordrecht. The Netherlands.

Daily G, Polasky S, Goldstein J, Kareiva P, Mooney H, Pejchar L, Ricketts T, Salzman J, Shallenberger R. 2009. Ecosystem services in decision making: Time to deliver. Frontiers in Ecology and the Environment 7: 21–28.

Delavan W, Epp DJ. 2001. "Benefits transfer: The case of nitrate contamination in Pennsylvania, Georgia, and Maine." In: The Economic Value of Water Quality. Northhampton, MA: Edward Elgar. 121–136.

Dickie M. 2003. Defensive behavior and damage cost methods. In: Champ P, Boyle K, Brown T, editors. 2003. A primer on nonmarket valuation. Dordrecht (The Netherlands): Kluwer Academic Publishers. p. 395–444.

Dionne G, Michaud P. 2002. Statistical analysis of value of life estimates using hedonic wage method. Working Paper 02-01. [cited 2010 Jan 22]: Available from: http://www.u-cergy.fr/IMG/2002-13Dionne.pdf.

Eduljee GH. 2000. Trends in risk assessment and risk management. The Science of the Total Environment 249(1): 13–23.

Eeckloudt L, Hammitt J. 2001. Background risks and the value of a statistical life. The Journal of Risk and Uncertainty 23: 261–279.

Farber S, Griner B. 2000. Using conjoint analysis to value ecosystem change. Environmental Science Technology 34: 1407–1412.

Field B, Field M. 2006. Environmental Economics. Boston (MA). McGraw-Hill. P 503.

Greenstone M, Gallagher J. 2005. Does hazardous waste matter? Evidence from the housing market and the superfund program. Washington (DC): National Bureau of Economic Research. Working Paper 11790.

Haab T, McConnell K. 2003. Valuing environmental and natural resources. Cheltenham (UK): Edward Elgar Publishing, Inc.

Hahn R, Tetlock P. 2007. Has economic analysis improved regulatory decisions? Washington (DC): AEI-Brookings Joint Center for Regulatory Studies. Brookings Working Paper 07–08.

Harrington W, Morgenstern R. 2004. Evaluating regulatory impact analyses. Washington (DC): Resources for the Future. Discussion Paper 04–04.

Jenkins R, Owens N, Wiggins L. 2001. Valuing reduced risks to children: The case of bicycle helmets. Contemporary Economics Policy 19: 397–408.

Khadam IM, Kaluarachchi JJ. 2003. Multi-criteria decision analysis with probabilistic risk assessment for the management of contaminated groundwater. Environmental Impact Review 23(4): 683–721.

Kubiszewsk I, Farley J, Costanza R. 2010. The production and allocation of information as a good that is enhanced with increased use. Ecological Economics 69(6): 1344–1354.

Matthews H, Lave L. 2000. Applications of environmental valuation for determining externality costs. Environmental Science and Technology 34(6): 1390–1395.

Mrozek J, Taylor L. 2002. What determines the value of life? A meta analysis. Journal of Policy Analysis and Management 21(1): 253–270.

Muradian R, Corbera E, Pascual U, Kosoy N, May PH. 2005. Reconciling theory and practice: An alternative conceptual framework for understanding payments for environmental services. Ecological Economics 69(6): 1202–1208.

Nelson E, Mendoza G, Tegetz J, Polasky S, Tallis H, Cameron R, Chan K, Daily G, Goldstein J, Kareiva P, Lonsdorf E, Naidoo R, Ricketts T, Shaw M. 2009. Modeling multiple ecosystem services, biodiversity conservation, commodity production, and tradeoff at landscape scales. Frontiers in Ecology and the Environment 7(1): 4–11.

Nijkamp P, Vindigni G, Nunes P. 2008. Economic valuation of biodiversity: A Comparative study. Ecological Economics 67(6): 217–232.

Nocera S, Bonato D, Telser H. 2002. The contingency of contingency valuation. How much are people willing to pay against Alzeimer's disease. International Journal of Health Care Finance and Economics 2(1): 219–240.

Pascual U, Muradian R, Rodríguez LC, Duraiappah A. 2005. Exploring the links between equity and efficiency in payments for environmental services: A conceptual approach. Ecological Economics 69(6): 1237–1244.

Plummer M. 2009. Assessing benefit transfer for the valuation of ecosystem services. Frontiers in Ecology and the Environment 7(1): 38–45.

Poggio L, Vrscaj B, Hepperle E, Schulin R, Marsan FA. 2008. Introducing a method of human health risk evaluation for planning and soil quality management of heavy metal-polluted soils-An example from Grugliasco (Italy). Landscape and Urban Planning 88: 64–71.

Pollard SJT, Davies GJ, Coley F, Lemon M. 2008. Better environmental decision making-Recent progress and future trends. Science of the Total Environment 400: 20–31.

Richmond A, Kaufmann R, Myneni R. 2007. Valuing ecosystem services: A shadow price for net primary production. Ecological Economics 64(2): 454–462.

Sabau GI. 2010. Know, live and let live: Towards a redefinition of the knowledge-based economy sustainable development nexus. Ecological Economics 69(6): 1192–1201.

Schaffner FC. 2006. Accelerated terrestrialization of a subtropical lagoon: The role of agency mismanagement. In: Cannizzaro P, editor. Proceedings of the 32nd Annual Conference on Ecosystems Restoration and Creation. Tampa, FL. p 92–110.

Shanmugam K. 2000. Valuation of life and injury risk. Environmental and Resource Economics 16(2): 379–389.

Shogren J, Stamland T. 2002. Skill and Value of Life. Journal of Political Economy 110(6): 1168–1173.

Spash C, Vatn A. 2006. Transferring environmental value estimate: Issues and Alternatives. Ecological Economics: 379–388.

Sunstein C. 2003. Lives, life-years, and willingness to pay. The Law School. The University of Chicago. Working Paper No. 03–5.

Taylor L. 2003. The hedonic method. In: Champ P, Boyle K, Brown T, editors. A primer on nonmarket valuation. Dordrecht (The Netherlands): Kluwer Academic Publishers. p. 331–394.

Tietenberg T, Lewis L. 2009. Environmental and natural resource economics. Boston (MA). Addison Wesley.

[USEPA] US Environmental Protection Agency. 2000. Risk characterization handbook. EPA 100-B-00-002.

[USEPA] US Environmental Protection Agency. 2001. Economic valuation of mortality risk reduction: assessing the state of the art for policy applications. In: National Center for Environmental Economic Workshop: Nov 6 -7, 2001. Silver Spring, Maryland.

[USEPA] US Environmental Protection Agency. 2006. Morbidity and mortality: How do we value the risk of illness and death?. In: National Center for Environmental Economic Workshop: April 10-12, 2006. Washington, DC, Maryland.

[USEPA] US Environmental Protection Agency. 2010. What is risk? What is a stressor?. [Internet]. [cited 2010 March 8]. Available from http://epa.gov/riskassessment/basicinformation.htm#risk.

Vatn A. 2010. An institutional analysis of payments for environmental services. Ecological Economics 69(6): 1245–1252.

Viscusi W. 2008. How to value a life. Valderbilt University. Working Paper Number 08-16.

Viscusi W, Aldy J. 2003. The Value of a statistical life: A critical review of market estimates throughout the word. The Journal of Risk and Uncertainty 27: 5–76.

Viscusi W, Aldy J. 2007. Labor market estimated of the senior discount for the value of statistical life. Journal of Environmental Economics and Management 53: 377–392.

Whitehead J. 2000. A practitioner's primer on contingency valuation. Greenville (NC). (Department of Economics). East Carolina University.

Chapter 6

Adjusted Genuine Savings and Human Development Index: A two dimensional indicator for sustainability

Holger Schlör, Wolfgang Fischer, Jürgen-Friedrich Hake and Eddie N. Laboy-Nieves

SUMMARY

The Brundtland Commission introduced a critical new dimension into our understanding of economic growth by raising the issue of sustainable development. Universal access to basic water services, for example, is one of the most fundamental conditions for sustainable human progress. To determine whether or not we are living in a sustainable way, information is needed about the ecological, economic and social conditions of the environment in which we live. In particular it is necessary to identify the components of the natural and social system and to define indicators that can provide essential and reliable information about these systems. Genuine Savings (GS), an indicator developed by the World Bank, is a sustainability indicator based on the concepts of national accounts that can provide comprehensive information about the system shaping sustainable development. GS measures the rate of savings of an economy taking into account investment in human capital, depletion of natural capital and damage caused by carbon dioxide emissions. However, the present GS of the World Bank does not cover all aspects of the Earth's system. In particular it does not consider water as a renewable resource. Water is a normal capital good that provides critical functions for natural and social systems for which no substitutes are available. Not considering water as a critical capital good leads to a loss of prosperity and not considering water in the GS results in a lack of information about the condition of the Earth's system. The aim of this chapter is to present a two-dimensional indicator that combines the modified Adjusted Genuine Savings (AGS) with the Human Development Index (HDI). The AGS/HDI considers water as the most important renewable resource, an asset to attain sustainability.

6.1 INTRODUCTION

Risk arises for any country from the ignorance of its true welfare state, covered by the three pillars of sustainability (ecological, social and economic issues). The control of risks is a method of "promoting the concept of sustainable development" and thereby protecting the welfare basis of the country (McQuaid 2000). Many indicators have attempted to capture the various dimensions of sustainability in terms of sub-components as well as the way these are combined or aggregated (Pillarisetti and van den Bergh 2010). Therefore, we use the two-dimensional indicator Adjusted Genuine Savings (AGS)/Human Development Index (HDI) to identify the risks associated with the sustainable development of the analysed countries. This indicator gives impetus to the scientific discussion on the measurement of risks, the configuration of an international indicator system for risks and sustainability measurement, and for the construction of a better risk management system. The indicator identifies risk areas for sustainable development and discloses those countries that are threatened by non-sustainable development.

The sub-indicators of both the AGS (raw material consumption, economic development and the educational level) and the HDI (life expectancy, education and economic development) help to describe the risks of each of the three dimensions of sustainability. Hence, a country's policy and society can access information on the entire social risk for a country and on the particular risks for the subsystems as well (Vandermoere 2008). Unlike the risk management of a company, credit risks or environmental risks, which focuses on single field factors, risk management in the context of sustainability takes a broader look at hazards which have been classified as sustainability relevant for the development of the country (McQuaid 2000). Our analysis attempts to systematically capture the sustainability risks of countries as well, in order to be consistent with (Cawdery and Marshall 1989) postulates.

In this chapter, we evaluated the sustainability risk approach described by Mehta (1997) and argued that good policies require the integration of risk assessment as a tool and sustainability as a principle. We applied the risk minimizing used to develop environmental policy (Groh 1992; Rammel and van den Bergh 2003). Our objective is to avoid risks by minimizing the approach applied by Rammel and van den Bergh (2003), who reported that sustainability minimizes risks ignored by conventional policy. We expanded the Mehta (1997) approach by adopting sustainability not only as a principle but also as a tool for AGS/HDI two-dimensional indicator.

6.2 SUSTAINABILITY CONCEPTS

Since the Brundtland Report of 1987, the United Nations Earth Summit in Rio de Janeiro in 1992 and the Johannesburg Conference in 2002, sustainable development has been set up as a model for social and political processes (WCED 1987; UN 1992, 2002). The concept and its implementation have been discussed extensively both by the science community as well as society (Beckerman 1994, Beckerman 2003, Daly 1995, Jacobs 1995, Victor 2006, Neumayer 2001, Mäler 2008, O'Riordan 2008). Although sustainable stems from the need to find alternatives for environmental deterioration and to reorient the interrelation between society and nature (Cunningham and Cunningham 2005), it is not the place where you are going, rather how you make the journey (Meadows 2008). Two sustainability concepts are currently being discussed, which define different rules for Meadow's (2008) journey: weak or strong sustainability. There is a consensus about the fundamentals of sustainable growth, namely the protection of the capital basis (e.g.

natural, social, and human capital) of society. But strong and weak sustainability differ in two factors, as pointed out by (Pearce et al. 1996):

- Unlike the weak sustainability concept, strong sustainability restricts the "substitutability between natural assets and other assets – human and manufactured assets."
- Strong sustainability "stresses 'discontinuity' and 'non-smoothness' in ecological systems"

The strong sustainability concept has its starting point in ecological imperatives, whereas its weak counterpart starts with the standard assumptions in economics (Pearce et al. 1996). The differences result from the importance of the individual capital goods in the various sustainability concepts. It should be noted that weak sustainability implies barriers to the consumption of non-renewable resources, since the profits obtained from the sales of resources according to Hotelling's Rule have to be invested in reproducible capital to keep the capital basis constant over time (Hotelling 1931).

Our analysis concentrates on water as a basic and critical capital good for life and for economic and social development. Human beings cannot live without water; they need at least 50 liters of water per day (Gleick 1993), and this minimum requirement cannot be replaced by man-made capital. Water is the most important critical natural capital good integrated into the AGS and the HDI indicators. Access to basic water services is one of the most fundamental conditions of human development and thus indispensable for sustainable development. We acknowledge that the substitutability of the various natural capital goods is limited and that the HDI considers the well-being of people in terms of capabilities. By embedding critical capital goods (such as water) in the genuine savings concept, and considering the well-being of people, our approach links weak and strong sustainability for a better interpretation of the concept. Therefore, we present a two-dimensional indicator which combines AGS and HDI.

6.3 MEASURING SUSTAINABILITY USING ADJUSTED GENUINE SAVINGS

6.3.1 Genuine Savings Indicator

The Genuine Savings Indicator (GSI) is a weak sustainability indicator based on the concept of national accounts (El Serafy 1991; Hamilton 1994). GSI measures the rate of savings of an economy taking into account investment in human capital, depletion of natural capital and damage caused by CO_2 emissions (Bolt et al. 2002). GSI tries to offer decision makers a simple indicator showing how sustainable their policy is (Hamilton 2000). The main objective of the GSI is to answer the question of whether the welfare of a society increases or decreases by analyzing its capital stock (Atkinson et al. 1997).

GSI represents the first numeric approximation of the degree to which a nation satisfies the Hartwick-Solow rule, often called weak sustainability (Hartwick 1977, 1978; Solow 1974a, 1974b, 1993; Rapp Nielsen 2010; Arbex and Perobelli 2010).

A nation which reinvests all of its profits exploiting non-renewable natural resources in the formation of human capital through its educational system would have imposed no net opportunity cost on the country's future citizens (Hamilton 2000). GSI records the changes in natural, man-made, and human capital, and evaluates the total change in economic assets providing an index of whether an economy is on a sustainable path (Hamilton and Bolt 2006). If the GSI is negative, the welfare of the country will decrease in the future and the development path of that country is no longer sustainable (Dasgupta and Mäler 2000). This means the quality of life only improves if genuine investment is positive (Dasgupta and Mäler 2001).

6.3.2 Adjusted Genuine Savings Indicator

The World Bank recommended that further research in the field of genuine savings should include other natural capital goods such as water, because it is a prominent element (Kunte et al. 1998). The Adjusted Genuine Savings Indicator (AGSI) does not consider water as the most important natural critical resource. However, water should be regarded as a natural capital good (WI 2008) in the context of the weak sustainability approach, and as an economic good in the tradition of the Dublin Principles (ICWE 1992). Fankhauser (1994) proposed a formula which considers water as the most important renewable resource, subjected to depletion (Equation 6.1):

$$\text{Water Rent} = (\text{ANR} - \text{AWW}) \times \text{water price} \times \text{rent rate} \tag{6.1}$$

where ANR stands for Annual Renewable Water, AWW = Annual Water Withdrawal, and the rent rate is determine with Equation 6.2 as follows:

$$\text{Rent Rate} = \frac{(\text{market price of water} - \text{production cost})}{\text{market price of water}} \tag{6.2}$$

The ANR concept relates to the annualized theoretical maximum volume of water resources available in a country, while the AWW refers to gross amount of water extracted from the resources. For the water rent equation, the production costs have to consider the principles of full cost recovery and the polluter pays principle (Stallworth 2003). The degree to which water prices cover costs can only be determined and assessed if all the expenses and expenditures, as well as all incomes, are disclosed. However, no reliable figures are available on the expenditure or the income side. In many countries, water prices and tariffs have long been used as instruments of social policy and for regional development. Subsidies have been used to reduce the real costs of water especially in the agriculture sector (OECD 1997), but it is essential to reveal who receives the subsidies (Dinar 2000; Boland and Whittington 2000; Komives et al. 2005). Water subsidies encourage users to overuse it, while pricing this resource tends to promotes its productive use and conservation (WI 2008).

A clean environment is a valuable good that has no real price unlike other resources and products. Supply and demand for a clean environment do not meet in the market place. This is unfortunate since if the environment had a price, it would be far easier for companies and households to incorporate environmental protection measures in their everyday economic decisions (WI 2008). Water prices have to reflect the principles of full cost recovery and the polluter pays principle (Stallworth 2003). Economic principle approaches (e.g. polluter pays principle), tools (e.g. cost effectiveness analysis) and economic instruments (e.g. water pricing) have to be considered for achieving environmental objectives for good water (Hansjürgens and Messner 2002). Operational, social and environmental costs have to be considered in water pricing to foster incentives for sustainable water uses (Kraemer and Piotrowski 1998). However, only Germany and the Netherlands have achieved a high degree of cost covering by determining the price of water.

6.3.3 Human Development Index

According to the United Nations Development Programme (UNDP), there are three factors to attain acceptable living standards: life expectancy, the opportunity to acquire knowledge, and access to natural resources (UNDP 1990, 2002). This organization defines political freedom, guaranteed human rights and self-respect as additional choices (UNDP 1990). Human development accounting should not concentrate only on traditional measures of economic progress (such as the gross national product per head), but rather in the development of national income accounting to people-centered policies (Haq 1995). The Human Development Index (HDI) was developed by the UNDP to define a new process for enlarging people's choices (UNDP 1990; Fukuda-Parr 2003). HDI brings a pluralist conception of progress to the exercise of development evaluation (Sen 2000), and capability approaches to provide the intellectual foundation for human development and for centered in participation, human well-being and freedom (Sen 1985; Nussbaum and Sen 1993; Anand and Sen 2000; Nussbaum 2007; Graymore et al. 2010).

Anand and Sen (1994) reported that human beings are the real end of all activities, and development must be centered on enhancing their achievements, freedoms, and capabilities. Sen (2000) defined well-being in terms of capabilities to achieve substantive freedoms of individuals to live the kind of life they have reason to value, including the choice to have education, a long life, and enough to eat. The advantage of the capabilities approach is that it is conceptually "eclectic", which means that one can fit further issues into it, as long as the issues are empirically operational, like the environment. Development in the sense of the capability approach is thereby seen as a method to remove obstacles for illiteracy, ill health, lack of access to resources, or negligible civil and political freedom (Fukuda-Parr 2003).

If progress is to be assessed by capability expansion, then it is meritorious to identify the factor to achieve it, because the sort of human potentials is infinite, so is the value people assign to each capability as well. The HDI includes three capabilities or dimensions: to be knowledgeable, to survive, and to enjoy a decent living standard (Anand and Sen 1994). The performance level of each one of these three dimensions ranges between 0 and 1, after applying Equation 6.3 (UNDP 2002):

$$\text{Dimension Index} = \frac{\text{actual value} - \text{minimum value}}{\text{maximum value} - \text{minimum value}} \qquad (6.3)$$

It has been assumed that achieving a respectable level of human development does not require unlimited income (UNDP) 2002). Hence, the HDI is computed as an arithmetic mean of three indexes: life expectancy (LEI), educational attainment (EAI), and real gross domestic product per capita adjusted for purchasing power parity (GDP), as depicted in Equation 6.4.

$$\text{HDI} = \tfrac{1}{3}[\text{LEI}] + \tfrac{1}{3}[\text{EAI}] + \tfrac{1}{3}[\text{GDP}] \tag{6.4}$$

The HDI then takes values between 0 and 1. The ranking order of any country depends on how near that country's HDI is to the maximum value. This framework has continued to be the keystone of the UNDP Annual Reports Series (UNDP 2009).

The HDI does not capture the development and the condition of the environmental capital of the individual countries, thus an extension of the HDI concerning environment and sustainable development issues is inevitable (Kelley 1991; McGillivray 1991; Desai 1995; Sagar and Najam 1998). Neumayer (2001) suggested combining the HDI with existing sustainable indicators like the Genuine Savings Index of the World Bank instead of modifying the HDI to create a new two-dimensional indicator to measure sustainable development, which is the focus of this chapter.

6.3.4 The AGS/HDI Two-Dimensional Indicator

Using 2000 data from the Adjusted Net Saving Data Center (WB 2000), the World Resources Report (UNDP 2000), and the Human Development Report (UNDP 2001), we developed the AGS/HDI Two-Dimensional Indicator, which results are depicted in Table 6.1. This AGS/HDS model demonstrated that 38 nations (nearly 900 million people) exhibit unsustainable conditions due to a negative Adjusted Genuine Savings (AGS). Fifteen nations were classified as least developed countries. In addition, 60 million people living in Bolivia, Central Africa, the Ivory Coast, Guinea, Laos, Madagascar, Nicaragua and Tajikistan were affected by unsustainable development if water depletion is considered after estimating the AGS (Figure 6.1). The analysis also revealed that unsustainable nations lost US $154 billion in

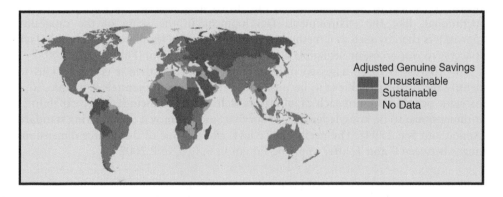

Figure 6.1 Sustainability level of nations according to their Adjusted Genuine Savings (AGS).

Table 6.1 Unsustainable nations in 2000.

Country	HDI Rank	Population in millions	GS in % GNI	AGS in % GNI	Capital lost by not considering water in US$
Angola	Low	13.134.000	−97,36	−102,36	−4.862.391.326
Amenia	Medium	3.803.000	−5,02	−5,51	−106.484.859
Azerbaijan	Medium	8.049.000	−49,45	−49,42	−2.431.654.928
Bolivia	Medium	8.328.700	2,87	−2,26	−182.337.911
Burundi	Low	6.807.000	−5,81	−6,59	−44.368.409
Cameroon	Medium	14.876.000	−0,54	−7,69	−637.128.822
Central African Republic	Low	3.717.000	5,90	−13,35	−127.057.113
Chad	Low	7.694.000	−0,58	−2,33	−32.609.920
Colombia	Medium	42.299.300	−3,77	−6,88	−5.423.927.957
Zaire	Low	50.948.000	−13,46	−38,65	−1.934.163.371
Congo	Low	3.018.000	−31,11	−48,60	−1,084.777.827
Ivory Coast	Low	16.013.000	0,80	−0,43	−37.237.669
Ecuador	Medium	12.646.000	−5,55	−9,75	−1.206.667.532
Ethiopia	Low	64.298.000	−7,27	−9,50	−601.580.286
Gabon	Medium	1.230.000	−37,60	−42,56	−1.804.213.789
Georgia	Medium	5.024.000	−6,14	−8,33	−253.461.258
Guinea	Low	7.415.000	2,21	−6,14	−180.037.156
Haiti	Low	7.959.000	−1,05	−1,37	−55.504.521
Iran	Medium	63.664.000	−12,49	−12,55	−13.212.344.528
Kazakhstan	Medium	14.869.000	−29,64	−30,05	−5.124.441.702
Kuwait	High	1.984.400	−8,43	−8,43	−3.769.147.956
Kyrgyzstan	Medium	4.915.000	−2,86	−6,62	−80.942.783
Laos	Low	5.279.000	10,12	−2,54	−42.414.640
Lebanon	Medium	4.328.000	−9,79	−2,02	−352.339.044
Madagascar	Low	15.523.000	1,23	−7,76	−295.080.281
Malawi	Low	10.311.000	−8,10	−9,11	−151.180.983
Nicaragua	Medium	5.071.000	5,89	−5,03	−106.221.258
Niger	Low	10.832.000	−6,27	−6,55	−118.405.964
Nigeria	Low	126.910.000	−31,80	−32,57	−11.960.057.974
Russia	Medium	145.555.008	−13,36	−15,31	−36.726.887.939
Saudi Arabia	Medium	20.723.150	−27,28	−27,27	−47.353.903.658
Sudan	Low	31.095.000	−6,19	−6,42	−626.077.050
Syria	Medium	16.189.000	−27,91	−27,89	−4.453.364.311
Tajikistan	Medium	6.170.000	5,15	−0,62	−5.810.053
Ukraine	Medium	49.501.000	−4,18	−4,31	−1.329.456.722
Uzbekistan	Medium	24.752.000	−61,81	−61,44	−4.555.941.356
Venezuela	Medium	24.170.000	−0.71	−1,54	−1.840.718.271
Yemen	Low	17.507.160	−18,23	−18,25	−1.347.907.308
Living unsustainable water not considered		809.091.018			−154.458.248.438
Living unsustainable water considered		876.607.718			
Addition people living unsustainable		*67.516.700*			

Source: Own calculation from 2010 based on data from UNDP 2000 and World Bank 2000/ GNI = Gross National Income

2000, whereas Russia and Saudi Arabia lost nearly US $85 billion, mainly due to vanishing oil and gas capital; they are responsible for more than half of the world capital loss in 2000. In Europe, only Russia and Ukraine were classified as unsustainable. These findings are consistent with the postulate that societies do not become wealthy first and then invest in water management (UNESCO 2003; WI 2008).

The AGS/HDI model showed that 38 countries are living in unsustainable conditions (Figure 6.2). From 53 highly developed countries, Kuwait is the only one assessed as unsustainable: it represents 0.2% of all the people living in those countries. The model identified 20 medium-developed countries (~476 million inhabitants) living under unsustainable conditions, whereas nearly 50% of the low-developed countries with a total population of nearly 400 million live under unsustainable conditions. It is predicted that in 2025 nearly 3.8 billion people will live under severe water stress conditions (Alcamo et al. 2000; UNESCO 2003). Therefore, countries have to find ways to manage critical resources such as water to pursue economic growth.

Living under unsustainable conditions is a problem for people in developing countries, mainly from Africa and Asia. Fifteen of the seventeen unsustainable (low HDI) countries are classified as underdeveloped, because of the depreciation of their natural capital stock and a lack of education, health services, and purchasing power. These countries did not or could not use their wealth of natural resources to build up new capital stock to ensure the three dimensions identified by (Anand and Sen 1994). Low HDI countries ought to avoid wasting natural capital rents from renewable and non-renewable resources for consumption; rents should lead to create new capital stock for future generations and thus ensure intergenerational justice. Nations living under unsustainable conditions (15% of the world's population) have to invest US $160 billion in order to reach a sustainable investment level. Thus, the current global economic situation jeopardizes future generation's potential for achieving sustainable development.

Laboy-Nieves (2009) pointed out that the human dimensions component of sustainability has become an integral part to achieve the rehabilitation of the environment; but the challenge will be to preserve ecosystem functions and to use

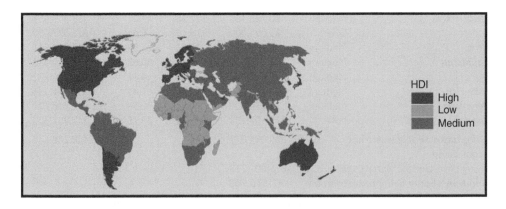

Figure 6.2 Sustainability level of nations according to their Human Development Index (HDI).

natural capital while simultaneously sustaining local communities. Research is required to evaluate and monitor sustainable development and to determine if all these indexes are able to sufficiently capture its multidimensional nature, and to identify and rank nations accordingly (Pillarisetti and van den Bergh 2010).

6.4 RISK EVALUATION AND SUSTAINABILITY

This study incorporated water as a central natural capital good given water's irreplaceable role in human lives. Water availability has a central meaning for sustainable risk management. Our findings show that including water could be a driver to change the risk evaluation for some countries, and could modify the "risk ranking" and thus the relative position in relation to other countries.

The measurement of sustainability avoids the risk of stimulating unsustainable socio-economic patterns, and sacrificing long-term stability for short-term "optimums and gains of efficiency" (Rammel and van den Bergh 2003). The often unconscious and careless risk behavior of today's generations leads to irreversible damages nowadays and also constitutes a heavy burden for future generations. We thereby confirm that risk assessment and risk management offer a way for an operationally sustainable development (Karlsson 2006). This is very valid for developing countries, especially for the underdeveloped ones. Information on critical resources offers an opportunity for the international community to use this knowledge to assess current and future welfare level within the range of a risk-reducing sustainability strategy.

This analysis is particularly significant for developing countries: they get information on their real welfare, which enables to protect natural resources more effectively, and thereby to improve their socioeconomic development perspective. The method can further be developed by taking into consideration water quality, supply, and expenses (social dimension of sustainability). The inclusion of water in the AGS/HDI indicator also opens the door for participative concepts of risk management, which should be included in the sustainability strategy (McDaniels et al. 1999).

In our view the AGS/HDI two-dimensional indicator has the advantage of the sum of individual indicators, enabling us to determine for each country the individual indicators responsible for its development. The AGS enables to determine whether the growth is caused by the consumption of fixed capital, or resource depletion such as water, or the damage caused by CO_2 emissions. Finally, we think it is necessary to include further critical natural capital goods, such as fisheries or the preservation of biodiversity, in the concept of genuine savings. The relevance of biodiversity to human health is also becoming a major international political issue. Moreover, a minimum requirement for specific critical capital stocks can also be considered. The annual quantity of water withdrawn, for example, should not exceed the annual water renewal rate. Thus, a decline in capital stocks below a certain level could be an indicator of an unsustainable state policy irrespective of the remaining capital stock. This procedure could represent a connection between the concept of strong and weak sustainability. Finally, the two-dimensional indicator can be used independently of the chosen sustainability paradigm. Hence, AGS/HDI provides a critical new dimension for the relative interpretation of economic development.

6.5 CONCLUDING REMARKS

Since the concept of sustainable development introduced a critical new dimension into the understanding of economic growth, the question whether we are living in a sustainable way is in the focus of national and international science and policy. Thus, information is needed about the ecological, economic and social conditions of the world. It is necessary to identify the components of the natural and social system and to define indicators that can provide reliable information about these systems. The World Bank developed the Genuine Savings (GS) indicator, to measure sustainability based on the rate of savings of an economy taking into account investment in human capital, depletion of natural capital and damage caused by CO_2. However, the GS does not consider water, which is a capital good that provides indispensible functions for natural and social systems. Overriding water in the GS leads to a loss of prosperity, a lack of information about the condition of the Earth's system, and impedes a proper analysis of societal risks. The AGS/HDI two-dimensional indicator herein analysed, shows a way to cope with that problem. The AGS considers water as the most important renewable resource asset, while the HDI delivers information about human quality of life. Including water in the calculation changes the evaluation of societal risk for some countries and lays a foundation for sustainable policies.

REFERENCES

Alcamo J, Henrichs T, Rösch T. 2000. World Water in 2025. Kassel, Center for Environmental Systems Research. 48 p.

Anand S, Sen A. 1994. Human development index: methodology and measurement. New York: UNDP. 25 p.

Anand S, Sen A. 2000. Human Development and Economic Sustainability. World Development 28(12): 2029–2049.

Arbex M, Perobelli FS. 2010. Solow meets Leontief: Economic growth and energy consumption. Energy Economics 32(1): 43–53.

Atkinson G, Duboung R, Hamilton K, Munasinghe M, Pearce D, Young C. 1997. Measuring sustainable development: macroeconomics and the environment. Cheltenham, UK: Edward-Elgar Publisher. 271 p.

Beckerman W. 1994. Sustainable development – is it a useful concept?. Environmental Values 3(3): 191–209.

Beckerman W. 2003. A poverty of reason. Sustainable Development and Economic Growth. Oakland, California: The Independent Institute. 112 p.

Boland JJ, Whittington D. 2000. The political economy and water tariff design in developing countries: Increasing block tariffs versus uniform price with rebate. In: Dinar A, editor. The political economy of water pricing reforms. New York: Oxford University Press: 215–236.

Bolt K., Matete M., Clemens M. 2002. Manual for calculating adjusted net savings. Environment Department. Washington: World Bank. 23 p.

Cawdery J, Marshall D. 1989. Sustainable development: A promising new avenue for transferring risk analysis technology to developing countries. Risk Analysis 9(2): 151–152.

Cunningham WP, Cunningham MA. 2005. Principles of Environmental Sciences. Minnesota (USA): McGraw Hill. 418 p.

Daly HE. 1995. On Wilfred Beckerman's critique of sustainable development. Environmental Values 4(1): 49.

Dasgupta P, Mäler K-G. 2000. Net national product, wealth, and social well-being. Environment and Development Economics 5(1): 69–93.

Dasgupta P, Mäler K-G. 2001. Wealth as a criterion for sustainable development. World Economics 2(3): 19–44.

Desai M. 1995. Human development: concepts and measurement. European Economic Review 35(2–3): 350.

Dinar A, editor. 2000. The political economy of water pricing reforms. New York: Oxford University Press. 416 p.

El Serafy S. 1991. The Environment as capital. In: Costanza R., editor. Ecological Economics, the Science and Management of Sustainability. New York: Columbia University Press. p 168–175.

Fankhauser S. 1994. The social costs of greenhouse-gas emissions – an expected value approach. Energy Journal 15(2): 157–184.

Fukuda-Parr S. 2003. The humand development paradigm: Operationalizing Sen's ideas on capabilities. Feminist Economics 9(2-3): 301–317.

Gleick PH, editor. 1993. Water in Crisis: A Guide to the World's Freshwater Resources, New York: Oxford University Press. 473 p.

Graymore MLM, Sipe NG, Rickson RE. 2010. Sustaining Human Carrying Capacity: A tool for regional sustainability assessment. Ecological Economics 69(3): 459–468.

Groh D. 1992. Strategien, Zeit und Ressourcen. Risikominimierung, Unterproduktivität und Mußepräferenz—die zentralen Kategorien von Subsistenzökonomien (Strategies, time and resources. Risk minimizing, under productivity and leisure preference. Central categories of subsistence economies). Anthropological dimension of history. Frankfurt: Suhrkamp. 314 p.

Hamilton K. 1994. Green adjustments to GDP. Resources Policy 20(3): 155–168.

Hamilton K. 2000. Sustaining economic welfare: Estimating changes in wealth per capital. In: World Bank (ed.) Policy Research Working Paper. Washington D.C.: World Bank. 28 p.

Hamilton K, Bolt K. 2006. Genuine saving as an indicator of sustainability. In: Atkinson G , Dietz S, Neumayer E, editors. Handbook of Sustainable Development. Cheltenham: Edward Elgar. p 292–306

Hansjürgens B, Messner F. 2002. Kostendeckende und verursacher-gerechte Wasserpreise in der EU-Wasserrahmenrichtlinie (Full cost recovery and polluter-pays principle based water prices in the EU water framework directive). Wasser and Boden 54(7/8): 66–70.

Haq M. 1995. Reflections on human development. New York: Oxford University Press. 288 p.

Hartwick JM. 1977. Intergenerational equity and the investing of rents from exhaustible resources. American Economic Review 67(5): 972.

Hartwick JM. 1978. Substitution among exhaustible resources and intergenerational equity. The Review of Economic Studies 45(2): 347–354.

Hotelling H. 1931. The economics of exhaustible resources. Journal of Political Economy 39(2): 137.

[ICWE] International Conference on Water and the Environment. 1992. The Dublin Statement on Water and Sustainable Development [Online]. Dublin: International Conference on Water and the Environment (ICWE). Available: http://www.gdrc.org/uem/water/dublin-statement.html [Accessed 10 November 2003].

Jacobs M. 1995. Sustainable development, capital substitution and economic humility: A response to Beckerman. Environmental Values 4(1): 57.

Karlsson M. 2006. Science and norms in policies for sustainable development: Assessing and managing risks of chemical substances and genetically modified organisms in the European Union. Regulatory Toxicology and Pharmacology 44(1): 49–56.

Kelley Ac. 1991. The human development index: 'Handle with care'. Population and Development Review 17(2): 315–324.

Komives K, Foster V, Halpern J, Wodon Q, Abdullah R. 2005. Water, electricity, and the poor: Who benefits from utility subsidies? Washington: World Bank. 283 p.

Kraemer A, Piotrowski R. 1998. Comparison of water prices in Europe – summary report, Berlin: Ecologic. 19 p.

Kunte A, Hamilton K, Dixon J, Clemens M. 1998. Estimating national wealth: Methodology and results. Washington: World Bank. 53 p.

Laboy-Nieves EN. 2008. Ética y Sustentabilidad Ambiental en Puerto Rico (Environmental Ethics and Sustainability in Puerto Rico). Actas del Foro Internacional de Recursos Hídricos. INDRHI: Dominican Republic. p 63–74.

Laboy-Nieves EN. 2009. Environmental management issues in Jobos Bay, Puerto Rico. In: Laboy-Nieves EN, Schaffner F, Abdelhadi AH, Goosen MFA, editors. Environmental management, sustainable development and human health. London: Taylor and Francis: 361–398.

Mäler KG. 2008. Sustainable development and resilience in ecosystems. Environmental and Resource Economics 39(1): 17–24.

McDaniels T, Gregory R, Fields D. 1999. Democratizing risk management: Successful public involvement in local water management decisions. Risk Analysis 19(3): 497–510.

McGillivray M. 1991. The human development index: yet another redundant composite development indicator? World Development 19(10): 1461–1468.

McQuaid J. 2000. The application of risk control concepts and experience to sustainable development. Process Safety and Environmental Protection 78(4): 262–269.

Meadows D. 2008. Social and technical innovations for sustainable development. International Sustainability Conference. Basel, Switzerland, August 21, 2008, slides available from: http://www.isc2008.ch/materials/Meadows.pdf.

Mehta MD. 1997. Risk assessment and sustainable development: Towards a concept of sustainable risk. Risk: Health, Safety & Environment 8: 137–154, available from: http://www.piercelaw.edu/risk/vol8/spring/mehta.htm#fn14.

Neumayer E. 2001. The human development index and sustainability – a constructive proposal. Ecological Economics 39(1): 101–114.

Nussbaum M. 2007. Human rights and human capabilities. Harvard Human Rights Journal 20(Spring): 21–24.

Nussbaum M, Sen A. 1993. Quality of Life. Oxford: Oxford University Press. 472 p.

O'Riordan T. 2008. Some reflections on the conditions for favouring integrated sustainability assessments. International Journal of Innovation and Sustainable Development 3(1/2): 153–162.

[OECD] Organisation for Economic Co-Operation and Development. 1997. Water subsidies and the environment. Paris: OECD Publications. 43 p.

Pearce D, Hamilton K, Atkinson G. 1996. Measuring sustainable development: progress on indicators. Environment and Development Economics 1(1): 85–101.

Pillarisetti JR, van den Bergh JCM. 2010. Sustainable nations: what do aggregate indexes tell us? Environmental Development and Sustainability 12: 49–62.

Rammel C, van den Bergh J. 2003. Evolutionary policies for sustainable development: adaptive flexibility and risk minimising. Ecological Economics 47(2–3): 121–133.

Rapp Nilsen, Heidi 2010. The joint discourse 'reflexive sustainable development' – From weak towards strong sustainable development. Ecological Economics 69(3): 495–501.

Sagar AD, Najam A. 1998. The human development index: a critical review. Ecological Economics 25(3): 249–264.

Sen A. 1985. Commodities and capabilities. Amsterdam: Elsevier. 104 p.

Sen A. 2000. A decade of human development. Journal of Human Development 117: 23.

Solow R. 1993. Sustainability: An economist's perspective. In: Dorfman R, Dorfman N, editors. Selected Readings in Environmental Economics. New York: Norton. p 179–187.

Solow RM. 1974a. The economics of resources or the resources of economics. American Economic Review 64(2): 1–14.

Solow RM. 1974b. Intergenerational equity and exhaustible resources. Review of Economic Studies 41(1): 29–45.

Stallworth H. 2003. Water and wastewater pricing: An informational overview. In: Management Oow (ed.). Washington DC: U.S. Environmental Protection Agency. 7 p.

[UN] United Nations. 1992. Report of the United Nations Conference on Environment and Development. New York: UN, documents available from http://www.un.org/esa/dsd/resources/res_docukeyconf_eartsumm.shtml.

[UN] United Nations. 2002. Report of the World Summit on Sustainable Development. New York: UN, documents available from http://www.un.org/esa/dsd/resources/res_docukeyconf.shtml.

[UNDP] United Nations Development Programme. 1990. Human Development Report 1990, New York: Oxford University Press. 122 p.

[UNDP] United Nations Development Programme. 2000. World Resources 2000–2001: People and ecosystems: The fraying web of life. Washington: World Resource Institute. 276 p.

[UNDP] United Nations Development Programme. 2001. Human Development Report 2001. New York: UNDP. 264 p.

[UNDP] United Nations Development Programme. 2002. Human Development Report 2002. New York: UNDP. 292 p.

[UNDP] United Nations Development Programme. 2009. Human Development Report 2009. New York: UNDP. 217 p.

[UNESCO] United Nations Educational Scientific and Cultural Organization. 2003. Water for people water for life. Oxford: UNESCO and Berghan Books. 576 p.

Vandermoere F. 2008. Hazard perception, risk perception, and the need for decontamination by residents exposed to soil pollution: The role of sustainability and the limits of expert knowledge. Risk Analysis 28(2): 387–398.

Victor DG. 2006. Recovering Sustainable Development. Foreign Affairs 85(1): 91–103.

[WB] World Bank. 2000. Adjusted Net Saving Data Center 2000. World Bank: New York, available from http://web.worldbank.org/WBSITE/EXTERNAL/TOPICS/ENVIRONMENT/EXTEEI/0,contentMDK:20502388~menuPK:1187778~pagePK:148956~piPK:216618~theSitePK:408050,00.html.

[WCED] World Commission on Environment and Development . 1987. Our Common Future Oxford. New York: Oxford University Press. 400 p.

[WI] Worldwatch Institute. 2008. State of the World 2008: Innovations for a sustainable economy, New York: WW Norton + Company. 260 p.

Rustic stove in Barahona, Dominican Republic (@Eddie N. Laboy-Nieves)

Management of surface and groundwater resources

Sediments runoff after rainstorm in Aguada, Puerto Rico (@Eddie N. Laboy-Nieves)

Chapter 7

Management, public opinion and research on Costa Rica's surface and groundwater resources

Peter Phillips

SUMMARY

Costa Rica has a reputation as an ecotourism destination and a haven of biodiversity. However, in terms of the management of its water resources, its reputation for progressive environmental management is seriously challenged. The country's water resources are managed by an array of government institutions and these are guided by legislation dating from 1942 long considered inadequate. Attempts to update the law have not yet met with success. A review of newsprint media during 2007–2008 revealed serious public concerns regarding government ability in managing water resources; the safety of potable water, ecosystem viability and especially surface and groundwater contamination. The major issues were: Central Valley urban area water quality concerns due to nearly inexistent wastewater treatment; northwest and central Pacific coast nearshore concerns due to poor wastewater management and an unreliable potable water supply; and Caribbean coastal plain concerns due to alleged groundwater contamination from pineapple cultivation. In an analysis of surface water quality at 56 sites in the Central Valley, home to nearly 60% of the nation's population, highly degraded river courses were observed and high levels of nutrients corroborated the absence of wastewater treatment. With chaotic tourist development along the NW Pacific coast, similar observations were made at 15 sites. The Costa Rican government and public recognize the danger to human and aquatic ecosystem health. However, balancing population and economic growth against the availability of economic resources will continue to challenge their ability to manage their surface and groundwater resources.

7.1 INTRODUCTION

Costa Rica has global reputation of being a tropical ecological paradise, with a developing tourism sector based on popular advertising slogans such as "ecotourism" and "sustainability". This country ranks number three among 163 nations evaluated for their environmental public health and ecosystem viability (Emerson et al. 2010). Is Costa Rica's reputation as an ecotourism mecca and a haven of biodiversity deserved? When viewed through the lens of the management of its water resources, its reputation for enlightened and progressive environmental management is seriously challenged.

With the publication of the Environmental Performance Index (Emerson et al. 2010), Costa Rican newspaper editorials spoke proudly of the country being in the big league of world conservation, and as a moment of enormous pride but that the recognition must give impulse to take on the big remaining challenges. Reference was made to the government initiative Peace with Nature (Paz con la Naturaleza), containing a series of environmental commitments at the national and international level, destined to serve as a guide to developing countries and to stimulate cooperation between them and the richest nations in promoting sustainable development. However, the study signaled a very weak aspect of Costa Rica's capacity to manage river contamination; this latter having reached an emergency level in some major watersheds, and increasing contamination of some hydrological sources (Anonymous 2008a).

Various sources criticize the state of the nation's water resources as equally, or more, deteriorated than its land resources (Astorga 2009; PEN 2009). What international pressures are there to impact these resources? As ecotourism has grown, some NW and Central Pacific coastal areas of Guanacaste and Puntarenas provinces also saw the installation of large-scale resort hotel developments. By the middle of the 1st decade of 2000, there was an increasing volume of evidence showing that tourism and vacation home development was overwhelming the Pacific coastal region natural resource base and threatening the southern Caribbean zone as well. Sectors of the Pacific coast are under siege by such rapacious development that, in addition to the loss of the biological diversity that has traditionally attracted visitors, reliable scientific data has been published showing the local potable water supply will not satisfy new development, salt water intrusion into the ground water and aquifers is occurring (Calderón Sánchez et al. 2002), the absence of comprehensive wastewater treatment is damaging surface waterways and nearshore areas and solid waste management is seriously deficient (Anonymous 2008b).

Aside from the coastal development impact, the greatest impact on the Costa Rican citizenry comes from disorderly urban development, deficient infrastructure, lack of planning and chaotic wastewater handling in the Central Valley, or specifically the metropolitan area of San José, where nearly 60% of the population resides (Reynolds Vargas and Richter 1994; Reynolds Vargas and Fraile 2002; Salas et al. 2002).

Human and ecosystem well-being require steady and sustainable environmental conditions. However, alterations to river hydrology considered beneficial to humans often have negative consequences for the environment and for biodiversity. Freshwater ecosystems are among the most endangered worldwide (Nel et al. 2008). In an analysis of river flows juxtaposed against climate change and water withdrawal needs, future episodes of flooding or storage will affect hundreds of millions of people, reducing biodiversity and increasing risk to humans. Projections are that all populated basins of the world will be impacted by discharge reductions and many will experience water stress and a loss of ecosystem services (Nelson et al. 2008; Palmer et al. 2008). Palmer et al. (2009) provide an overview of expected impacts on river ecosystems due lack of management as well as the anticipated impacts due to climate change. To correct this situation, new water resource strategies need include more efficient use and pollution reduction. Additionally, worldwide more than one billion people live in watersheds that will likely need to be managed to accommodate anticipated climate change alterations in hydrology. An increasing area of research to combat these trends is to begin to conceptualize freshwaters in terms of global water systems thus improving their management from anthropogenic impacts. This involves identifying the ecological and policy implications of changes to global water systems, establishing international programs to understand and resolve major social and environmental issues arising from those changes, and developing broad-based mitigation or restoration techniques. Achieving these goals is paramount for maintaining human health as well as for the freshwater ecosystems upon which we depend (Naiman and Dudgeon 2010). Chapagaina and Hoekstrab (2008) take the assessment and management of freshwater to another dimension by incorporating the concept of virtual water and its movement via global trade. A country's water supply is incorporated into the products is produces during manufacturing or in agricultural production and thus that water used is exported. Countries have a net gain or loss in water via virtual water movement. Most countries of the Americas, including Costa Rica, have a net water export.

Arthington et al. (2010) claim that biogeochemical processes and biodiverse aquatic communities' regulation of freshwater quantity and quality is not well understood by the water resources managers. The establishment and enforcement of stream and river flow is critical to sustain environmental services. What the authors' term environmental flow is considered the most critical property to ultimately regulate aquatic ecosystems and meet the UN Millennium Ecosystem Assessment goals for wetlands and water. The greatest challenge is to manage streams, rivers and watersheds for multiple uses so as to satisfy ecosystem and human welfare needs.

Watershed urbanization, the driver behind most of the deterioration in Costa Rican streams and rivers, impacts waterways in three ways: loss of geomorphological complexity; reduced value to society in that streams become less attractive to the public for recreational use and; ecological simplification in terms of biodiversity and loss of stream environmental services as, for example, a site of nutrient reduction (Bernhardt and Palmer 2007). In terms of the latter, Craig et al. (2008) made a series of recommendations to reduce the load of nitrogen into streams by identifying those sites where the contribution is greatest and where measures may be taken to reduce the load in the riparian zone or the stream channel. The challenges that Costa Rica confronts are not unlike challenges shared by countries worldwide where human demand is greatly exceeding the capacity of surface and groundwater sources to sustain the level of withdrawal and the pollution loads they are subjected to (Falkenmark 1990; Gleick 1998; Postel 2000). In Costa Rica, there are, however, notable environmental protection successes. The implementation of a program of payment for environmental services is successfully promoting the concept of sustainable land stewardship as documented by an analysis of the program's impact in the Virilla River subwatershed (Miranda et al. 2003). This program operates within the National Fund for Forestry Financing. That agency reported that in 20 years over 10% of the national territory has been reforested. Recovery is due to environmental regulation as well as the decline of cattle farming. In 1960, 70% of the country had forest cover. By 1986, only 40% of the country was forested. This was reversed in the 1990s and by 1997 forest cover had increased to 42%, in 2000 to 47% and in 2005 to 51%. Two other factors support recovery; environmental awareness and the development of an ecotourist industry that uses forests. The expansion of national parks and other reserves and the substitution of the agricultural economic model for an industrial one aided this process as well (Anonymous 2006a). A recent strategy that may be the most prominent tool for preserving significant tracts of land in hydrologically-sensitive watersheds is the Payment for Environmental Services (PES) scheme in Costa Rica. This is one aspect of the greater economic theory of eco-system services covering all ecosystems, not just water (Fisher et al. 2008). Costa Rica has the most developed PES program (Pagiola et al. 2005; Zbinden and Lee 2005; Pagiola 2008). Developed to preserve the landscape for a variety of reasons, the greatest significance in Costa Rica is aimed toward conserving water quality and quantity. Although natural resource preservation is the primary goal, poverty alle-viation is suggested to be a corollary goal as well since upper watershed regions tend to have less fertile soils, be more damaged by deforestation and inhabited by the poor (Pagiola et al. 2005; Nelson and Chomitz 2007). This latter is not considered significant in Costa Rica where approximately 200,000 hectares are enrolled in the PES scheme. However, it has been suggested for Guatemala (Nelson and Chomitz 2007). Specifically, in Heredia, Costa Rica, an analysis of the PES scheme was found to be more likely effective when it tackled water quality problems, since there is less technical uncertainty and less divergence between public expectations and scientific evidence on the relationship between land use and water quality (Kosaya et al. 2007).

Natural resources management in general and water resources management in particular, are currently undergoing a major paradigm shift. Management practices have largely been developed and implemented by experts using technical means based on designing systems that can be predicted and controlled. In recent years, stakeholder involvement has gained increasing importance. Collaborative governance is considered to be more appropriate for integrated and adaptive management regimes needed to cope with the complexity of social-ecological systems (Pahl-Wostl et al. 2007). With ecotourism, or nature tourism, being a driver for environmental protection in Costa Rica, an overlooked component of the array of natural resources is water. Water is the ultimate repository of polluting human activities. It is analogous to the human body's circulatory system; it captures, collects, circulates and drains fallout from the atmosphere, runoff from the landscape (natural and that altered by human activities) and discharges from pipes carrying liquid residues laden with human and industrial effluents.

This chapter outlines and summarizes nationwide concerns over the state of the nation's surface and groundwater resources, and identifies major topics of this worry. It is based on reviews of newsprint media articles (mostly anonymous) featuring water resources, evaluations of government management of water resources, and a presentation of two examples of water quality research conducted during the same time period.

7.2 MANAGEMENT

Watershed management and regulations on water use: A total of 17 entities administer water resources for the country, although to a great extent the public institution, the National Institute of Aqueducts and Drainage (AyA), manages water resources. In recent years, the legislature has taken on a highly debated plan to revise and update the Water Law of 1942 as the Law of Conservation, Management and Use of Water Resources which will regulate water use, designate conservation areas and establish penalties. New legislation was proposed in 2001 and in 2005 a unified proposal emerged. Legislators still contend that the project is far from being a law that provides answers to the dilemma of water utilization in the country (Anonymous 2007a). Specifically regarding land management, regulations outlined in a 1995 Environmental Law and those regarding procedures for environmental impact assessments do not delineate what is environmental damage and the appropriate sanctions against those who incur damage. The building boom that has expanded across the country, and the associated impact on water resources, underlines the limitations of institutions, municipalities and the National Groundwater, Irrigation and Drainage System Agency (Anonymous 2008c).

Water supply and potability: Due to its geographic location and humid tropical climate with precipitation oscillating between 1,300–7,500 mm per year, Costa Rica is positioned among countries with the greatest freshwater resources in the world. It is third in water resources in Central America, with 112 km^3, and first in per capita water with 29,579 m^3 per person per year (Gleick 2003). The potable water supply amounts to 31 m^3 per person (Mora Alvarado 2004); however Costa Ricans use four times more water per day (200 liters) than the recommended consumption levels established by the World Health Organization (50 liters). Ninety-seven percent of Costa Ricans have access to water for their use from more than 4,000 wells, springs and surface water sources. However, potable water, defined as treatment by chlorination, is available to only 83% of the population (Anonymous 2008d). This leaves more than 180,000 people lacking this safe resource. Of these 4,000 sources, 79% are managed by Rural Aqueduct Administration Committees but only 20% receive disinfection (Mora Alvarado and Portuguez 2000). Greatest problems exist in communities of less than 10,000 inhabitants. AyA chlorinates 98% of the water in its system and has the goal of achieving 100% potability in all aqueducts of the country by 2015 (Anonymous 2007b).

With much of the nation's problem of fecal contamination now managed by chlorination, a newer concern is that potable water supply sources are in danger of chemical contamination, such as from fuels and agrochemicals and there is no system in place to detect the presence of these toxic substances. For example, in 2001, carbolina was detected in a Central Valley water treatment plant, in 2002 there was hydrocarbon contamination in another plant and in 2004, toluene and other substances were detected in the water supplied to the Caribbean coastal community of Limón. One of the most serious cases was detected in 2004 in a Central Valley aquifer stratum, contaminated with 20,000 liters of hydrocarbons from leaking gasoline station tanks. These leaks were near the Colima Superior aquifer which supplies 1.5 million residents with water (Anonymous 2007c).

Wastewater management and water pollution: Costa Rica only treats 5% of its waste water while 69% drains to septics and the remainder is sent directly to rivers and streams. Therefore, a significant percentage of its surface water supply for human consumption is highly vulnerable to contamination (UNDP 2006). According to the Interamerican Development Bank, that puts Costa Rica as number 5 in Latin America on a list with the worst levels of wastewater treatment. On average across Latin American, 14% of waste water is treated (Anonymous 2007d).

The Ministry of Environment, Energy and Telecommunications is currently initiating a fee collection system to be used in a program of watershed rescue and community training for wastewater management. A realistic time frame to expect a primary treatment plant for removal of solids for 1.6 million Central Valley users is 2011 (Anonymous 2007e).

Solid waste management and water pollution: Mishandling of solid wastes impacts water resources and Costa Rica produces 11,000 tons of garbage daily. The Ministry of Health estimates that 300 tons of waste daily ends up in ditches, rivers or empty lots. Poor management of domestic solid waste causes the emergence of diseases, groundwater contamination with toxic materials and drains choked with garbage during the rainy season (Anonymous 2007f).

Hydropower: In Costa Rica, hydropower accounts for approximately 80% of electrical production. Considering the growing demand for power, the country has a seeming advantage because it can generate most of its power needs via hydro plants. However, the installation of these plants has a huge impact on freshwater aquatic and land resources especially when impoundments permanently flood land (Anderson et al. 2006).

7.3 CENTRAL VALLEY VIRILLA RIVER SUBWATERSHED

Land use: The Virilla River subwatershed includes most of the Central Valley (Figure 7.1). The soils are of volcanic origin and are a product of geomorphological erosion. The land is 32% pasture, 25% urban, 17% under coffee cultivation and 9% protected forests. There was a 10% increase in urban coverage between 1996 and 2006. The topography is very steep resulting in high stream velocity that can lead to eroded stream banks and damaging floods (CENAT 2007).

The Central Valley's population doubled in recent years to approximately 2.4 million living on 4% of the national territory (2,044 km^2); this represents nearly 60% of country's population. The outcome has been generalized urban disorder, deficient wastewater and drainage systems and poor solid waste management. Land for expanding development threatens the ring of forests and pastures that still remain. According to the Ministry of Housing, this is a result of the lack of planning in the transition from an agricultural society to one that prefers to live in the city. This population growth has resulted in more river pollution (Anonymous 2006b).

Water supply: Reports indicate that between 25 and 50% of water consumed in the Central Valley is wasted. This problem originates in deteriorated network pipes with approximately 1,000 leaks per month reported. With $40 million from the Central American Economic Integration Bank, AyA plans distribution network rehabilitation and improvements in potable water treatment plants (Anonymous 2007g).

Surface and groundwater contamination: In a Ministry of Health study on fecal coliform contamination of river discharges into the Pacific and Caribbean coasts, estuaries and beaches, lakes and reservoirs and potable water supply sources, the Grande de Tárcoles River (including the Virilla subwatershed), draining the Central Valley, was the most polluted (Mora Alvarado 2004) and estimates are that it is the most contaminated river in Central America. There are central wastewater treatment plants in Costa Rica only outside of the Central Valley.

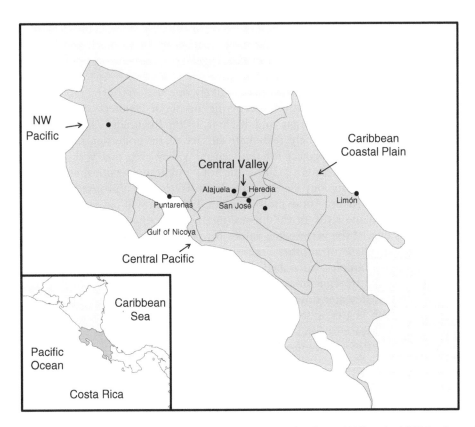

Figure 7.1 Location of the regions mentioned in the study: the Central Valley, the NW Pacific and Central Pacific coast and the Caribbean coastal plain.

Central Valley rivers receive an estimated 3,000 liters per second of untreated waste water and have been described by AyA as "open-air sewers". Sixty-seven percent of San José homes have a septic tank that doesn't function and only 3.5% of area homes have some type of sanitary drain to collect these waters. The leaky metro-politan area sanitary drainage system has four collectors extending 86 km and connections linked to the storm water drainage often overload the system. In addition to liquid wastes, residents dump garbage into storm drains. The National Light and Power Company shuts down its Virilla River subwatershed hydroelectric plant turbines weekly to collect hundreds of tons of garbage (tires, closets, dressers, bed frames, tables, millions of plastic bottles and bags) accumulated in reservoirs. This loss of generating power results in an annual revenue loss of more than $3 million. The Costa Rican Electricity Institute (ICE) collected 32 tons during six months in 2007 at its plant on the Virilla River. The municipality of San José targets $2 million annually and 30 staff to maintain drains (Anonymous 2007d).

Some 4,000 families live along the margins of contaminated rivers in squatter settlements in the San José area and suffer from poor hygienic conditions. The risk to their health includes landslides, flooding and illnesses. According to the Ministry of Health, cutaneous infections produced by water-borne bacteria such as

Streptococcus or *Staphylococcus* are common. Dengue is an added risk as well as diarrhea from *Giardia intestinalis*, an intestinal parasite, due to consumption of dirty river water. Bacterial gastroenteritis, an inflammation of the stomach and intestines, is brought on by the presence of *Salmonella*, *Shigella* or *Staphylococcus* (Anonymous 2008e).

This absence of wastewater treatment is likely one reason for the reported trend in increasing nitrate concentrations occurring in major aquifers serving Central Valley residents (Reynolds Vargas and Richter 1994; Reynolds Vargas and Fraile 2002; Salas et al. 2002). To help remediate surface water pollution, the Legislative Assembly has approved a project for the Central Valley area that will cover needs until 2025. During the decade of 2000, the four public universities, supported by the National High Technology Center, have been developing a project to rescue the Grande de Tárcoles River (Anonymous 2004a). This plan will cover the upper, mid and lower areas of the river. In the most contaminated zone of the Tárcoles, fecal coliform levels exceed the permissible values for potable water up to 373,333 times. This is an initial phase of another ambitious project that consists of cleaning the entire Grande de Tárcoles watershed. This public works project encompasses the rehabilitation of the aforementioned four principal collectors and network expansion to handle total wastewater flow. The project is receiving financing by the Japanese Cooperation Bank. Later a secondary treatment process (to eliminate biological residues such as disease-causing bacteria) plant construction will begin in 2012 and be operational in 2015 (Anonymous 2007b).

7.4 NORTHWEST AND CENTRAL PACIFIC COAST

Land use: Between 2005 and 2006, construction in the beach tourist areas of the NW Pacific grew 69.5% and 44.3% in the Central Pacific region, while in the rest of the country it grew 31%. To control coastal development, an Executive Power decree was issued with the aim of regulating construction development by prohibiting buildings greater than three stories in zones located next to beaches. The measure also includes land to four kilometers inland. The decree applies to those Guanacaste counties whose municipalities have not produced regulatory development plans and is intended to be in place for four years from 2008. Only 36 of 89 local governments in Costa Rica have regulatory plans, 52 have environmental offices but 16 of them have only one employee. The XIII State of the Nation report signaled that coastal populations are reproducing the chaotic patterns of the Central Valley as if Costa Ricans and foreigners tired of disorder and filth, went to the beach and constructed equally dirty and disorderly zones in those areas. Consequences are apparent: mangrove damage, lack of potable water, salt intrusion in wells and fecal contamination along the beaches. Whereas the National Technical Secretariat of the Environment contends that the country has sufficient laws; it doesn't have the ability to enforce them (Anonymous 2008f).

Water supply: If water extraction from a coastal aquifer is greater than that resupplied by rainfall, it runs the risk of depletion and contamination with salt water. In Costa Rica there are at least 20,000 wells, a third are illegal. In spite of scarce data, SETENA knows there are aquifers at risk in the communities along the NW

Pacific coast. The majority of applications to install wells are aimed at supplying the tourism industry where consumption per person per day reaches 450 liters. In 2003, University of Costa Rica researchers found the Playa Tamarindo aquifer has an annual estimated recharge of 6,900,000 m³, but 6,400,000 m³ were being extracted therefore indicating that exploitation was nearing maximum capacity. Water shortages have reached the point where local residents, whose consumption is minimal, have had service cut by the Costa Rican Tourism Institute so tourists in luxury hotels have enough water for drinking, bathing and swimming (Anonymous 2008g). Developers, hoteliers and tourist complex owners are contributing to the construction of an aqueduct to solve the water shortage that is affecting their projects. Once the work is completed, the aqueduct will be handed over to AyA since the plan is to benefit all inhabitants (Anonymous 2008h).

Surface and groundwater contamination: In the NW Pacific region, other water supply threats, in addition to over-extraction of water for consumption, include fecal contamination. Nearly all the towns that have sprouted where tourists congregate, the lack public sewage systems means responsibility falls on private citizens. There are small wastewater treatment plants at inland urban areas that were sized for populations of the 1970s. There is evidence that coastal zone ground water near popular tourist beaches suffer from fecal contamination, high nitrate concentration and salt water intrusion (Calderón Sánchez et al. 2002).

The two notable wastewater contamination events during the time period of this 2007–2008 research were the closure of a large resort hotel complex, for the first time ever, by the Ministry of Health due to direct untreated wastewater discharges into estuary of the Gulf of Papagayo. As an interim measure, the hotel trucked waste to an inland area for dumping. The incident spread over weeks and was the subject of intense investigative reporting. An opinion editorial clearly outlined the importance of the environment for tourism development by stating that the environment is the essential element of the product the country offers to those who visit. It noted that principal attractions aren't colonial architecture, archaeological ruins or world-class museums. What really distinguishes Costa Rica and generates the growing tourist flow, and which has in turn emerged as the principal source of foreign exchange earnings, is nature. The environment is as much a public good as an indispensable part of the country's tourism product (Anonymous 2008b). This contamination has had other impacts on the ecology of the nearshore area. A five-year study by the environmental group MarViva showed that fertilizer and wastewater nutrient runoff from development has been causing a bloom of the algae, *Caulerpa sertularioides*, and is clouding the water and killing corals (Anonymous 2007h).

The second major incident was centered on Playa Tamarindo. An AyA study found 11 sites along the beach with excrement. Samples were collected from pipes or streamlets located very close to businesses. One of the sites showed 460,000 coliforms and another 3.1 million per 100 ml of water, confirming that the discharge pipes are dumping waste water. As an incentive for communities to organize and care for water resources they strive to be awarded the Ecological Blue Flag (BAE), a program that has been in existence since 1996. For tourists, to visit a beach with the blue flag is a guarantee of sea water quality and fresh water quality apt for human consumption. Playa Tamarindo, along with seven other beaches in the country, lost their BAE designation in 2007 (Anonymous 2008i).

NW Pacific water quality problems also spread to inland areas. The Morote River, discharging to the Gulf of Nicoya, is contaminated due to liquid leachate wastes emanating from the community of Nicoya's garbage dump as confirmed by a National Autonomous University study (Anonymous 2008j). Another example is the plan by AyA to use Lake Arenal, a 1,570 million m^3 reservoir, as a future supply of water directed to NW Pacific coast residents. Water draining into Lake Arenal has elevated levels of fecal contamination from dairies. The lake's water still is potable because the contaminants are diluted by the huge water volume. However, after water descends through ICE electrical generating plants to the Santa Rosa and Magdalena Rivers, high fecal coliform counts are detected. Officials report that septic tank trucks, not having anywhere to dump their waste, will discharge to the rivers and irrigation canals (Anonymous 2007i).

In the Central Pacific Region, microbiological analysis by AyA shows that the sea water at heavily-developed Playa Jacó is apt for swimming. However, four tributaries draining to the beach are highly contaminated with feces. In 11 years the mean at these sites has oscillated between 23,130 and 58,974 fecal coliforms. If the tendency continues, AyA predicts that between 5 and 10 years the beach will be ruined (Anonymous 2007j).

Yet another regional water quality impact is the incidents of red tide. Red tide is an increase in the concentration of marine microorganisms, dinoflagellates, that may produce toxins. Red tide incidents are often related to nutrient-rich waters being deposited by rivers into nearshore areas. The dinoflagellates are consumed by mollusks. Mollusks are filters for the toxin, storing it in great quantities. Upon consuming mollusks, intoxication may occur and be manifested by numbness in the mouth, lips and extremities (hands and feet) to vomiting, diarrhea and abdominal pain. If the intoxication is severe, death may result. In Costa Rica, the majority of the mollusks ingested that sicken humans come from the zone of Guanacaste and Tárcoles, Puntarenas (Anonymous 2006c). Tárcoles is the final repository of pollution traveling down the Virilla and Gran de Tárcoles river system draining the Central Valley to the Gulf of Nicoya. Scientists also concur in the XI State of the Nation report that heavy metals and radioactive elements dumped in the Central Valley reach the gulf which is also the largest biologically-productive area of the country. For example, radioactive cesium has been detected in the mouth of the Grande de Tárcoles River and radioactive potassium has been found in the mouth of the nearby Tivives River. Both of these elements are used to irradiate foods or medicines; their appearance in coastal waters would indicate that Central Valley industries are also dumping their radioactive wastes to the watershed (Anonymous 2006d).

7.5 CARIBBEAN COASTAL PLAIN

Surface and groundwater contamination: AyA is investing nearly $113 million to carry out various aqueduct projects in which they aim to guarantee a potable water supply for the country for the next 20 years. Work will be conducted in the port city of Limón and in the south of the province the plan is to construct a drainage system. Work began in 2008 and will last five years (Anonymous 2007b).

However, the greatest media attention toward the Caribbean Region has been directed at pineapple cultivation. Costa Rica is the world's largest exporter of pineapple. It provides 80% of all exports of the fruit to the US. The value climbed 195% in 5 years from $159 to $470 million. In 2000, there were 12,500 ha cultivated and this has risen to 40,000 ha. This puts pineapple in fourth place of 3,800 export products, only exceeded by microprocessors, bananas and serum infusion equipment. Crop expansion has occurred where former cattle pasture lands and root crops are located. There are an estimated 1,000 small and medium producers contributing 4% of total production; remaining production is among large producers. The increase in cultivation has brought with it increasing complaints of alleged water contamination and soil destruction. Complaints have brought about the prohibition of water consumption from aqueducts in several rural population centers by the Ministry of Health based on reports of the presence of Bromacil, Diuron and Triadimefon in well water. These pesticides are used on pineapple farms. At one point, 6,000 residents were being provided with potable water via cistern trucks. The Environmental Tribunal discovered that many pineapple plantations do not abide by environmental regulations. Three principal problems were found; alleged contamination of surface tributaries, invasion of riparian protected zones and changes in soil use as a consequence of cutting trees to extend areas for cultivation. Claims against the plantations have been placed by various community activist groups. As a final note, a plan for an additional 1,500 hectares of pineapple plantations on nearby mountain slopes in an aquifer recharge zone created concern that agricultural contaminants will leach into drinking water aquifers. Throughout this controversial period, concern that potable water was being contaminated with chemicals had various Limón province communities confronting pineapple companies. In response, the companies, via certified laboratory analyses, responded that none of the 21 pesticides used by growers was detectable in water (Anonymous 2008k).

7.6 PUBLIC OPINION

Costa Rican public opinion regarding the state of their surface and groundwater resources is documented among the articles used here to review management practices. The newsprint media's content is categorized by major topic and geographic region in order to focus importance on greatest public concerns.

Review of newsprint media: Concern and interest on the topic of water resources among the Costa Rican public was gauged by reviewing news media articles and focusing on the approximate time period of 2007–2008. To organize them, they were summarized among four major categories; policy, supply, contamination and ecology. Additionally, these were organized around the most impacted geographic regions; Central Valley, NW Pacific, Central Pacific and the Caribbean Drainage. The Caribbean coastal plain is contained within the Caribbean drainage and all other areas are within the Pacific drainage (Table 7.1). Of the 78 articles reviewed, over two-thirds were concerned with water contamination. Approximately one-half of the articles focused on the NW Pacific Region. Media articles focusing on the Central Valley dealt with urban development impact on water resources from runoff, lack of centralized sewage treatment and septic system failure. For the NW Pacific, although several articles covered the Tempisque basin draining to the Gulf of Nicoya and contamination of Lake Arenal, most articles focused on Playa Tamarindo and

Table 7.1 Number of newsprint media written on the topic of water resources in Costa Rica principally during 2007–2008; used to gauge public interest, concern and perception of water quality.

Topic	Central Valley	NW Pacific	Central Pacific	Caribbean Drainage
Policy	2	2	0	1
Supply	3	6	2	3
Contamination	14	31	3	7
Ecology	1	1	0	2
Total	20	40	5	13

the Gulf of Papagayo fecal coliform pollution and depletion of groundwater and aquifers due to excessive and unregulated well-drilling and extraction. The few Central Pacific articles covered water supply shortages and fecal coliform pollution as well as metal and pesticide pollution from the Grande de Tárcoles River draining the Central Valley into the Gulf of Nicoya. All articles from the Caribbean drainage region emanated from possible groundwater contamination due to agrochemicals applied to pineapple farms.

7.7 RESEARCH

Virilla River subwatershed: Fifty-six sites located on 22 streams were sampled around the eastern and central region of the Central Valley, all within the Virilla River subwatershed. These eventually drained to the Virilla River which downstream joins the Grande River to form the Grande de Tárcoles River. This severely polluted and degraded waterway discharges to the Pacific Ocean at the Gulf of Nicoya. Some sites were located in forested high-elevation regions surrounding San José. Others were located within the centers of San José and two other major Central Valley population centers, Heredia and Alajuela. Several sites were in rural areas of lower-elevation regions west of the major population centers. A sweep of these sites was conducted in the rainy season (September to November 2007) and another in the dry season (January to March 2008).

Mean dissolved oxygen (DO) levels were adequate to support aquatic fauna in all watershed regions (>8 mg/L). These DO levels were consistently higher than those observed by the author in the Yaque del Norte watershed in Dominican Republic (DR) (Phillips et al. 2007) and may be due to overall more rapid flow and flushing aspects of Central Valley streams and rivers compared to other regions of high input of raw sewage in developing countries (Thorne et al. 2000, Daniel et al. 2002). Even during the dry season, turbidity was high in the urban watershed area (122 NTU), a situation similar to that found in areas of high sediment load in Dominican Republic. Considering the absence of rainfall and consequent erosion runoff, this turbidity was possibly partially due to wastewater input. The sediment load associated with high turbidity can be severely damaging to aquatic life; clogging fish gills, smothering eggs or small benthic life stages (Davies-Colley and Smith 2001). Water pH was satisfactory in all areas (>7.8). Specific conductivity was highest in the dry season in the urban area (0.265 ms/cm), typical of contaminated water, although in areas with high wastewater input in Santiago, DR, a city of comparable size to San José, mean specific conductivity exceeded 1 ms/cm (Phillips et al 2007) and similarly high values were reported from Brazil (Daniel et al 2002). In Costa Rica, the rainy season water volumes likely dilute the constituents that contribute to the water's conductivity (0.197). Overall turbidity is higher in the rainy season.

Sampling for nutrients (nitrogen and phosphorus) was only conducted in the dry season. Background nutrient concentrations have been reported in the US for nitrogen ranging from 0.02 to 0.5 mg/L and for phosphorus from 0.006 to 0.08 mg/L (Smith et al. 2003). While Costa Rica does not have specific nutrient standards for surface waters, a general standard in the US is that total nitrogen (a measure of both inorganic and organically-bound nitrogen forms) should not exceed 1.5 mg/L. Total phosphorus should not exceed 0.06 mg/L. Mean total nitrogen in the urban watershed sites was 1.53 mg/L and for total phosphorus 0.67 mg/L. Mean nitrogen concentrations were significantly lower in all rural localities, however phosphorus concentrations were consistently high. Collectively, all these forms of nitrogen and phosphorus likely have their origin in urban areas from raw waste water. In highland areas of the upper rural watershed, nutrient origin may be from dairy farm input. Thus, although mean nutrient concentrations were the highest in the urban area, the upper rural watershed showed higher values than the lower rural watershed areas. Comparable nutrient analyses in DR showed total nitrogen concentration higher than 6 mg/L in the center of Santiago while total phosphorus only exceeded 0.2 mg/L in tributaries carrying a high wastewater load (Phillips et al. 2007).

NW Pacific: Fifteen sites were sampled during the rainy season (November 2007) along the NW Pacific coast in the area of the Gulf of Papagayo southward. All sites were in small streams discharging onto beaches. A similar sampling sweep was conducted in the dry season (March 2008) however all but four sites were completely dry.

The detection of high specific conductivity values (mean 0.427 ms/cm) as well as the high nutrient values (total nitrogen at 1.84 mg/L in the rainy season; total phosphorus at 0.19 or 0.18 mg/L dry and rainy seasons) indicates presence of contamination. This is likely due to wastewater input, a case overwhelmingly supported by newsprint media information in the same time period.

7.8 RISK ASSESSMENT

While critical of Costa Rica, is this situation really any different than from any other country? Does it manage its water resources better or worse? Can CR acquire a balance between supply and demand and environmental protection? Costa Rica has an admirable record of good governance and management of forest resources but it has not sufficiently addressed how to manage water resources. The same fresh water management challenges confound countries with the most developed economies in the world (Jackson et al. 2001, Karr 1991). Regionally across Latin America, the most water resource rich area of the world, regional shortages in supply, negative impacts from pollutants, deforestation, damming and mining are common and increasing as human population increases pressure on these resources (Hillstrom and Hillstrom 2004).

Infrastructure deficiencies: The major challenge for the country's water resources is the handling of wastewater effluent. An emerging challenge, from both the country's growing industrial activity as well as the discovery of formerly unknown foci of historical contamination, is that of organic and metal, and possibly low-level radioactive, wastes in surface, nearshore coastal and ground waters. These contamination problems are present due to near total absence of infrastructure such as comprehensive wastewater treatment plants to handle an increasing volume of waste water, a lack of clear and updated regulatory policy to guide development and a process to correct and remediate waste production and historical accumulations of contaminants (Miranda et al. 2003).

Disease and public health: In concert with the lack of infrastructure is the generalized degradation of stream and river riparian zones. Throughout the country, destabilization of stream and river banks is visually apparent. In mountainous terrain subject to high rainfall, periodic high velocity and high volume water is the norm in the rainy season. The lack of forested or, at least, diversely vegetated buffers enhances the instability of the riparian stream and river banks. Generalized deforestation over decades has exacerbated the problem. Although during the last 15 to 20 years, forest cover has significantly increased in Costa Rica, this doesn't mean specifically that stream and river edges are stabilized or stabilizing. Added to this problem, in any urbanized area, no matter how small the settlement, it is common to observe squatter dwellings on the edge of waterways; and in the metropolitan area these evolve into significantly-sized communities. Since these settlements grow outside of any type of regulatory framework and are only slowly able to acquire municipal services, the management of waste is lacking and is up to the individual. In these cases, waste water is normally diverted directly to the adjacent stream or river. In rainy season periods, flooding is common along these high velocity waterways. It is not uncommon that entire neighborhoods are destroyed, with some loss of life. In many rural areas of the country similar challenges to provide safe water is common (Welsh 2006). However, the more chronic problem is that of having created foci for disease vectors to flourish. This is recognized by the Ministry of Health but is an exceedingly difficult problem to address.

Coastal resources and ecology threat: The minimal information available reports on the damage to coastal resources such as corals and fisheries from wastewater input. The addition of nitrogen and phosphorus promotes eutrophication (Carpenter et al. 1998) along with a change in composition of marine flora and fauna. From a practical economic prospective, abundant, diverse and stable coastal resources provide fishermen and their families with a livelihood. The tourist economy requires the same.

Tourism development and the health risk to this sector: Costa Rica has developed not only ecotourism but beach tourism and is heavily promoting it (Honey 2008). The inability of beach communities and individual resort developments to manage their waste water was more than apparent and egregious during the course of this study. Costa Rica's reputation has been harmed by overdevelopment of key beach areas; it is not uncommon that visitors bathe in waters possessing high fecal coliform bacteria populations. This will clearly be a challenge in the future.

All of these present and future water resource management issues will continue to challenge the country's resources. The risk to human health and the environment will not be easily managed. The solution should require a broad-based comprehensive adaptive management plan involving all stakeholders to be successful (Gregory et al. 2006) such as large-scale environmental restoration projects progressively occurring in developed areas such as the Everglades in Florida (Redfield 2000).

7.9 CONCLUSION

This review of newsprint media, assessment of public opinion and presentation of water quality research shows the precarious and threatened state of Costa Rica's surface and groundwater resources. At this time, knowledge of the issue among government officials and the general public is pervasive but the challenges of deficient infrastructure, haphazard development, neglect and multiple demands on economic resources makes the issues, of not only correcting and remediating existing historical accumulated problems of surface and groundwater resources, but the challenge of going forward and managing water resources in a sustainable fashion, formidable. However, Costa Rica has a history of positive achievements in the greater arena of governance that would indicate that the country does certainly have the capability to address all of its environmental problems including that of sustainably managing its water resources. For example, the government's Peace with Nature initiative is a comprehensive global policy statement outlining ambitious goals for insuring perpetual safeguarding of all aspects of the nation's environment. With this clear statement, a guiding policy is now set out as a standard to follow and will hopefully guide policy-makers and be a motivation for the general public to demand sound stewardship of all the country's resources including water.

ACKNOWLEDGEMENTS

The author is grateful to Winthrop University for granting a sabbatical to conduct this research. The Department of Biology generously provided equipment and travel support. The Costa Rican Universidad Nacional Autónoma kindly provided a visiting professor appointment and the Escuela de Ciencias Biológicas provided physical space and personnel support to conduct the research. The Escuela de Ciencias Ambientales also provided personnel, laboratory and field support. Christopher Storie, Winthrop University geographer, prepared the map of Costa Rica.

REFERENCES

Anderson EP, Pringle CM, Rojas M. 2006. Transforming tropical rivers: an environmental perspective on hydropower development in Costa Rica. Aquatic Conservation: Marine and Freshwater Ecosystems 16: 679–693.

[Anonymous]. 2004a Jan 24. Plan para rescatar la cuenca del río Tárcoles [Plan to restore the Tárcoles River watershed]. La Nación (San José, Costa Rica). (On-line Ed). Spanish.

[Anonymous]. 2006a Nov 30. País recupera bosque perdido [The country recovers lost forest]. La Nación (San José, Costa Rica). (On-line Ed). Spanish.

[Anonymous]. 2006b Nov 29. Población metropolitana se duplicó en 11 años [Metropolitan population duplicated in 11 years]. La Nación (San José, Costa Rica). (On-line Ed). Spanish.

[Anonymous]. 2006c Jan 9. Persiste alerta por mareas rojas [Warning persists for red tides]. La Nación (San José, Costa Rica). (On-line Ed). Spanish.

[Anonymous]. 2006d Jan 9. Metales pesados y radioactivos amenazan biología en el Pacífico [Heavy metal and radioactive elements threaten the biology of the Pacific]. La Nación (San José, Costa Rica). (On-line Ed). Spanish.

[Anonymous]. 2007a Sep 22. Retrocede en Congreso plan para regular uso del agua [Regulatory plan for water use recedes into the background in Congress]. La Nación (San José, Costa Rica). (On-line Ed). Spanish.

[Anonymous]. 2007b Sep 22. AyA impulsa nuevas obras [AyA pushes new work]. (On-line Ed). Spanish.

[Anonymous]. 2007 c Apr 12. Agua potable en riesgo de contaminación química [Potable water at risk of chemical contamination]. La Nación (San José, Costa Rica). (On-line Ed). Spanish.

[Anonymous]. 2007d Dec 30. Millones de litros de aguas negras caen sin tratar a ríos [Millions of liters of sewer water dumped without treatment to rivers]. La Nación (San José, Costa Rica). 2007 Dec 30 (On-line Ed). Spanish.

[Anonymous]. 2007e Mar 27. Cobro por contaminar ríos lleva tres años atrasado [Fee for contaminating rivers is three years behind]. La Nación (San José, Costa Rica). (On-line Ed). Spanish.

[Anonymous]. 2007f Dec 31. Mayoría de municipios lanza la basura a cielo abierto [Majority of municipalities place garbage in open-air dumps. La Nación (San José, Costa Rica). (On-line Ed). Spanish.

[Anonymous]. 2007g Dec 25. AyA busca nuevas fuentes de agua [AyA looking for new sources of water]. La Nación (San José, Costa Rica). (On-line Ed). Spanish.

[Anonymous]. 2007h Aug 7. Pacific coral reefs endangered by algae. Tico Times (San José, Costa Rica). (On-line Ed

[Anonymous]. 2007i Oct 2. Agua de 'la bajura' depende de 'la altura' [Water in the 'lowlands' depends on the 'highlands']. La Nación (San José, Costa Rica). (On-line Ed). Spanish.

[Anonymous]. 2007j Sep 10. Playa Jacó en peligro de ser insalubre dentro de 5 años [Jacó Beach in danger of being unhealthy in 5 years]. La Nación (San José, Costa Rica). (On-line Ed). Spanish.

[Anonymous]. 2008a Jan 28. Distinción y compromiso [Distinction and commitment]. La Nación (San José, Costa Rica). (On-line Ed). Spanish.

[Anonymous]. 2008b Feb 10. El ambiente, bien turístico [The environment, a public good]. La Nación (San José, Costa Rica). (On-line Ed). Spanish.

[Anonymous]. 2008c. Normas ambientales no se adecúan al desarrollo actual [Environmental regulations don't work for current development]. La Nación (San José, Costa Rica). (On-line Ed). Spanish.

[Anonymous] 2008d. 180.000 personas reciben agua sin tratamiento de cloro [180,000 people receive water without chlorination]. La Nación (San José, Costa Rica). (On-line Ed). Spanish.

[Anonymous]. 2008e Mar 10. Vecinos de ríos expuestos a mal olor, ratas, zancudos y diarreas [Residents near rivers exposed to bad odor, rats, mosquitoes and diarrea]. La Nación (San José, Costa Rica). (On-line Ed). Spanish.

[Anonymous]. 2008f Apr 10. Decreto regulará desarrollo inmobiliario en Guanacaste [Decree will regulate building development in Guanacaste]. La Nación (San José, Costa Rica). (On-line Ed). Spanish.

[Anonymous]. 2008g Feb 8. AyA exige restringir permisos para construir pozos de agua [AyA insists upon restricting permits to dig wells]. La Nación (San José, Costa Rica). (On-line Ed). Spanish.

[Anonymous]. 2008h Feb 19. Empresarios tras dinero para acueducto en Jacó [Businessmen seeking Money for aqueduct in Jacó]. La Nación (San José, Costa Rica). (On-line Ed). Spanish.

[Anonymous]. 2008i Mar 25. Ocho playas pierden bandera azul ecológica [Eight beaches lose blue ecological flag]. La Nación (San José, Costa Rica). (On-line Ed). Spanish.

[Anonymous]. 2008j Jan 29. Nicoya sigue llena de basura [Nicoya continues to be filled with garbage]. La Nación (San José, Costa Rica). (On-line Ed). Spanish.

[Anonymous]. 2008k Apr 15. k 26 fincas piñeras en Limón [Environmental Court investigates 26 pineapple farms. La Nación (San José, Costa Rica). (On-line Ed). Spanish.

Arthington AH, Naiman RJ, McClain ME, Nilsson C. 2010. Preserving the biodiversity and ecological services of rivers: new challenges and research opportunities. Freshwater Biology 55:1–16.

Astorga, Y. 2009. Situación del recurso hídrico. Ponencia preparada para el XV Informe Estado de la Nación [State of the hydrological resource]. Programa Estado de la Nación, San José, Costa Rica.

Bernhardt ES, Palmer MA. 2007. Restoring streams in an urbanizing world. Freshwater Biology 52: 738–751.

Calderón Sánchez H, H Madrigal Solís, J Reynolds Vargas. 2002. Contaminación química y microbiológica del agua subterránea en la zona costera de Guanacaste [Chemical and microbiological contamination of ground water in the coastal zone of Guanacaste]. In: Reynolds Vargas J, editor. Manejo sostenible de las aguas subterráneas: Un reto para el futuro. Editorial UNED, Costa Rica. p. 33–48.

Carpenter SR, Caraco NF, Correll DL, Howarth RW, Sharpley AN, Smith VH. 1998. Nonpoint pollution of surface waters with phosphorus and nitrogen. Ecological Applications 8(3): 559–568.

CENAT. 2007. Diagnóstico preliminar de la calidad de las aguas superficiales de las micro-cuencas que componen la subcuenca del río Virilla [Preliminary diagnosis of surface water quality of the microwatersheds that comprise the Virilla River subwatershed], presented by UNA Laboratorio de Análisis Ambiental and Ministerio de Salud, 10 Aug 2007.

Chapagaina AK, Hoekstrab AY. 2008. The global component of freshwater demand and supply: an assessment of virtual water flows between nations as a result of trade in agricultural and industrial products. Water International 33(1): 19–32.

Craig LS, Palmer MA, Richardson DC, Filoso S, Bernhardt ES, Bledsoe BP, Doyle MW, Groffman PM, Hassett BA, Kaushal SS, Mayer PM, Smith SM, Wilcock PR. 2008. Stream restoration strategies for reducing river nitrogen loads. Frontiers in Ecology and the Environment 6: 529–538.

Daniel MHB, Montebello AA, Bernardes, MC, Ometto JPHB, DeCamargo PB, Krusche AV, Ballester MV, Victoria RL, Martinelli LA. 2002. Effects of urban sewage on dissolved oxygen, dissolved inorganic and organic carbon, and electrical conductivity of small streams along a gradient of urbanization in the Piracicaba River basin. Water, Air, and Soil Pollution 136: 189–206.

Davies-Colley RJ, Smith DG 2001. Turbidity, suspended sediment, and water clarity: A review. Journal of the American Water Resources Association 37: 1085–1101.

Emerson J, Esty DC, Levy MA, Kim CH, Mara V, de Sherbinin A, Srebotnjak T. 2010. 2010 Environmental Performance Index. New Haven (CT): Yale Center for Environmental Law and Policy.

Falkenmark M. 1990. Global water issues confronting humanity. Journal of Peace Research 27(2): 177–190.

Fisher B, Turner K, Zylstra M, Brouwer R, de Groot R, Farber S, Ferraro P, Green R, Hadley D, Harlow J, Jefferiss P, Kirkby C, Morling P, Mowatt S, Naidoo R, Paavola J, Strassburg B, Yu D, Balmford A. 2008. Ecosystem services and economic theory: Integration for policy-relevant research. Ecological Applications 18(8): 2050–2067.

Gleick PH. 1998. Water in crisis: Paths to sustainable water use. Ecological Applications 8(3): 571–579.

Gleick, PH. 2003. The World's Water 2002–2003: The Biennial Report on Freshwater Resources. Island Press, Washington, DC.

Gregory R, Ohlson D, Arvai J. 2006. Deconstructing adaptive management: Criteria for applications to environmental management. Ecological Applications 16(6): 2411–2425.

Hillstrom K, Hillstrom LC. 2004. Latin America and the Caribbean: A Continental Overview of Environmental Issues. ABC-CLIO's The World's Environment Series. Santa Barbara (CA). p 131–153.

Honey, M. 2008. Ecotourism and Sustainable Development: Who Owns Paradise? 2nd Edition. Island Press, Washington, DC.

Jackson RB, Carpenter SR, Dahm, CN, McKnight DM, Naiman RJ, Postel, SL, Running SW. 2001. Water in a changing world. Ecological Applications 11(4): 1027–1045.

Karr JR. 1991. Biological integrity: a long-neglected aspect of water resource management. Ecological Applications 1(1): 66–84.

Kosoya N, Martinez-Tuna M, Muradian R, Martinez-Aliera J. 2007. Payments for environmental services in watersheds: Insights from a comparative study of three cases in Central America. Ecological Economics 61: 446–455.

Miranda M, Porras IT, Moreno ML. 2003. The social impacts of payments for environmental services in Costa Rica. A quantitative field survey and analysis of the Virilla watershed. International Institute for Environment and Development, London.

Mora Alvarado D. 2004. Calidad microbiológica de las aguas superficiales en Costa Rica [Microbiological quality of the surface waters of Costa Rica]. Revista Costarricense de Salud Pública Rev. 13(24): unnumbered pages.

Mora Alvarado D, Portuguez CF. 2000. Diagnóstico de la cobertura y calidad del agua para consumo humano en Costa Rica a principios del año 2000 [Diagnosis of the coverage and quality of water for human consumption in Costa Rica at the beginning of the year 2000]. Revista Costarricense de Salud Pública 9(16): unnumbered pages.

Naiman RJ, Dudgeon D. 2010. Global alteration of freshwaters: influences on human and environmental well-being. Available from: SpringerLink Ecological Research DOI 10.1007/s11284-010-0693-3: http://www.springerlink.com/content/6551113837q7n725/

Nel JL, Roux DJ, Abell R, Ashton PJ, Cowling RM, Higgins JV, Thieme M, Viers JH. 2008. Progress and challenges in freshwater conservation planning. Aquatic Conservation: Marine and Freshwater Ecosystems 19(4): 474–485.

Nelson A, Chomitz KM. 2007. The forest-hydrology-poverty nexus in Central America: An heuristic analysis. Environment, Development and Sustainability 9: 369–385.

Nelson KC, Palmer MA, Pizzuto JE, Moglen GE, Angermeier PL, Hilderbrand RH, Dettinger M, Hayhoe K. 2008. Forecasting the combined effects of urbanization and climate change on stream ecosystems: from impacts to management options. Journal of Applied Ecology 46(1):154–163.

Pagiola S. 2008. Payments for environmental services in Costa Rica. Ecological Economics 65: 712–724.

Pagiola S, Arcneas A, Platais G. 2005. Can payments for environmental services help reduce poverty? An exploration of the issues and the evidence to date from Latin America. World Development 33(2): 237–253.

Pahl-Wostl C, Craps M, Dewulf A, Mostert E, Tabara D, Taillieu T. 2007. Social learning and water resources management. Ecology and Society 12(2): 5. Available from: http://www.ecologyandsociety.org/

Palmer MA, Reidy Liermann CA, Nilsson C, Flörke M, Alcamo J, Lake PS, Bond N. 2008. Climate change and the world's river basins: anticipating management options. Frontiers in Ecology and the Environment 6(2): 81–89.

Palmer, MA, Lettenmaier DP, LeRoy Poff N, Postel SL, Richter B, Warner R. 2009. Climate change and river ecosystems: Protection and adaptation options. Environmental Management 44(6): 1053–1068.

[PEN] Programa Estado de la Nación. 2009. Accessed 20 Feb 2010: http://www.estadonacion.or.cr/pdf/Resumen_armonia.pdf

Phillips P, Russell FA, Turner J. 2007. The effect of non-point source runoff and urban sewage on the Yaque del Norte River, Dominican Republic. International Journal of Environment and Pollution 31(3/4):244–266.

Postel SL. 2000. Entering an era of water scarcity: The challenges ahead. Ecological Applications 10(4): 941–948.

Redfield, GW. 2000. Ecological research for aquatic science and environmental restoration in south Florida. Ecological Applications 10(4): 990–1005.

Reynolds Vargas J, Richter DD. 1994. Nitrate in groundwaters of the Central Valley, Costa Rica. Environment International 21: 71–79.

Reynolds Vargas J, Fraile J. 2002. Presente y futuro de las aguas subterráneas en el Valle Central [Present and future of Central Valley ground water]. In: Reynolds Vargas J, editor. Manejo sostenible de las aguas subterráneas: Un reto para el futuro. Editorial UNED, Costa Rica. p 19–32.

Salas R, Bornemisza E, Zapata F, Chaves V, Rivera A. 2002. Absorción del fertilizante nitrogenado por la planta de café y su influencia sobre la contaminación de las aguas subterráneas [Absorption of nitrogen fertilizer by coffee plants and its influence on ground water contamination]. In: Reynolds Vargas J, editor. Manejo sostenible de las aguas subterráneas: Un reto para el futuro. Editorial UNED, Costa Rica. p 89–103.

Smith RA, Alexander RB, Schwarz GE. 2003. Natural background concentrations of nutrients in streams and rivers of the conterminous United States. Environmental Science and Technology 37(14): 3039–3047.

Thorne RSJ, Williams WP, Gordon C. 2000. The macroinvertebrates of a polluted stream in Ghana. Journal of Freshwater Ecology 15(2): 209–217.

[UNDP] United Nations Development Programme. 2006. Human Development Report 2006 – Beyond Scarcity: Power, Poverty and the World Water Crisis, was included in the World Human Development Report 2006 of UNPD.

Welsh K. 2006. Assessing access to potable water in rural communities in Costa Rica. Tropical Resources Bulletin 25: 66–71.

Zbinden S, Lee D. 2005. Paying for environmental services: an analysis of participation in Costa Rica's PSA program. World Development 33(2): 255–272.

Comerío reservoir, Puerto Rico (@Eddie N. Laboy-Nieves)

Chapter 8

Stormwater management practices: The case of Port-au-Prince, Haiti

Farah A. Dorval, Bernard Chocat, Evens Emmanuel and Gislain Lipeme-Kouyi

SUMMARY

Many developing countries are characterized by unstable weather situations: harsh climates with extreme seasonal and annual rainfall variations. Significant degradations have been observed in those countries, basically due to intensification of poor water management practices and an inadequate infrastructure management. Recurrent catastrophic floods have huge social and economic effects that the important investments deriving from several institutions remain insufficient. Those countries currently face persisting and new water management system challenges. The improvement of management practices in developing countries must be based on experience in the collection of long term data series, the use of extended numerical tools, and the elaboration of realistic methods. This chapter focuses on Port-au-Prince, Haiti, to illustrate hydrological processes on urban catchment based on measurements and hydrological models. Discussions highlight the importance of hydrological data collection activities which enable to further understand the catchment behavior. These measurements will help to set the relevant hydraulic design parameters and stormwater management strategy appropriate for climate changes. Emphasis is given on the need of integration of actors involved in stormwater drainage. Hence, this method would allow having an accurate idea on phenomena related to the water resources management (frequency of floods, droughts, and other weather hazards which impact urban drainage management). At last, some suggestions are presented to adapt this methodology in developping countries.

8.1 INTRODUCTION

Many developing countries are characterized by unstable weather situations and increasing urbanization (harsh climates with extreme seasonal and annual rainfall variations, anarchic construction). Those main characteristics cause significant degradations basically due to intensification of poor water management practices and inadequate disaster management infrastructure (Silveira 2002). Recurrent catastrophic floods have huge social and economic effects that the important investments deriving from several institutions remain insufficient. Those countries currently face persisting and new water management system challenges (Myrza 2003). Waterways and receiving waters nearby urban areas are often adversely affected by urban stormwater runoff. The degree and type of impact varies from a city to another one, but it is often significant relative to other sources of pollution and environmental degradation. Urbanization affects water quality, water quantity, biological resources, public health, and the aesthetic appearance of urban waterways (Figure 8.1). They have to deal with insufficient drainage network when their most important goal is to protect urban areas from flooding and pollution.

Large analytical tools have been developed, from simple conceptual models of entire system to very detailed physically based system models for complex hydraulic and physical processes in sewage systems (Ashley et al. 1998). Researches in urban hydrology field constitute to develop new concepts for urban stormwater drainage management whose main objectives are to:

– Reduce stormwater discharges and volumes (reduction and attenuation of flow within drainage system by source control techniques and use of alternatives techniques such as retention ponds, infiltration trenches, porous pavements with reservoir structures, soak away and grass filter strips or swales).
– Decrease pollutants loads which are discharged in natural environment (treatment of collected surface water before discharge).

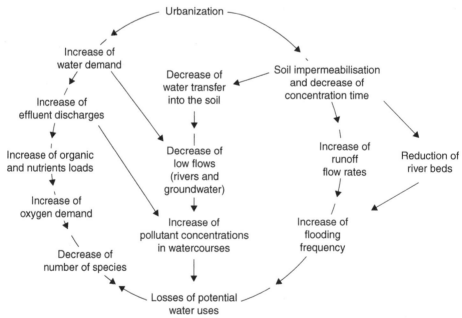

Figure 8.1 Impacts of urbanization on aquatic environments (adopted from Chocat 1997 by Bertrand-Krajewski et al. 2000).

Those technical tools and principles for urban drainage control are well known in developed countries, but their direct application in developing countries has not been successful, because they required adaptations to the condition of each country (Azzou et al. 1994). While developed countries are concerned with fines and specifics problems in stormwater management, developing countries show undeveloped drainage equipment adding to socioeconomic constraints. Due to the rapid expansion of urban areas, existing urban drainage systems become inefficient and inept solid waste management worsen complexity of the situation.

There are many difficulties to effectively implement a sustainable urban stormwater approach in developing countries. The interactions between local developments, sub-catchment drainage and natural watercourses are complex, and must be considered as a whole in order to coordinate and maximize the benefit of measures to control runoff while achieving environmentally acceptable solutions (Maskell et al. 1992). Then, the improvements of management practices must be based on learning the collection of long term data series, the use of extended numerical tools, and the elaboration of realistic methods, taking into account legal and socioeconomic conditions of urban areas (Tayler and Parkinson 2005).

Port-au-Prince, the capital city of Haiti, is considered as a typical example where drainage systems are relatively poor and hardly well-managed. The population density living in urban area related to paving brings significant changes in land surface physical properties such as decrease of infiltration. As a result of this situation we can also observe: fast runoff with high peak flow and significant pollution problems. Sediments and waste accumulated in dry weather period are carried later by rain flow and discharged in adjacent environment. The situation is aggravated because the country is located in a zone exposed to natural disasters as hurricanes and seismic activity. The country is subject to violent tropical storms events which are at the origin of several inundations (2008: Fay, Gustav, Hanna, Ike; 2004: Jeanne; 1998: Georges; 1994: Gordon; 1963: Flora; 1954: Hazel). The city is also located at twenty kilometers to the Enriquillo Fault. Over the past three centuries, Port-au-Prince had registered majors' earthquake in 1751, 1770 and recently in January 2010, a violent earthquake of magnitude 7.3 on Richler level that almost destroyed the City. Unfortunately, risk management had concentrated their efforts on preventing hurricanes than earthquakes which are frequent less. However, consequences of those natural disasters are severe and are worsened by poverty, deforestation, bad quality of constructions, political instability and low level of education.

Actually, if scientists cannot predict the occurrence of these natural disasters, they can at least help to prevent their effects by a better knowledge resulting on the physical phenomena studies and researches which allow transfer of this knowledge towards the structures in charge with planning. This chapter proposes to describe the complexity of urban drainage systems in cities of developing countries by taking as typical example Port-au-Prince city and some guidelines for a sustainable management for urban stormwater will be also presented. This information will be useful for municipalities in developing countries, such as those at Port-au-Prince, for better drainage systems management in order to protect the environment against flooding and pollution.

8.2 URBAN DRAINAGE SYSTEMS IN DEVELOPING COUNTRIES

In developing countries, urban drainage systems management becomes a big challenge for municipalities. In some cities, many efforts have been done to attenuate the peak of stormwater runoff (Parkinson and Mark 2005). In the poorest countries, the installation of drainage devices remains sometimes more and more complicated because of the their low financial and economic situations and low technical capacities of local water managers to arrive to a sustainable solution (Mechler 2004). Most of local governments in developing countries have built several programs to manage urban stormwater runoff by constructing detention ponds or infiltration trenches. Amount of money have been spent for the improvement of those programs which usually named best management practices. Sometimes, their application from developed to developing countries are suggested to major discussions related to the context of the countries (Rochon et al. 2010).

In developed countries, management practices are based on local input such as historical rainfall-runoff data, surface water data, drainage network data, performer local personnel for maintenance or management of system and technology. Drainage plans are generally insert in ambitious improvement master plan including important and expensive investment in drainage network and complex technologies (Butler and Davies 2004). Therefore, transfer of only tools as software and technology, without those local supports, often product a complete failure of primary objectives of improvement master plan in developing countries. Maksimovic et al. (1993) summarized contemporary urban drainage problems for humid tropical areas (Table 8.1). This overview is relevant for developing countries and can be assessed as a base to establish appropriate technologies that will match the local needs of each country.

For example, in Porto Alegre (Brazil), Goldenfum (2007) reported that the main difficulties identified for urban drainage were a bad balance of knowledge among the many involved actors, and the inexistence of adequate institutional arrangements (lack of integrated action among the different sectors results in inundation problems and pollution, with costs paid by the public sector). In Tanzania Gondwe (1990) noted that, the lack of co-ordination between the different authorities concerned with water management is largely to blame for failures of the water systems. Maksimovic (1993) pointed out that the lack of reliable and accurate data is often one of the major obstacles for the correct application of appropriate technologies in modern urban drainage systems.

Effective urban stormwater management cannot only employ structural practices. In Praia (Cape Verde), Sabino et al. (1999), concluded that watershed management failed due to insufficient involvement of beneficiaries during project implementation. Specific recommendations on how to take into account non-structural strategies for mitigating the effect of floods on city residents and urban infrastructure are presented in Okoko (2008). A successful program was implemented in Santo-Domingo (Chavez 2002), where an NGO built an urban drainage plan integrated approach toward local socio-economic and environmental surveys.

Opening new lines for cooperation convey and dissemination of experiences, and new knowledge gained during full-scale experiments is the first step for transferring technologies. Such transmission of management practices should be less concentrated on traditional solutions; it should promote new sustainable system

Table 8.1 Major differences between humid tropics and regions with temperate climate pertinent to urban storm drainage (Maksimovic 1993).

Differences	Effects
Meteorological conditions in large metropolitan area	• conditions for formation development and profile of storm are different
Storm characteristics	• design criteria have to be established
• Season, duration, depth, aerial distribution, temporal distribution	• response of the system different
	• hurricanes can occur
Vegetation cover and seasonal variation	• separate modules for infiltration and more rigorous calibration
Building practice retention capacities	• initial loss vary
	• surface depression different
	• surface runoff altered
Interaction with solid wastes	• carrying capacity of receiving waters reduced
	• pollution hazard increased
Existing system, inadequate or outdated	• need for integrated approach
	• low priority
Economical potentials low	• need for establishment of pilot study areas and continuous training
• lack of qualified staff and reliable data	• design is unreliable
• lack of all kinds of data	• need for systematic and organized data collection
	• appropriate methods have to be sought
Cultural, educational, traditional, hygienic conditions	• appropriate priority measures have to be invested
Uncontrolled construction, poor soil protection after deforestation cause severe sedimentation problems	• sedimentation has to be prevented
Public awareness very low priority in investment in these project low as well	• public awareness needs to be raised

solutions affordable to less developed countries, considering local traditions and participation of stakeholders in the decision-making process.

8.3 THE CRISIS OF PORT-AU-PRINCE

The metropolitan city of Port-au-Prince is the principal urban settling of Haiti, because of its population density (324 hab/km²) and urban services (IHSI 2003). For two decades, the economic crisis has had considerable impacts on urbanization and land use. This particular situation has facilitated flimsy districts emergence and population increase by rural depopulation towards the capital and the other neighboring urban centers, and considerable issues related to deterioration of sanitary infrastructure (Emmanuel 1997; OPS/OMS 1998). The City has grown at an

Figure 8.2 Structural destruction after the January 2010 earthquake.

exponential rate but without necessary funds to extend and rehabilitate existing drainage systems. The sewer network is not consistent with the extraordinary city expansion and generally stops at the first residential area. Consequently, this wild and uncontrolled urbanization has a strong pressure on the weak sewer infrastructure (Baptiste 2002).

After the earthquake in January 2010 other major's issues have been noticed and increased the existing list of problems. The poor quality of materials used in construction in Haiti had weakened the structure of the buildings making them less resistant to seism. Most of the open channels are full of debris and concretes resulting to buildings collapses (Figure 8.2A). The cracks observed in the streets may reveal underground structural damages on the drainage system (Figure 8.2B). Until now there's no official report on actual state off the drainage system of Port-au-Prince.

8.4 DRAINAGE SERVICE: CONDITIONS AND MANAGEMENT

In Port-au-Prince, flooding and pollution problems are the most imperative needs to be solved including:

– Remedial solutions for existing drainage system dysfunctions.
– Preventives measures for future trend of city expansion.

Primary methods of rainfall drainage encountered in Port-au-Prince are construction of underground sewer system or one or two open canals in one or both sides of streets. The open canals are built parallel with the streets and their slopes. This "formal" drainage system is completed by a gully network. "Formal" and "informal" drainage systems are constituted of fast transportation of polluted waters (rainwater; wastewater) directly (without any treatment) to receiving water. This kind of urban drainage management faces different levels of difficulties related in Leger (2002). We propose to resume them in two main problems: the lack of drainage infrastructure and the ineffectiveness of master plan for urban drainage.

8.5 LACK OF DRAINAGE INFRASTRUCTURE

Historically, development of the drainage infrastructures in Port-au-Prince was made in the first urbanized zone. The drainage system continues in mountain and recent urbanized zone by natural channels. As shown in Figure 8.3, the drainage infrastructures do not follow the expansion of urbanization; most of the system is almost localized only in the part of the City which corresponds to the pre 1980 urban areas.

In addition to standard and regular urbanization, sewer system could be subject to maintenance problem (mainly due to cost of rehabilitation, sedimentation or misunderstanding of their function (Figure 8.4A); with reduced sections the hydraulic capacity of the drain or the open channel decrease and cause significant

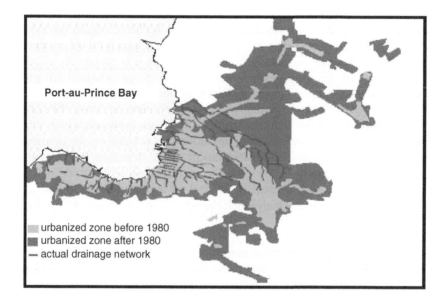

Port-au-Prince Bay

urbanized zone before 1980
urbanized zone after 1980
— actual drainage network

Figure 8.3 Port-au-Prince drainage network and urbanization expansion (Baptiste 2002).

Figure 8.4 Typical flooding and pollution problem in Port-au-Prince: (A) solid wastes and sediments disposal into open channel; (B) anarchic building on floodplains of natural storm drainage channel.

implication on the drainage system performance. Disturbance and flooding at each rainfall event became the consequences of this matter. This situation will get to worse while adding a high population density and anarchic urbanization (Figure 8.4B). Then, flooding constitutes a risk for urban population at each rainy event. The poorest habitants are the more vulnerable, they prefer the benefits to leave close employment opportunities in urban area to disadvantages related to flood risk which is often accepted as a fact of life.

8.6 ABSENCE OF MASTER PLAN FOR URBAN DRAINAGE

As the Country is prone to violent rain, the action plans by the Ministries of Public Work are reduced mainly to actions on the damages caused by the last worse weather or paralleling to actions aiming the increase of drainage infrastructures for the zones most touched by the effects of the floods. While acting on the symptoms by neglecting the fundamental causes of the problem, the interventions on the system of drainage do not have any effect on the long term. Funds continuously allocated to actions (increasing the capacity of the system, drains, or open channels cleaning) do not bring satisfactory solutions to the problems of floods which become increasingly alarming as the progression of the degradation phenomena occur on the catchment area. The resolution of these problems must pass by an approach which integrates the various aspects of the problem and must utilize knowledge on the constraints to the development of a sustainable master plan of urban drainage. In Port-au-Prince, those problems related firstly to theoretical constraints while designing new system or analyzing the existing one, because of the lack of hydrological data; and secondly to socio-economic constraints which block the operational performance of the drainage network.

- Social-economical and political constraints

Socio-economical and political constraints are the most specific and perhaps the most important point to put the emphasis in drainage project development. To be successful during the project execution, sociocultural behavior and political stability of the area need to be taken into account (Manuel-Navarrete et al. 2007). Indeed, the execution of drainage plan is broken down on the long term. However each government has unfortunately their own policies and priorities which will not be systematically followed by the next in place. Socio-economic constraints concern especially the low-income population whom sanitary issues are careless. Principal actions of this population in the drainage process during wet weather period can be resumed as followed:

- Runoff generation

The catchment area of Port-au-Prince is dealing with a serious process of deforestation (Paskett and Phyloctete 1990; Dolisca et al. 2007). Populations who are still living in wooded area are forced due to their economic situation to cut the remaining trees which are transformed into charcoal (activity which is considered as one of the last activity able to generate profits). Such behaviors only accentuated the degradations and facilitated appearance of new runoff and erosions phenomena which are more violent and very more frequent. These flows come into the network with streaming, alluvia and solid wastes.

– **Diminution of drains and channel capacity**

Solid waste management has quickly become one of the main problems on the drainage system (Figure 8.4). Solid wastes are accumulated in the streets, gutters, the river beds and even at Port-au-Prince Bay (Bouchon 2000). In one hand, they decrease the hydraulic capacity of drainage installations and, on the other hand they block the inlet to the storm water drainage system (Figure 8.5B) and thus accentuate the risk of flooding (Bras et al. 2009).

– **Vandalism**

Poverty remained as the main cause of vandalisms registered on the drainage network accessories (Figure 8.5A). The drainage devices such as manhole cover and grids are in most cases removed and used as construction materials. These inlets without covers constitute the entrance point in the network for solid wastes and sediments which can completely block water circulation (Figure 8.5B). As a result of this phenomenon, the network cannot fulfill any more the functions for which it was built.

Public administrators often have to handle decisional problems about urban hydrology design and analysis consideration which require for their solution the evaluation and the correlation of complex and large data about set. Among these data, the most relevant are climatologic (rainfall hazards) and hydrological conditions (catchment area, housing density, drainage system, groundwater table).

Such planning must be based on local data. Thus, gathering of reliable and adequate hydrological data is an important task of urban hydrologists. The return frequency of flooding is a crucial parameter which influences the design of urban drainage system and it is relies directly to rainfall data along with watershed

Figure 8.5 (A) manhole without cover; (B) blocked inlet to the stormwater drainage system.

characteristics (intensity and duration). Rain data requirements for urban applications need a high temporal and spatial resolution. Ideally a long data series (at least two times longer than the return period of flooding for which drainage infrastructure had been designed) with 1 min time resolution, $1 km^2$ spatial resolution, and time synchronization errors below 1 min should be available. For some applications and for regular climate these requirements might be relaxed (Shilling 1991; Chocat 1997; Berne et al. 2004).

In Port-au-Prince, hydrological data are mainly quantitative such as rainfall, level and discharge; information about water quality are rare and there are no records or observed performances of the drainage system in term of flooding experienced. Due to the anarchic urbanization there are no archives or maps describing the construction of building or infrastructure. A good knowledge of base describing land use and topography is fundamental for the computation of runoff and to facilitate the evaluation of the existing drainage system and to propose designs or alternatives for new system. Without this, investments are unsustainable and the benefits illusory. Unfortunately, most of the hydraulic infrastructure was designed using hydrologic design parameters that were established some decades ago, based on assumptions about climatic and geographical conditions that are now out of date. This scheme produces erratic values for design of drainage system and sometimes causes some great damage to other structures rely to the catchment (Figure 8.6).

This deficiency of hydrologic data is mainly due to:

– lack of funds for hydrologic records in urban areas,
– lack of habilitated person (in the administration responsible of drainage management) to deal with hydrologic network, and
– the rapid change of the environment which required continuous update of data base.

However, efforts have to be done in the research sector in order to provided deciders on data reliable to actual and future risk for urban drainage system performance.

Figure 8.6 Impact of flooding in urban catchment: (A) bridge removed due to high peak of flow; (B) street flooding.

8.7 RISK MANAGEMENT

The first step in the development for a storm drainage plan consists on the preparation of a project base map. In order to develop solutions more effectively, global and regional pressures must be recognised and used to drive the design and management processes of urban drainage systems. Therefore, a lot of research and development is still necessary to develop practical approaches, local understanding of the dominant physical processes and alternatives to unquestioning transfer of technologies and design parameters (Parkinson and Mark 2005). One good example from a facility in a developed country is the Experimental Observatory for Urban Hydrology Program in Lyon (France). It was built by a partnership of a multidisciplinary team representing universities, research laboratories, engineers, and the community (OTHU 2009). The Program is based on long term approaches to understand and interconnect natural sciences with hydraulics, social science and economy. The main objectives are to provide results, better knowledge and methodologies to assess sustainability of urban water system such as modeling, decision-making tools, definition of objectives, and metrology. Information is taken from the catchments by the implementation of *in situ* continuous measurements. Reliable knowledge about the short-and long-term behavior of urban water systems is necessary to evaluate indicators and criteria which will be used in various methodologies for assessing sustainability (Bertrand-Krajewski et al. 2000).

From this base, a planning process in developing countries should identify the watershed areas and subareas, the land use and cover types, the soil types, the existing drainage patterns, the topographic features, the measure and understanding of drainage system trough monitoring and applied research (Figure 8.7). This database will then be supplemented with environmental factors, economic factors, legislation and regulation of investments from government and nongovernment sector. This integrated approach will involves identifying the main complex problems which will be eventually decomposed into a series of simpler minor problems. They will could be solved sequentially in short and long term issue by an optimization technique consist to determine optimum solution for a given set of decisions required for each goal identified.

The frame proposed and described in Figure 8.7 is an useful integrated development plan applicable for urban drainage management in Port-au-Prince. Beside the fact that priorities is now accentuated on providing basic needs, questions on how to rebuild sustainable the city should be pose. Lessons have to be learned from passed errors and future urban network must be designed by tacking into account the inevitability of strong hurricanes or earthquakes. It's the chance to build durable and sustainable.

The experience of Port-au-Prince with natural disasters reveals a need of better understanding of the vulnerability of Haiti to be devastated by a plethora of hazards. Hence, emphasis should be placed on monitoring and data collection on natural and anthropogenic phenomena that inflict the Haitian natural environment and the welfare of residents. Solutions are available, but depend either on rigorous observations or qualitative evaluations to clear the presently increasing uncertainties of urban drainage management. Changes should not be limited to applied technologies but also in education systems, aid programs, social habits, policies, and willingness

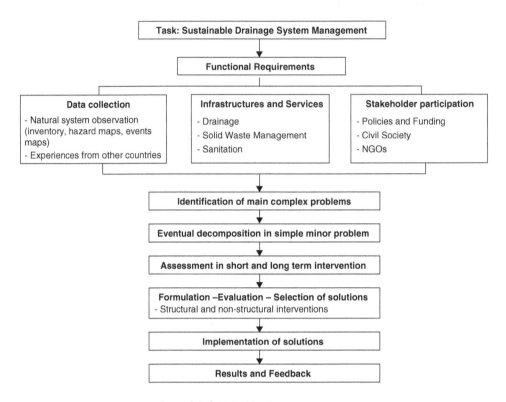

Figure 8.7 Sustainable plan management.

to structure and manage a sustainable Caribbean society. This could be a revolution that could take decades, because policies to obtain national or international funds to invest in short term goals focus on repairing damages, but not on preventing them. After the earthquake, apparently new pathways are being traced to invest in information which could give support to better projects formulations and management decisions. Holistic management ought to integrate other urban services involved in the perennial drainage system and to consider local population needs and its social economical conditions.

8.8 CONCLUSION

This chapter has highlighted the problem of urban drainage in developing countries with an evaluation of the situation of Port-au-Prince. It appears that in developing countries the failing of drainage systems is mainly due to a top-down approach with limited involvements of stakeholders and to unknowing the origin and cause of flooding. Traditional approaches limited to upgrading or rehabilitating urban drainage by simply increasing the capacity of the system, show inadequacy and have to be changed. Global and regional pressures must be recognized and used to drive the design and management processes of urban water systems as well as hydrological parameters. Coordination should be made with regulatory agencies or others that have interests in drainage matters. Actor's involvement may come from any level of government (federal, state, or local). However the ultimate responsibility of properly maintaining the drainage systems rests with the developing countries themselves.

REFERENCES

Ashley R, Hvitved-Jacobsen T, Bertrand-Krajewski JL. 1999. *Quo vadis* sewer process modelling. Water Science and Technology 39(9): 9–22.

Azzou Y, Alfakih E, Barraud S, Cres FN. 1994. Techniques Alternatives d'assainissement pluvial. Choix, conception, réalisation et entretien (Alternative techniques in urban storm drainage. Choice, design, realization and maintenance). Paris: Tec & Doc Lavoisier. 371 p.

Baptiste M. 2002. Development of a Drainage Network in Port-au-Prince [Internet]. Washington (DC). [cited 2010 Apr. 27] ; 3 p. Available from: http://www.aag.org/sustainable/gallery/projects/baptist.pdf.

Berne A, Delrieu G, Creutin JD, Obled C. 2004. Temporal and spatial resolution of rainfall measurements required for urban hydrology. Journal of Hydrology 209: 166–179.

Bertrand-Krajewski JL, Barraud S, Chocat B. 2000. Need for improved methodologies and measurements for sustainable management of urban water systems. Environmental Impact Assessment Review 20(3): 323–331.

Bouchon C. 2000. Diagnostic écologique des écosystèmes et des ressources marines côtières de la République d'Haïti, la région de Port-au-Prince à Saint-Marc (Ecological diagnosis of the ecosystems and the coastal marine resources of the Republic of Haiti, Saint-Marc: area of Port-au-Prince). From the Center of Studies Applied to the Antilles and Guyana Natural Environments. Report. S.L. : S.N. 20 p.

Bras A, Berdier C, Emmanuel E, Zimmerman M. 2009. Problems and current practices of solid waste management in Port-au-Prince (Haiti). Waste Management 29: 2907–2910.

Butler D, Davies JW. 2004. Urban drainage, second edition. London: Spon Press. 549 p.

Chavez R. 2002. Barrios el café, la mina and hermanas Mirabal. In Santo Domingo: A best practice in urban environmental rehabilitation. Washington, DC: World Bank Thematic Group on Services to the Urban Poor. Urban notes on Upgrading Experiences No 4.

Chocat B. 1997. Encyclopédie de l'Hydrologie Urbaine et de l'Assainissement (Encyclopedia of Urban Hydrology and Urban Drainage). Paris : Tec & Doc Lavoisier. 1124 p.

Dolisca F, McDaniel J, Teeter LD, Jolly CM. 2007. Land tenure, population pressure, and deforestation in Haiti: the case of Forêt des Pins Reserve. Journal of Forest Economics 13: 277–289.

Emmanuel E. 1997. Water in Haiti, resources and management: market imperfections and distorsions. Proceedings of the Water Environment Federation 70th Annual Conference and Exposition; October 18–22; Chicago, Illinois U.S.A. p. 393–406.

Gondwe E. 1990. Water management in urban areas of a developing country. Proceedings of the Hydrological Processes and Water Management in Urban Areas; April; Duisberg, Germany. p. 337–340.

Goldenfum JA, Tassi R, Meller A, Allasia DG, Da Silveira AL. 2007. Challenges for the sustainable urban stormwater management in developing countries: from basic education to technical and institutional issues. Proceeding in 6th International Conference on Sustainable Techniques and Strategies for Urban Water Management Novatech: 25–28 june; Lyon, France. p. 357–364.

[IHSI] Institut Haïtien de Statistique et d'Informatique (Haiti). 2003. Enquête sur les conditions de vie en Haïti (Inquire into the living conditions in Haiti) (ECVH-2001), Port-au-Prince: Vol. 1. 640 p.

Léger RJ. 2002. Shared management of the infrastructures of drainage and sewerage: Case of the upstream/downstream municipalities (Pétion-Ville/Port-au-Prince). Proceeding in International Conference on integrated water management in Haiti; 26–28 june; Port-au-Prince, Haiti. p. 136–152.

Manuel-Navarrete D, Gómez JJ, Gallopín G. 2007. Syndromes of sustainability of development for assessing the vulnerability of coupled human–environmental systems: The case of hydrometeorological disasters in Central America and the Caribbean. Global Environmental Change 17: 207–217.

Maksimovic C, Todorovic Z, Braga BPF. 1993. Urban drainage problems in the humid tropics. Proceedings of the International Conference on Hydrology of Warm Humid Regions; july; Yokohama, Japan. p 377–401.

Maskell AD, Sheriff JDF, Leonard OJ. 1992. Scope for control of urban runoff. London. General Technical Reports N:123–124.

Mechler R. 2004. Natural disaster risk management and financing disaster losses in developing countries. Germany: Verlag Versicherungswirtschaft. 312 p.

Myrza MMQ. 2003. Climate change and extreme weather events: can developing countries adapt? Climate Policy 3: 233–248.

Okoko E. 2008. The urban stormwater crisis and the way out: Empirical evidences from Ondo Town, Nigeria. Medwell Journals, The Social Sciences 3(2): 148–156.

PAHO/WHO. Pan American Health Organization of the World Health Organization. 1998. Analyse du secteur Eau Potable et Assainissement. Haiti: Comité National Interministériel (Analyzes of drinking water and sanitation sectors. Haiti: Interdepartmental National Committee). General Technical Report No. 21. 12 p.

OTHU. Research Field Observatory in Urban drainage. [Internet]. [updated 12 03 2009]. France; [cited 2010 04 27]. Available from: http://www.graie.org/othu/.

Parkison J, Mark O. 2005. Urban stormwater management in developing countries. London: IWA. 218 p.

Paskett CJ, Philoctete CE. 1990. Soil conservation in Haiti. Journal of Soil and Water Conservation 45(4): 457–459.

Rochon GL, Niyogi D, Fall S, Quansah JE, Biehl L, Araya B, Maringanti C, Valcarcel AT, Rakotomalala L, Rochon HS, Mbongo BH, Thiam T. 2010. Best management practices for corporate, academic and governmental transfer of sustainable technologies to developing countries. Clean Technologies and Environmental Policy 12: 19–30.

Sabino AA, Querido AL, Sousa MI. 1999. Flood management in Cape Verde: the case study of Praia. Urban Water 1(2): 161–166.

Schilling, W. 1991. Rainfall data for urban hydrology: what do we need? Atmospheric Research 27(1–3): 5–21.

Silveira ALL. 2002. Problems of modern urban drainage developing countries. Water Science and Technology 45(7): 31–40.

Tayler K, Parkinson J. 2005. Strategic planning for urban sanitation – a 21st century development priority? Journal of Water Policy 7: 569–580.

Chapter 9

Impact of surface runoff on the aquifers of Port-au-Prince, Haiti

Urbain Fifi, Thierry Winiarski and Evens Emmanuel

SUMMARY

The infiltration of urban stormwater runoff is considered as one of the main factors of the deterioration of soils and groundwater quality. This situation is very critical in developing countries where urban surface runoff is seriously contaminated by organic and inorganic pollutants. Informations, reported in the literature, showed that the subjacent groundwater of several cities in developing countries is extremely contaminated by pollutants, particularly pathogenic microorganisms and heavy metals. In addition to metals or microbiological contaminations, groundwater from coastal aquifers of developing countries is also exposed to seawater contamination. This situation can not merely lead, in term of water quality, to the scarcity of groundwater resources, but also to risks for human health consumers. Indeed, in developing countries, drinking water supply by the groundwater resources is generally privileged to catchment site of water surface. In the specific geographic context of developing countries, there is an important challenge of public health related to the possibility of the appearance of chemical risks due to the contamination of groundwater resources by chemical substances present in urban surface runoff. In order to understand and to manage these risks, this chapter aims to examine groundwater problems in developing countries related to the infiltration of surface runoff, by analyzing the plain of Cul-de-sac aquifer which contributes approximately to 50% of the water supply of urban community of Port-au-Prince, Haiti.

9.1 INTRODUCTION

Environmental management in developing countries has become today a great challenge for municipalities. Cities are generally characterized by a considerable urban explosion and an accumulation of socio-economical and political problems. Their urban area offers the best social security benefits, but paradoxically, it is the enclave for unhealthy conditions and social effervescence. Developing countries know the meaning of a sprawl resulting by increasing inner cities called "shanty town". This phenomenon is the product of several combined factors: increasing demography, accelerated and unplanned urbanization and property of urban population fringe. Consequently, poor and hardly well-managed stormwater drainage systems and pollution sources have become a major environmental challenge for these countries. Stormwater are generally mixed with others wastewater such as domestic wastewater, leachates, waste oils and hospital effluents. These point sources containing biodegradable organic matter, inorganic and organic chemicals, toxic substances and disease causing agents are frequently discharged into soils, oceans, rivers, lakes, or wetlands without treatment. Technical alternatives such as infiltration basins, retention ponds, porous pavements with reservoir structures, soakaways and grass filter strips or swales are practically non-existent. However, urban stormwater infiltration is considered as one of the main factors of the deterioration of soils and groundwater quality.

Approximately 35% of the world population depends on groundwater resources (600 to 700 km^3 per year), mostly from shallow aquifers to supply their water needs (UNEP 2002). Although groundwater can be distributed with little treatment, the degradation of its quality is a function of wastewater discharging and the overexploitation of this resource. In developing countries, inhabitants use wells or sources in contact with groundwater without any perimeter of protection, as their only means of drinking water supply. The consumption of these waters represents a sanitary risk for urban population with the appearance of epidemic pathogen that typically affects children at lower ages. Every year, more than five million deaths are recorded in developing countries due to diseases from water consumption (UNEP 2002). Nearly 25 to 35% of cholera cases are attributed to environmental factors (Smith et al. 1999). In developing countries, this percentage is about 18% of which 7% would be attributed to the insufficiency of the network water supply, 4% to air pollution inside residences, 3% to vectors of diseases, 2% to urban atmospheric pollution and 1% to agro-industrial waste (UNEP 2002). In sub-Saharan Africa, the proportion is even higher (26.5%), mainly because of insufficiency and purification of water supply consumption and others vectors of diseases (UNEP 2002).

In addition to bacterial or salt pollutions, concentrations in various pollutants (heavy metals, organic compounds) have been detected in urban groundwater from developing countries: Madras in India (Howard and Beck 1993), Cairo in Egypt (Soltan 1998), and Port-au-Prince in Haiti (Emmanuel et al. 2009b). This chapter aims to examine groundwater problems in developing countries related to the infiltration of surface runoff, by analyzing the plain of Cul-de-sac aquifer which contributes approximately to 50% of the water supply of urban community of Port-au-Prince, Haiti.

9.2 SURFACE RUNOFF GENERATION

Rainwater, throughout its way from the atmosphere to its release into urban drainage systems, washes the urban surface (roofs, roads, parks) which it crosses. When it arrives on the city, waters is mixed to dried wastewater and loaded with pollutants, and become *"urban stormwater runoff"*. A part of these waters goes to rivers and the sea, while the other remains in soils and infiltrates gradually to groundwater. The Figure 9.1 illustrates the mechanisms of surface runoff production in developing countries.

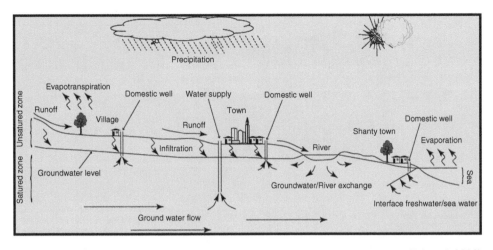

Figure 9.1 Schematic diagram of the urban hydrologic cycle in developing countries (Fifi et al. 2008).

Runoff production mechanisms are influenced by several factors such as: the condition of the soil surface and its vegetative cover, the soil texture, and the antecedent soil moisture content (Manfreda 2008). These characteristics of soil are usually modified by urbanization. This phenomenon substantially increases the volume and peak flow of surface runoff, by creating more impervious surfaces such as pavement and buildings that do not allow percolation of the water down through the soil to the aquifer. The cumulative effect of increased runoff volumes and localized peaks then contributes to flooding of downstream areas. In developing countries, flooding usually cause various devastating damages and a degradable life framework permanently in cities. For example, in the Asia-Pacific region, from 1991 to 2000, natural disasters caused more than 550,000 deaths which represent 83% of the world total (IFRC 2001), particularly by floods in Bangladesh, China and India (Mirza and Ericksen 1996). From 1971 to 2000, China reported more than 300 natural disasters with a death toll exceeding 311,000; India about 300 disasters (~120,000 deaths); Philippines 300 (34,000 deaths); Indonesia 200 (~15,000 deaths); and Bangladesh 181 disasters with more than 250,000 deaths (UNEP 2002). By 2004, the cities of Gonaïves and Port-de-Paix (Haiti) were flooded by a tropical storm, causing 2,000 deaths and more than 100,000 people left homeless (PAHO 2007). Frequent flooding is generally accompanied by violent epidemics associated to polluted water consumption and others vectors of diseases (Kim 2000).

9.2.1 Sources and loads of pollutants

Excessive pollutant loading from storm runoff has been identified as an important component of nonpoint source pollution in urban areas in both developed and developing countries (Eriksson et al. 2007; Zhang et al. 2009). The pollutant loading varies with land uses, human activities inside cities, automobile emissions, solid wastes incineration, and chemical incomes from industrial processes. Although developing countries are characterized by a lower industrialization, pollution from

Figure 9.2 Scenarios of wastewater and solid wastes discharging in Haiti. (a) wastewater stagnation in Cité Soleil (Smeets 2008); (b) paint effluents discharging containing heavy metals to Bizoton Stream (Joseph 2009); (c) wastewater stagnation and solid waste deposit on a street of Port-au-Prince (Coiffier and Théodat 2005a); and (d) obstruction of a stormwater drainage system by solid wastes (Coiffier and Théodat 2005b).

urban activities is aggravated by the poor solid waste and wastewater management and city planning, as illustrated in Figure 9.2 for Port-au-Prince, Haiti.

Urban runoff pollutants are many and varied depending on the land uses and pollutant sources present in an urban area (Table 9.1). Typically loadings of urban pollutants are greatest from industrial and commercial areas, roads and freeways, and higher density residential areas. Although sources of specific pollutants may vary widely in urban areas, motor vehicles are recognized to be a major source of pollutants, contributing oils, greases, hydrocarbons, and toxic metals (Jang et al. 2010). However, there is a general assumption that the major sources of these contaminants particularly metals, are known although the actual proportions contributed by each source to total catchment stormwater loads are uncertain.

In developed countries, many studies are focused to urban stormwater characterization, for instance the USA Nationwide Urban Runoff Program (USEPA 1983); french campaigns (Deutsch and Hémain 1984); CIPEL Studies in Switzerland (Rossi et al. 1996); OTHU projects at Lyon (Graie 2005); OPUR studies at Paris (Kafi-Benyahia et al. 2005); and Fuchs et al. (2004) studies in Germany. However, the implementation of a characterization or data acquisition systems for urban stormwater remains also very complex in developing countries. Wondimu (2000) discussed that there are two main factors to explain this complexity: (i) available means badly managed: corruption, tracing of models and parachuted methods and, (ii) political and institutional

Table 9.1 Pollution sources of urban stormwater runoff (USEPA 1999).

Contaminant	Sources
Sediment	Streets , lawns, driveways, roads, construction activities, atmospheric deposition, drainage channel erosion
Pesticides and herbicides	Residential lawns and gardens, roadsides, utility right-of-ways, commercial and industrial landscaped areas, soil wash-off
Organic materials	Residential lawns and gardens, commercial landscaping, animal wastes
Metals	Automobiles, bridges, atmospheric deposition, industrial areas, soil erosion, corroding metal surfaces, combustion processes
Oil, grease and hydrocarbons	Roads, driveways, parking lots, vehicle maintenance areas, gas stations, illicit dumping to storm drains
Bacteria and viruses	Lawns, roads, leaky sanitary sewed lines, sanitary sewer cross-connections, animal waste, septic systems
Nitrogen and phosphorus	Lawn fertilizers, atmospheric deposition, automobile exhaust, soil erosion, animal waste, detergents

Table 9.2 Heavy metals concentrations range from urban stormwater in some developing countries.

Country	Parameter				
	pH	Pb (mg/l)	Cu (mg/l)	Cd (mg/l)	Zn (mg/l)
India[a]	[5.8–6.5]	[0.022–0.067]	[0.009–0.116]	[0.0012–0.0024]	[0.006–0.151]
China	[6.7–7.6][d]	[0.003–0.286][b]	[0–0.005][b]	[0.003–0.023][b]	[0.008–0.185][c]
		[0.001–0.015][c]	[0.0014–0.025][c]		
Iran[d]	[6.9–7.6]	[0.018–0.558]	–	–	[0.015–2.386]
Egypt[e]	–	[[0.17–0.35]	[0.26–2.37]	[1.9–5.9]	[0.35–2.3]
Haiti (Port-au-Prince)	[6.8–7.1][f]	[0.20–0.25][f] [> 1.67][g]	[0.07–0.14][f]	[0.0015–0.0016][f] [0.011][h]	[0.44–0.63][f]
		[0.012–0.015][h]			

[a] (Rattan et al. 2005)
[b] (Zhu et al. 2004)
[c] (Huang et al. 2007)
[d] (Taebia and Droste 2004)
[e] (Hamad 1993)
[f] (Angerville 2009)
[g] (Carré 1997)
[h] (Emmanuel et al. 2009a)

instability, combined with financial constraints and the poor scientific and technical capacity of many local actors. Studies from some developing countries focused mainly on the heavy metal determination in waters from urban sewer systems (Table 9.2).

9.2.2 Surface runoff infiltration and its impacts on aquifers

The principal environmental issues associated with runoff are the impacts to surface water, groundwater and soil through transport of water pollutants to these systems. Ultimately these consequences translate into human health risk, ecosystem disturbance and aesthetic impact to water resources. In developing countries, characterized by a high population density and insufficiency or malfunctioning drainage systems, subjacent groundwater reservoirs may be affected by surface runoff infiltration. According to Foppen (2002), in poor urban areas where drainage systems are not enough developed, the risks of groundwater pollution by wastewater is crucial. In mostly developing countries, wastewater is usually a mix of domestic and industrial wastewater and stormwater (Qadir et al. 2010) which constitutes main sources of groundwater recharge (Figure 9.3). This recharge is also important to replenish groundwater supplies that we use for drinking, farming, industry, horticulture, and irrigating gardens and open spaces. However, the sources and the pathways for groundwater recharge in urban areas are more numerous and complex than in rural environments. The recharge of rural aquifers is accomplished by precipitation, rivers and others surface water bodies, inter-aquifers flows, and irrigation (Lerner 1990). In some urban areas, the increase of groundwater recharge depends generally on three primary sources: stormwater, wastewater and water supply and sewer leakage (Lerner 2002).

Wastewater and stormwater infiltration (Figure 9.3) are considered not only the main sources of recharge of aquifers but may be constributed also to the deterioration

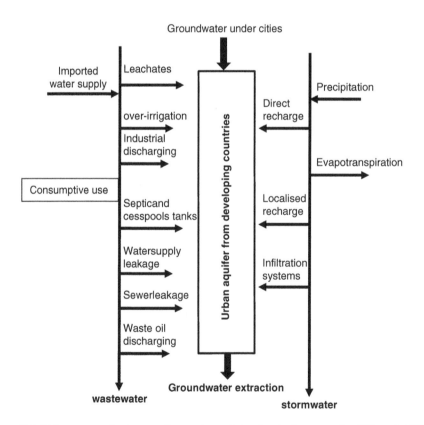

Figure 9.3 Main sources of urban aquifers recharge from developing countries (Fifi et al. 2010).

of groundwater quality. For example, in Bilawayo (Zimbabwe), it was shown that 27% of wells were contaminated with coliform bacteria, which was thought to be caused by leaking sewers (Mangore and Taigbenu 2004). Poorly-constructed sanitary well seals have been suggested as the cause of elevated pathogen contamination in the Patiño Aquifer in Asuncion, Paraguay, where 70% of drilled wells are contaminated with fecal coliforms (Hirata et al. 2002). Numerous studies have also found wastewater infiltration to be relatively free of chemical contaminants such as heavy metals. High concentrations of these pollutants have been detected in some urban groundwater samples in developing countries (Table 9.3). As showed in Table 9.3, the maximum concentrations of lead and cadmium measured in groundwater from urban areas are higher than threshold values fixed by the World Health Organization (WHO 2004). Higher heavy metal concentrations are detected in groundwater samples collected nearby industrial installations and landfill sites. However, heavy metals are generally considered as a threat toward human health and ecosystems because of their potentially high toxicity (Qin et al. 2006). Some of them are toxic even if their concentration is very low and their toxicity increases with accumulation in water and soils (Bhattacharyya and Gupta 2007). Canfield et al. (2003) found a decrease of 4.6 intelligence quotient (IQ) points corresponding to every 10 μg/dl increase in blood Pb level. Lanphear et al. (2000) also found that deficits in cognitive and academic skills associated with lead exposure occurred at blood lead concentrations lower than 5 μg/dl.

9.3 CASE STUDY FROM PORT-AU-PRINCE, HAITI

Located between 18° and 20°6′ Northern latitude and between 71°20′ and 74°30′ Western longitude, Haiti divides with Dominican Republic "the island of Hispaniola" which is the second biggest island of the Caribbean. By 2005, its total population was estimated at 8,763,588 inhabitants (42.2% in urban area) and its population density was 324 inhabitants per Km^2 (IHSI 2003). Port-au-Prince accounts 37% of the total population of Haiti that is approximately (IHSI 2003). This city was settled at the bottom of the Gulf of "La Gonâve", in the south border of Plain of Cul-de-sac and in the north catchment area of the "Massif de la Selle" piedmont. The main municipalities which constitute urban community of Port-au-Prince are: Port-au-Prince, Delmas, Pétion-ville, Croix-des-bouquets, Gressier and Carrefour.

Haiti is exposed to a considerable ecological imbalance, characterized by catastrophic flooding associated to torrential rains and hurricanes, devastating earthquakes, and deforestation. Other problems, resulting from this imbalance include: land use forming the immediate perimeter of headwaters and wells, wetlands draining, arable soils erosion, the decrease of the headwaters flow and groundwater, seawater intrusion, sewers obstruction and fecal pollution (Saade 2006). However, Haiti has a considerable water potential of which only approximately 10% of water resources available are exploited. The aquifers potential is estimated at approximately 56 billion m^3, distributed in 48 billion m^3 of continuous aquifers, the remainder being discontinuous (Saade 2006). Aquifers exploitation is currently very weak, except at the plain of Cul-de-sac aquifer which is overexploitated (Figure 9.4). Then, Port-au-Prince city is not exposed yet to the situation known as "water stress" because the resource is still available but it is badly managed and exploited. In the

Table 9.3 Heavy metals concentrations range detected from urban groundwater in developing countries.

City	Country	Pb (µg/l)	Cu (µg/l)	Cd (µg/l)	Zn (µg/l)	References
Hyderabad	India	[20.4–82.3]	[6.2–13.0]	[0.1–0.4]	[19.9–171.7]	(Satyanarayanan et al. 2007)
Madras	India	[1.7–25.4]	[6.3–170.2]	[0.01–4.8]	[19.6–6179.6]	(Ramesh et al. 1995)[a]
North India	India	[8.0–106.4]	[1.0–68.0]	[0.0–5.0]	[5.0–5500.0]	(Singh et al. 2005)[b]
Western Delhi	India	[22.0–41.0]	[7.0–9.0]	[1.1–1.8]	[3.0–38.0]	(Rattan et al. 2005)
Northern Jordan	Jordan	[0.0–42.0]	–	[0.0–6.0]	[0.0–34.0]	(Abu-Rukah and Al-Kofahi 2001)[c]
Khozestan	Iran	–	[3.0–1200.0]	[0.0–150.0]	[65.0–510.0]	(Nouri et al. 2006)
Tiaret	Algeria	[30.0–340.0]	[30.0–150.0]	–	[530.0–2016.0]	(Mokhtaria et al. 2007)[d]
Mohammedia	Morocco	[5.4–23.3]	[1.3–29.2]	[0.1–2.1]	[156.3–256.3]	(Sherghini et al. 2003)
Epworth	Zimbabwe	–	[10.0–610.3]	–	[30.0–4320.0]	(Zingoni et al. 2005)
The valley of Rift	Ethiopia	[0.023–46.0]	[0.26–27.0]	[0.0–6.41]	[1.1–5140.0]	(Reimann et al. 2003)
Vihovići	Bosnia	[6.0–8.0]	[70.0–160.0]	[16.0–42.0]	[50.0–200.0]	(Calò and Parise 2009)
Hattar	Pakistan	[1.0–2340.0]	–	[1.0–210.0]	[7.0–1340.0]	(Manzoor et al. 2006)[e]
Ibb	Yemen	[142.0–283.0]	[107.0–9611.0]	[9.5–189.0]	[85–2550.0]	(Al Sabahi et al. 2009)[f]
Southern Nigeria	Nigeria	[0.61–14.3]	[0.23–53.7]	[0.06–1.07]	[8.60–1653]	(Asubiojo et al. 1997)
Benin	Nigeria	[30.0–200.0]	–	[20.0–230.0]	[980.0–8500.0]	(Erah et al. 2002)
Aswan and Kom Ombo	Egypt	[11.0–29.0]	[11.0–29.0]	–	[200.0–1400.0]	(Soltan 1998)[g]
Port-au-Prince	Haiti	[10.0–40.0]	–	–	–	(Emmanuel et al. 2009b)
WHO* threshold values		10	2000	3	3000	(WHO 2004)

a Concentrations measured on monsoon summer
b alluvial aquifer of Gangetic plain, North India
c groundwater under El-Akader landfill site–north Jordan
d groundwater under landfill site of Tiaret
e Trace metals from thee textile effluents site in relation to soil and groundwater
f Groundwater under the landfill site of the Ibb city (Yemen)
g Samples from 10 wells along the distance between Aswan and Kom Ombo cities (Egypt)
* World Health Organization

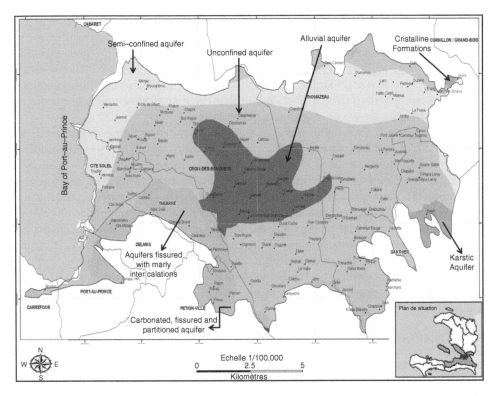

Figure 9.4 Aquifers systems of plain of Cul-de-sac (CNIGS conception–modified by Fifi et al. 2010).

30 next years, it is possible that urban pollution caused by the human activities will oblige water managers to give up currently exploited headwaters.

9.3.1 Groundwater contamination

Groundwater resources at Port-au-Prince are vulnerable to contamination related to polluted water infiltration such as leachates, cesspools and septic tanks, stormwater runoff, waste oil discharging, over-irrigation and industrial discharging (refer to Figure 9.3). These sources of groundwater recharge may contain organic and inorganic compounds which can be in dissolved and colloidal forms or associated to particles. Previous researches showed an impact of waters quality due to urban contaminants. The presence of lead (up to 1670 μg/l) has been reported were measured in wastewater discharged by paint factories in Port-au-Prince (Carré 1997). Lead concentrations ranging from 40 μg/l to 90 μg/l were measured in the drinking water of Port-au-Prince (Emmanuel et al. 2009b). Some concentrations in chloroform (118 μg/l); dibromochloromethan (118 μg/l) and faecal coliforms (700 NPP/100 ml) have been also detected in some wells from plain of Cul-de-sac aquifer (Emmanuel et al. 2009b). These values are largely higher than threshold values recommended for drinking water (WHO 2004). Therefore, Fifi et al. (2009) studied three sites at Cul-de-sac plain (refer to Figure 9.4) to assess the soil reactivity towards heavy metals. In their study, simple mathematical relationships were carried out afer equilibring initial concentations of heavy metals with some soil samples. They concluded that heavy metals transfer into soil and groundwater is governed by diversified physicochemical mechanisms and results have proved that cadmium may pose groundwater ressources at Port-au-Prince than lead and copper. Therefore, Gomes et al. (2001) point out that soil loading of Cd, Ni, and Zn appeared to be of greater environmental concern than Cr, Cu, and Pb and that the first group could accumulate in the tissue of plants grown on sludge-treated plots.

In addition to bacterial and metal contaminations, it was found that aquifers in Haiti are also exposed to seawater pollution. In the northeast of Haiti (Ouanaminthe), concentrations in chlorides [318.66–810.89 mg/l], higher than WHO standard [250 mg/l], were measured in some wells and inhabitants in this area develops an high blood pressure probably due to water consumption. However, groundwater quality at Port-au-Prince become more and more vulnerable to contamination related to the earthquake of 2010 January 12. After this terrible disaster, many mass graves were dug in Titanyen (in the vicinity of the plain of Cul-de-sac aquifer) where about 20,000 cadavers were deposited without any preliminary geological investigation. These actions are present concerns about sanitary and epidemiological risks related to microorganisms necrotic activities in and the infiltration of viruses, bacteria and pollutants through soil cracks.

9.3.2 Risk management of groundwater contamination

In the USA, Europe, Japan and other developed countries, Environmental Risk Analysis (ERA) has rapidly become not only a scientific framework for analysing problems of environmental protection and remediation but also a tool for setting standards and formulating guidelines in modern environmental policies. Indeed, ERA deals mainly with the evaluation of uncertainties in order to ensure reliability in a broad range of environmental issues, including utilization of natural resources (both in terms of quantity and quality), ecological preservation and public health considerations (Ganoulis and Simpson 2006). This environmental analysing tool may be also applied in developing countries, such as Haiti, for a sustainable groundwater management. Then, the risk management of groundwater ressources is defined through assessments of groundwater vulnerablity, aquifer potential and source protection areas. In this context, five key stages for a decision-making framework for groundwater management in cities of developing countries (such as Port-au-Prince City) must be useful: (1) investigation of groundwater management issues; (2) identification and quantification of water users and uses; (3) Knowledge of aquifers behaviour; (4) planning and implementation, and (5) monotoring and evaluation.

9.4 IDENTIFICATION OF AQUIFERS MANAGEMENT ISSUES

The primary management issues are related to the impacts of surface runoff infiltration on groundwater ressources. These key issues are mainly associated to a poor management of watershed, city planning, drainage systems and urban environment. Pollution from urban environment is widely recognized as one of the most serious challenge to the sustainable management of groundwater resources (FAO 2003). For example, at Port-au-Prince, the quality of groundwater ressource is generally under a considerable potential of contamination especially in karstic areas, which are characterized by infiltration of intensive polluted surface runoff. The development of efficient stategies for groundwater management in developing countries requires also an assessment of groundwater protection including: the protection of recharge points (rivers, lakes, ponds, etc.), piezometers, wells and springs.

9.5 IDENTIFICATION AND QUANTIFICATION OF WATER USERS AND USES

The identification and quantification of water users and uses is one the main processes to establish the groundwater balances. An estimation of water use for irrigation, domestic activities, industries and others activities is utmost important to ensure the availablility of the ressource. The availability of information on water use will be necessary to the reliability of the technical assessments (for example modelling the system behaviour) and definition of user provisions. Therefore, there are two main user provision issues to resolve: (i) the proportion of the available consumptive water that is to be extracted from the groundwater resource; and (ii) the allocation of the water amongst the various user groups, including urban and domestic, industrial, mining, stock, irrigation, environmental and downstream requirements.

9.6 KNOWLEDGE OF AQUIFERS BEHAVIOUR

The integration of groundwater knowledge at the assessments stage allows the development and testing of use strategies such as aquifer storage and recovery, protection of ecosystems and opportunistic cycling of the use of the resource within wet and dry periods. Knowledge generation aims to provide an understanding of how the groundwater system works which provides the information required to develop the conceptual models of the hydrogeological and hydrological system. In the case of plain of cul-de-sac aquifer, typical collection data such as recharge, permeability, porosity, bulk density, groundwater flux and diffusion is utmost important for modelling. The utilisation of integrated models will enable not only to create vulnerablity maps, but also to facilitate and optimise the decision-making process reating to problems of use/water management/environmental protection. Various mathematical descriptions have been developed in this context to understand the aquifers behaviour. Some of them are focused to groundwater flow, contaminant transport and diffsion though porous media have been developped for the governing relationships and solution of the relationships subject to various initial boundary conditions. MODFLOW is recognized as the world standard groundwater flow model. This software may be useful to understand both groundwater flow and pollutants transfer in aquifers in order to ensure a sustainable groundwater management at Port-au-Prince.

9.7 PLANNING AND IMPLEMENTATION

The groundwater management plan is intended to provide a flexible, adaptive plan for achieving the overall goal that groundwater will continue to be a reliable, safe, efficient, and cost-effective water supply. The main elements of the plan are the operating rules, a policy framework and a monitoring and evaluation strategy. In some countries, governments have introduced legislation to regulate groundwater development and to constrain activities that might compromise groundwater availability and quality. This water legislation offers usually considerable advantages, since it provides a legal basis for the effective and sustainable management of groundwater. Therefore, the implementation of the water legislation may include river basin management plans and action plans to secure lakes, watercourses and coastal aquifers which may contribute to groundwater recharge. This implementation of water legislation must be focused also on groundwater protection and sustainable groundwater exploitation.

9.8 MONITORING AND EVALUATION

Monitoring and evaluation programs are an essential tool for checking on the status of the groundwater ressources, assessing the impact of human activities on groundwater, and evaluating long-term groundwater trends associated with the aquifers maganement issues. These programs aim to: (a) identify monitoring targets to assess performance; (b) optimise monitoring network/data requirements; (c) identify data gaps or weak data areas where improved information and knowledge is required, and (d) review the performance and identify any need for further investigation/monitoring/modelling.

9.9 CONCLUSION

This chapter has fosuced on aquifers contamination in developing countries related to surface runoof infiltration by analysing Plain of cul-de-sac aquifer which contributes approximately 50% of the water supply of urban community of Port-au-Prince. It has showed that aquifers in developing countries are seriously vunerable to pollutants from surface runoff such as heavy metals. Concentrations range of these metal pollutants has been detected in some groundwater samples from developping countries and some of them are higher than threshold values fixed by WHO. The situation is also critical at Port-au-Prince where the quality of groundwater ressource is generally under a considerable potential of contamination especially in karstic areas. Indeed, if the situation remains unchanged, Port-au-Prince will face serious difficulties of water supply due to a poor management groundwater ressources to ensure the durability of the ressource. The decision framework developped in this chapter provides tools to municipalities in Haiti for analysing environmental risk management for a sustainable aquifers management.

REFERENCES

Abu-Rukah Y, Al-Kofahi O. 2001. The assessment of the effect of landfill leachate on groundwater quality–a case study. El-Akader landfill site–north Jordan. Journal of Arid Environments 49: 615–630.

Al Sabahi E, Abdul Rahim S, Zuhairi WYW, Al Nozaily F, Alshaebi F. 2009. The character-istics of leachate and groundwater pollution at municipal solid waste landfill of Ibb City, Yemen. American Journal of Environmental Sciences 5(3): 256–266.

Angerville R. 2009. Ecotoxicological risks assessment related to urban stormwater discharges in waterways: application to a french city and a Haitian city. Ph.D. Thesis. Lyon: INSA de Lyon. 485 p.

Asubiojo OI, Nkono NA, Ogunsua AO, Oluwole AF, Ward NI, Akanle OA,Spyrou NM. 1997. Trace elements in drinking and groundwater samples in southern Nigeria. The Science of the Total Environment 208: 1–8.

Bhattacharyya KG, Gupta SS. 2007. Adsorptive accumulation of Cd(II), Co(II), Cu(II), Pb(II), and Ni(II) from water on montmorillonite: Influence of acid activation. Journal of Colloid and Interface Science 310: 411–424.

Calò F, Parise M. 2009. Waste management and problems of groundwater pollution in karst environments in the context of a post-conflict scenario: The case of Mostar (Bosnia Herzegovina). Habitat International 33: 63–72.

Canfield RL, Henderson CR, Cory-Slechta DA, Cox C, Jusko TA, Lanphear BP. 2003. Intellectual impairment in children with blood lead concentrations below 10 µg per decili-ter. New England Journal of Medicine 348: 1517–1526.

Carré JC. 1997. Impact study of paintings lacquers and varnished on the environment and health. Port-au-Prince: Pan American Heath Organization–World Heath Organization–Ministry of Environment in Haiti. 50 p.

Coiffier N, Théodat JM. 2005a. Haïti. Port-au-Prince: Electrification of a street and accumu-lation of alluvia and solids waste [Internet]. [cited 2010 Feb 18]. Available from: http://energigeo.veille.inist.fr/images/448.jpg

Coiffier N, Théodat JM. 2005b. Haïti. The river of "Bois de Chêne" carries out plastic bottles and emerges in a shanty town of Port-au-Prince [internet]. [cited 2010 Feb 18]. Available from: http://energigeo.veille.inist.fr/images/448.jpg

Deutsch JC, Hémain JC. 1984. Main results of the French national programme of urban run-off quality measurement. Proceedings of the Third International Conference on Urban Storm Drainage, Göteborg, Sweden. Volume 3: 939–946.

Emmanuel E, Lacour J, Balthazard-Accou K, Joseph O. 2009a. Ecological hazard Assessment of the effects of heavy metals and nutrients contained in urban effluents on the bay ecosys-tems of Port-au-Prince (Haiti). Aqua-Lac 1(1): 19–28.

Emmanuel E, Pierre MG, Perrodin Y. 2009b. Groundwater contamination by microbiologi-cal and chemical substances released from hospital wastewater: Health risk assessment for drinking water consumers. Environment International 35: 718–726.

Erah PO, Akujieze CN, Oteze GE. 2002. The Quality of Groundwater in Benin City: A baseline study on inorganic chemicals and microbial contaminants of health impor-tance in boreholes and open wells. Tropical Journal of Pharmaceutical Research 1(2): 75–82.

Eriksson E, Baun A, Scholes L, Ledin A, Ahlman S, Revitt M, Noutsopoulos C, Mikkelsen PS. 2007. Selected Stormwater Priority Pollutants–a European Perspective. Science of the Total Environment 383: 41–51.

[FAO] Food and Agriculture Organization of the United Nations. 2003. Groundwater man-agement: the search for practical approaches–Water Reports 25 [Internet]. Rome: Natural ressources and environment [cited 2010 May 12]; 55 p. Available from: ftp://ftp.fao.org/agl/aglw/docs/wr25e.pdf.

Fifi U, Winiarski T, Emmanuel E. 2010. Urban groundwater quality in developing countries: Case study of Plain of Cul-de-sac aquifer of Port-au-Prince. In: Thierry Verdel and Jacques Bourguois, Editors. International conference on "water, waste and sustainable develop-ment", 2010 March 28–31, Alexandria, Egypt. CD-ROM.

Fifi U, Winiarski T, Emmanuel E. 2009. Sorption Mechanisms Studies of Pb (II) Cd (II) and Cu (II) into Soil of Port-au-Prince. Journal of the International Hydrological Programme for Latin America and Caribbean, Aqua-Lac 1 (2): 164–171.

Fifi U, Winiarski T, Emmanuel E. 2008. Study of the sorption mechanisms of Pb(II), Cd(II) and Cu(II) into soil of Port-au-Prince [Poster]. In: Abstracts of Third Annual Conference of the International Center for Environmental and Sustainable Development Studies (CIEMADeS). Puerto Rico, Universidad del Turabo. Puerto Rico 2008 December 11–13.

Foppen JWA. 2002. Impact of high-strength wastewater infiltration on groundwater quality and drinking water supply: the case of Sana'a, Yemen. Journal of Hydrology 263(1–4): 198–216.

Fuchs S, Brombach H, Weiß G. 2004. New database on urban runoff pollution. Proceeding of NOVATECH'2004–5th International Conference on Sustainable Techniques and Strategies in Urban Water Management, 2004 June 6–10. Lyon, France. p. 599–606.

Ganoulis J, Simpson L. 2006. Environmental risk assessment and management: promoting security in the Middle East and the mediterranean region. In: Morel B, Linkov I, editors. Environmental Security and Environmental Management: The Role of Risk Assessment, Springer, p. 245–253.

Gomes PC, Fontes MPF, Silva AG, Mendonça E, Netto AR. 2001. Selectivity sequence and competitive adsorption of heavy metals by Brazilian soils. Soil Science Society of America journal 65: 1115–1121.

Graie M. 2005. Research Field Observatory in Urban drainage. [Internet]. [updated 12 03 2009]. France; [cited 2010 Feb 12]. Available from: http://www.graie.org/othu/.

Hamad TMH. 1993. Sewage effluent used for irrigation and its impact on soil environment in some developing African countries. PhD Thesis. Cairo: University of Cairo, Institute of African Research and Studies, Department of Natural Ressources. 142 p.

Hirata R, Ferrari L, Ferreira L, Pede M. 2002. Groundwater exploitation of hydrographic watershed of the Alto Tiete: A chronicle of a crisis fortold. Boletín Geológico Minero (Mining Geology Bulletin) 113(3): 273- 282.

Howard KWF, Beck PJ. 1993. Hydrogeochemical implications of groundwater contamination by road de-icing chemicals. Journal of Contaminant Hydrology 12: 245–268.

Huang JL, Du PF, Ao CT, Lei MH, Zhao DQ, Ho MH, Wang ZS. 2007. Characterization of surface runoff from a subtropics urban catchment. Journal of Environmental Sciences 19: 148–152.

[IFRC] International Federation of the Red Cross and Red Crescent Societies. 2001. World Disasters Report 2001: Focus on Recovery [Internet]. [cited 2010 Feb 18]; 240 p. Available from: http://www.ifrc.org/publicat/wdr2001/ .

[IHSI] Institut Haïtien de Statistique et d'Informatique. 2003. 4th general census of the population and the habitat [Internet]. [cited 2010 Feb 10]. Available from: http://www.ihsi.ht/recensement.htm .

Jang YC, Jain P, Tolaymat T, Dubey B, Singh S, Townsend T. 2010. Characterization of roadway stormwater system residuals for reuse and disposal options. Science of the Total Environment 408:1878–1887.

Joseph O. 2009. Study of the potential use of Haitian agricultural by-products for the wastewater treatment by biosorption technique. PhD Thesis. Lyon: INSA de Lyon, 206 p.

Kafi-Benyahia M, Gromaire MC, Chebbo G. 2005. Spatial variability of characteristics of urban wet weather pollution in combined sewers. Water Science and Technology 52(3): 53–62.

Kim S. 2000. Southern Africa Swamped by Rains. Disaster News Network [internet]. [cited 2010 Feb 14]. Available from: http://www.disasternews.net/news/article.php?articleid=154.

Lanphear BP, Dietrich K, Auinger P, Cox C. 2000. Cognitive deficits associated with blood lead concentrations B 10 mg/dL in US children and adolescents. Public Health Reports 115(6):521–529.

Lerner DN. 1990. Groundwater recharge in urban areas. Atmospheric Environment 24: 29–33.

Lerner DN. 2002. Identifying and quantifying urban recharge: a review. Hydrogeology Journal 10: 143–152.

Manfreda S. 2008. Runoff generation dynamics within a humid river basin. Natural Hazards and Earth System Sciences 8: 1349–1357.

Mangore E, Taigbenu A. 2004. Land-use impacts on the quality of groundwater in Bulawayo. Water SA 30(4): 453–464.

Manzoor S, Shah MH, Shaheen N, Khalique A, Jaffar M. 2006. Multivariate of trace metals in textile effluents in relation to soil and groundwater. Journal of Hazardous Materials A137: 31–37.

Mirza MQ, Ericksen NJ. 1996. Impact of Water Control Projects on Fisheries Resources in Bangladesh. Environmental Management 20(4): 527–539.

Mokhtaria MM, Eddine BB, Larbi D, Azzedine H, Rabah L. 2007. Characteristics of the public landfil of Tiaret City and its impact on groundwater quality. Courrier du Savoir 8: 93–99.

Nouri J, Mahvi AH, Babaei AA, Jahed GR, Ahmadpour E. 2006. Investigation of metals in groundwater. Pakistan Journal of Biological Sciences 9(3): 377–384.

[PAHO] Pan American Health Organization. 2007. Haiti: Health in the Americas 2007 – Volume II [Internet]. [cited 2010 may 10]. Available from: http://www.paho.org/hia/archivosvol2/paisesing/Haiti%20English.pdf

Qadir M, Wichelns D, Raschid-Sally L, McCornick PG, Drechsel P, Bahri A, Minhas PS. 2010. The challenges of wastewater irrigation in developing countries. Agricultural Water Management 97: 561–568.

Qin F, Wen B, Shan XQ, Xie YN, Liu T, Zhang SZ, Khan SU. 2006. Mechanisms of competitive adsorption of Pb, Cu, and Cd on peat. Environmental Pollution 144: 669–680.

Ramesh R, Purvaja GR, Raveendra VI. 1995. The problem of groundwater pollution: a case study from Madras city, India. Man's Influence on Freshwater Ecosystems and Water Use (Proceedings of a Boulder Symposium, July 1995). IAHS Publication. p. 147–157.

Rattan RK, Datta SP, Chhonka PK, Suribabu K, Singh AK. 2005. Long-term impact of irrigation with sewage effluents on heavy metal content in soils, crops and groundwater–a case study. Agriculture, Ecosystems and Environment 109: 310–322.

Reimann C, Bjorvatn K, Frengstad B, Melaku Z, Tekle-Haimanot R, Siewers U. 2003. Drinking water quality in the Ethiopian section of the East African Rift Valley I–data and health aspects. The Science of the Total Environment 311: 65–80.

Rossi L, Fischer Y, Froehlich JM, Krayenbühl L.1996. Contamination study of stormwater runoff in urban area. Department of Rural Engineering, Hydrology and Land Improvement Lab., Swiss Federal Institute of Technology, Laussane. 72 p.

Saade L. 2006. Act together for an effective management of the drinking water services and sanitation in Haiti. United Nations, Economic Commission for Latin America and the Caribbean. Mexique. 44 p.

Satyanarayanan M, Balaram V, Al Hussin MS, Al Jemaili MAR, Rao TG, Mathur R, Dasaram B, Ramesh SL. 2007. Assessment of groundwater quality in a structurally deformed granitic terrain in Hyderabad, India. Environmental Monitoring and Assessment 131(1–3): 117–127.

Sherghini A, Fekhaoui M, El Abidi A, Tahri L, Bouissi M, Zaid AH. 2003. Metal contamination of the ground water in Mohammedia (Morocco). Cahiers d'études et de recherches francophones 13(3): 177–182.

Singh KP, Malik A, Singh VK, Mohan D, Sinha S. 2005. Chemometric analysis of groundwater quality data of alluvial aquifer of Gangetic plain, North India. Analytica Chimica Acta 550: 82–91.

Smeets A. 2008. International photo competition "UNICEF- Photo of the Year". [Internet]. [cited 2010 Feb 18]. Available from: http://www.alicesmeets.com/ .

Soltan ME.1998. Characterisation, classification, and evaluation of some ground water samples in upper Egypt. Chemosphere 37(4): 735–745.

Taebia A, Droste RL. 2004. Pollution loads in urban runoff and sanitary wastewater. Science of the Total Environment 327: 175–184.

[UNEP] United Nations Environment Programme. 2002. Global Environment Outlook 3 (GEO-3): Past, Present and Future Perspectives [Internet]. London: Earthscan Publications Ltd. [cited 2009 November 15]; 446 p.

Available from: http://www.grida.no/publications/other/geo3/?src=/geo/geo3/french/.

[USEPA] United States Environmental Protection Agency.1999. Preliminary data summary of urban storm water best management practices [Internet]. Washington D.C: Ingineering and Analysis Division/Office of Science and Technology.[cited 2006 october 20]; 448 p. Available from : http://www.epa.gov/waterscience/guide/stormwater/ .

[USEPA]. United States Environmental Protection Agency. 1998. Guidelines for Ecological Risk Assessment [internet]. Washington D.C: Environmental Protection Agency, [cited 2010 Feb 17]; 188 p. Available from: http://cfpub.epa.gov/ncea/cfm/recordisplay. cfm?deid=12460

[USEPA] United States Environmental Protection Agency. 1983. Results of the Nationwide Urban Runoff Program: Volume 1–Final report [Internet]. Washington DC: Water Planning Division. [cited 2008 April 9]; 186 p. Available from: http://www4.ncsu.edu/~rcborden/CE481/Stormwater_Refs/NURP_Results_Vol_1.pdf

[WHO] World Heath Organization. 2004. Guidelines for drinking water quality: Volume 1 recommendations, 3rd Edition [Internet]. Geneva. [cited 2009 Sept 22]. 540 p. Available from: http://www.who.int/water_sanitation_health/dwq/GDWQ2004web.pdf.

Wondimu A. 2000. Sustainable urban storm water management by urban upace management and subsidiarity: The case of Addis Ababa (Ethiopia). Ph.D. Thesis. Lyon: INSA de Lyon. 394 p.

Zhang W, Keller AA, Yue D, Wang X. 2009. Management of urban road runoff containing PAHs: Probabilistic modeling and its application in Beijing, China. Journal of the American Water Resources Association 45(4): 1009–1018.

Zhu LZ, Chen BL, Wang J, Shen HX. 2004. Pollution Survey of Polycyclic Aromatic Hydrocarbons in Surface Water of Hangzhou China. Chemosphere 54: 1085–1095.

Zingoni E, Love D, Magadza C, Moyce W, Musiwa K. 2005. Effects of a semi-formal urban settlement on groundwater quality Epworth (Zimbabwe): Case study and groundwater quality zoning. Physics and Chemistry of the Earth 30: 680–688.

Conservation of Land, Air, Water and Cultural Assets

Osaín sculpture at the Caguas Cultural and Botanical Garden, Puerto Rico
(@Eddie N. Laboy-Nieves)

Chapter 10

Ethnoecological restoration of deforested and agricultural tropical lands for Mesoamerica

Silvia del Amo Rodríguez, José María Ramos Prado and María del Carmen Vergara Tenorio

SUMMARY

In order to stop and reverse deforestation and land use changes, a series of management integral actions must be undertaken, which will help to direct rural communities on the path towards sustainability. However, they are required to include not only ecological elements of biodiversity conservation and ecological restoration, but also the social and economic counterpart: production systems, management policies, marketing strategies and effective funding schemes. In this chapter we present a framework called "biocultural resource management" based on 20 years of academic and practical experiences in the field of ethnoecological restoration and resource management, in tropical Mexico. We propose a feasible way to develop productive conservative strategies and community projects in rural tropical areas. The consideration of actual land use, local socio-economic problems and the expectations of the people involved, emphasizing cultural factors, are essential for better planning and more sustainable decisions. In the context of the current tropical areas, interdisciplinary and participatory methods are the most viable way to achieve biocultural resources conservation. Practical instruments for working with communities are landscape management plans that include productive, conservation and restoration projects considering human settlements, which represent sustainable alternatives for communities to develop sustainable societies.

10.1 INTRODUCTION TO A CRISIS

The subject of crisis has become reiterative over the last decade. Crisis exists and affects us everywhere, specially the environment. Attali (1982) defines crisis as "*a long and hard rewriting and thinking about on two world visions. The first one has proved incapable of solving problems and it is necessary to abandon; and the second one, that poses innovative alternatives, and conducts us towards a better balance among people and an improved relationship between humans and nature.*" This definition calls us to entail new meanings to previous actions, to analyze wise and wrong decisions and to assess mistakes with a critical perspective. These are key elements to solve problems, such as environmental degradation and to accept that a crisis always involves losses between human beings and nature and transformation of ethical, cultural and scientific aspects. Therefore, science must entail a commitment to the environment and its preservation and the formulation of a new or several new paradigms must be questioned by ethics, through profound inquiries, so we will create what Morin (1984) calls *Science avec conscience (Science with consciousness)*. We need a new scientific and social vision that would seek a renewed emphasis on the creation of sustainable communities (Capra 1996). According to Novo (2003), the new paradigm should reject indiscriminate domination and exploitation of nature, and embrace more balanced principles, which entail the abandonment of anthropocentrism of the last centuries and a new understanding of the human-environment relationship. These premises imply fostering values and attitudes for a better communication with nature, and the acknowledgement of it as a *subject of rights*. However, this also means that nature's rights cannot occur until human's communication is free of domination (Habermas 1984).

The millennium shift offers an excellent opportunity to open a space for discussion and to look back at the historic and political events that define us nowadays. The analysis of the world situation renders considerable inequities, broken promises, ecosystems loss, an increasing air, water and soil depletion, the demands of people who do not have access to resources and an overwhelming poverty. All these situations call for urgent solutions to the pending problems. One of the answers to a better relationship between society and nature is landscape management, as an instrument for local policy development and social inclusion. This approach guarantees that landscape management as an useful instrument for land use and for building sustainable societies. Furthermore, landscape management allows thinking about what is sustainability and its differences at the local, regional and global level.

Quiroga-Martínez (2003) explains that sustainability comprises the need of conserving our ecological and cultural heritage emphasizing the importance of assessing this need from a South-North perspective. In her view, developing countries would have a better understanding of their crises and options to plan for sustainability; considering social, cultural and economic rationalities, as well as, local richness and ecological heritage. Quiroga-Martínez (2003) implies that there are as many different sustainable societies, as ways of living exist, and calls for a redistribution of natural heritage instead of distribution of wealth, consistent with the ideas of Constanza et al. (1997) and Daly (1992, 1997). These authors agreed that the *minimal necessary condition to achieve sustainability is the maintenance or the increase of the total current natural heritage*. Redistribution of natural heritage means fair access to natural resources and energy sources, as well as, acknowledgement of the world's limited carrying capacity and responsibility for waste production (Boulding 1966). This process can be the foundation for a new paradigm centered on citizens to construct a sustainable society and a common welfare (Ralston 1997). In our projects, this sustainability conception is related to what we call biocultural resources management (Del Amo-Rodríguez et al. 2010).

10.2 THE NEED FOR AN INTERDISCIPLINARY APPROACH AND PEOPLE'S PARTICIPATION

For real ecological heritage redistribution we need to introduce ethics, and inter- and trans-discipline approaches (Naveh 2004). The respect for biological diversity prevails as a criterion when people understand the intrinsic value of natural resources and the need to preserve them for life maintenance. Also, respect for nature goes along with respect for cultural diversity to foster human development in the world (Novo 2003), and it is a key element in resource management (Del Amo-Rodríguez et al. 2010).

An excellent example that considers ethics and a trans-discipline approach is landscape ethnoecology, which is a combination of anthropology, ecology and culture methods. This new field could help developing a new scientific environmental paradigm with its own conceptual and instrumental systems; which means, transcending current advances in natural resources management and integrate the cultural and ethical dimensions to generate new solutions for sustainable societies. It is clear that without people's participation it will be difficult to advance into integral and sustainable solutions for environmental problems. Therefore, appropriate tools for landscape ethno ecology are action research and participatory methods to encourage shared responsibility for using internal and external resources and exchanging goods and services in a community framework (Geilfus 2008).

Participatory methods have been used since the 70's in different countries and popularized in Latin America by Freire (1970, 1990). The aim of these methods is to foster social participation and inclusiveness. One of the results while applying this methodology is a process for empowerment and reevaluation of social values (Bessette 2004). Empowerment is crucial to include groups of people usually excluded of any kind of decision. UNRISD (2002) indicates that social participation is *the organized efforts to increase the control over the resources and the social movements of those that have been traditionally excluded of control.* Then, a conscious citizen that uses resources in a communitarian context is the key for change (Carlsson and Berkes 2005). Examples of this change are the endogenous projects of Bolivia and Sri Lanka where the communities direct their own projects and flourish as sustainable societies (Haverlook et al. 2002; Rist 2002). Both of these experiences integrated the indigenous perspective (the comprehension of the peasant's inside world), and a trans-disciplinary approach. People's participation should involve respect for plurality, cultural ethnicity, and ways of interacting with nature (Haverkort and Rist 2004). There are three ethical guidelines described by Callicott (1998), to achieve this participation: 1) a philosophical criticism to modernity, its world view and its ways of relating to nature; 2) new ways of representing and relating to nature based on principles; and 3) attention to other cultures for new and non-western perspectives for conceiving and inhabiting nature.

10.3 FOREST CONSERVATION AND ETHNOECOLOGICAL RESTORATION

Tropical forests have suffered from intense deforestation and depletion as a result of anthropogenic activities. Great areas of forests that once contained a vast diversity, have been transformed into agricultural areas, secondary forests, degraded abandoned fields and isolated plots of forest. There are approximately 4,000 million hectares of forest left in the world, which represents 30% of the earth's terrestrial surface (FAO 2007). The deforestation rate has increased up to 13 million of hectares per year: Nine out of the ten countries, that own more than the 80% of the primary forests of the world, have lost at least 1% of its surface from the 2000 to 2005 (FAO 2007). Indonesia is at the head of the list (13%), followed by Mexico (6%), Papua New Guinea (5%) and Brazil (4%). In Mexico, from 1990 to 2000, the forest areas decreased at an annual rate of 0.52%, whereas in 2000–2005 declined to 0.40% (FAO 2007). This reduction of forest also implies the loss of biodiversity due to the risk of extinction of species and the loss of the maintenance and sustainment of the productivity of the planet (Myers 1984).

In Mexico forest lands have been substituted by agricultural, ranching and unsustainable forest activities for private consumption and market supplies. The development approach, based on an increased production-consumption, has satisfied urban elite ignoring the need of public policies in rural areas. The Country has lost or degraded forests, soil and water resources, and millenary indigenous management practices developed for sustainable living strategies (Toledo-Manzur 1992; Del Amo-Rodríguez et al. 2008 a, b). The loss of biological and cultural diversity is one of the major threats for Mexico and the whole planet, as the extinction of these resources is irreversible. The cumulative effect of the simplification of ecosystems and agroecosystems puts biodiversity, agrodiversity and human habitats at great risk (Myers 1984). For this reason, it is urgent to develop suitable mechanisms for recovering native species and for achieving ecological and cultural restoration. In this context, we believe that it is necessary to establish ethnoecological restoration strategies, using diversification methods, stratification, and rotation of useful and/or commercial native species (Mizrahi et al. 1997). Restoration could be implemented in familiar orchards, milpas, traditional and modern agricultural systems, cattle systems, forest plantations, as well as, in green corridors of urban and industrial regions.

Martínez-Ramos and García-Orth (2007) suggested that ecological principles should be used for efficient technologies to recover degraded forest. The same authors have developed a conceptual scheme that considers availability of native seedlings and the degree of environmental perturbation for vegetation regeneration in degraded forests. For regeneration to occur it is necessary to: 1) have a good biological knowledge of the native species for transplanting, as well as, knowledge of their ecological behavior in degraded atmospheres, and 2) to generate growing, transplanting and nurturing techniques and protocols, that will increase survival plant rates and decrease related expenses. New perspectives, such as ethnoecological ecology, should be introduced. For instance, sequential models are formulated using a delimited spacial and a time scale, which is implicit most of the times (Giampietro 2005; Ramos-Martín and Giampietro 2005; Roth 2004). The development of new models with a predictive capacity must incorporate spatial and time scales explicitly. In this way, the relevance of these models for restoration will significantly increase (Vega-Peña 2005).

10.4 BASIC PRINCIPLES OF ECOLOGICAL RESTORATION IN THE TWENTIETH CENTURY

According to Covington et al. (1998) restoration implies returning a damaged ecosystem to its previous condition or to its historical path of development, i.e., the recovery of the integrity and stability of a degraded or destroyed ecosystem, in terms of its structural and functional characteristics. The process of restoration is slow, accomplished by the reduction of anthropogenic disturbances, the elimination of exotic species, the rehabilitation of soil and aquifers, the remediation of contamination in soil and water, and the reintroduction of native species (Hambler 2004).

Restoration can be considered the science of the twenty first century, as conservation was for the previous century. The ecological restoration of degraded or damaged landscapes is a relatively new and experimental practical discipline. Also, the study of the processes and mechanisms that allow and limit the important factors for successful ecological restoration is new. Another important characteristic of restoration is its interdisciplinary approach, since it involves the study of degradation, destruction processes, holistic-systemic approaches and strategies for the recovery of ecosystems (Covington et al. 1998; Choi et al. 2008).

Vázquez-Yanes and Orozco-Segovia (1994) delimit the basic concepts and principles in ecological restoration as follows: a) a fundamentalist vision that considers restoration as a return to the previous existing conditions in the original natural communities; b) a practical approach that combines productive activities with environmental services; and c) landscape restoration that focuses on damaged landscape. The application of these three principles highly depends on the degree and extension of the perturbation, on the initial state of the forest, the soil degradation, the desired result, the time frame and the financial constraints and community participation (Chazdon 2008). The ethnoecological restoration considers productive activities and environmental services based on traditional resource management.

The importance of the conservation of flora and fauna in damaged areas such as secondary forests has gained recognition considering that under management conditions abundant species can recover quickly, and the diversity and characteristic species composition present in mature stages can be accelerated by means of its enrichment with clusters of key species (Mizrahi et al. 1997; Ramos-Prado and Del Amo-Rodríguez 1992, Del Amo-Rodríguez and Ramos-Prado 1993). In degraded soil areas like pasture fields, the rehabilitation through leguminous species can promote secondary succession and by means of the further introduction of native arboreal species, to partially reestablish the structure and functioning of the original forest ecosystems. In areas where agriculture has been less intensive and where use we can find patches of forest and agents for fauna dispersion that can assure the diversity of seed rain, natural regeneration is a less expensive option (Dunn 2004).

The diagnosis of tropical forest fragments, secondary forest and traditional agroforestry systems reveal information on the structure and the composition of the original ecosystem and its management history (Ramos-Prado and Del Amo-Rodríguez 1992; Del Amo-Rodríguez and Ramos-Prado 1993). The combination of ecological and cultural information, allows to draw conclusions on the ecosystem conservation state and to draw up its historical and successional trajectory. New strategies of ecological restoration in highly degraded or destroyed ecosystems could be generated through ecological information and the use of predictive successional models (Ramos-Prado et al. 1996; Martínez-Ramos and García-Orth 2007). The emulation of this sequential process during restoration, aids the development of the ecosystem towards states that are more coherent with its historical and evolutionary trajectory.

In the context of our research, we have established natural ecosystems and degraded tropical ecosystems as objects of study. Our main goal is to restore ecosystems using an ethnoecological approach (Ramos-Prado et al. 2004). The main objective is to transform degraded areas into productive successional systems throughout diversification. We focus on native species that can satisfy consumption needs, generate complementary income through commercialization and provide ecological and environmental services for the well-being of the local population. Our starting point is the diagnosis and the analysis of the degraded area. This diagnosis comprises an evaluation of the structure and the dynamics of the damaged ecosystem, as well as, the assessment of the degradation and destruction causes and factors. We use Márquez-Huitzil (2005) approach, who recommends five steps to design an ecological restoration project: 1) eliminate the perturbation source, 2) mitigate the effects produced by the perturbation source, 3) return the system to similar conditions that were present in a previous successional stage, 4) reincorporate original biotic or abiotic components into the system and 5) iteratively monitor and modify the restoration processes, directing the successional process in correspondence with its objective. It is important to mention that we frequently faced social, ecological and political limitations.

Through our work we have come to the conclusion that ethnoecological restoration, which entails the consideration of cultural aspects, constitutes the ecological basis for human survival. Cultural practices and ecological processes should reinforce themselves mutually. In the field of action research, hypotheses on the evolutionary and historical trajectory of cultural resource management by ethnic groups that have been coexisting with ecosystems for millennia can be drawn (Del Amo-Rodríguez et al. 2008 a, b). That is the case of the majority of ethnic groups and native cultures in Mexico, where Ramos-Prado et al. (2004) have proposed ethnoecological restoration to design productive projects and programs that involve local communities.

10.5 COMMUNITY DEVELOPMENT THROUGH SOCIAL CAPITAL

Restoration processes require a social and an individual cultural context and in this case, it is necessary to define alternatives to solve the problems in tropical areas. One of the most important aspects of an ethnoecological approach is to build and facilitate the creation of social capital within the communities the local communities that look for restoration. Bourdieu (1986) defines social capital as the sum of potential resources that are bound to a more or less institutionalized network of people, who recognize themselves as part of it and that can form friendship bonds. Other authors like Coleman (1988, 1990), consider that social capital consists of the relations established between the members of a social organization based on trust, collaboration, mutual aid and social norms. This group of individuals seeks to reach a specific objective and social capital facilitates the individual or collective action towards this goal. Robert Putman (1993, 1995a, 1995b) refers to social capital as the collective value of the social networks and the opportunities that arise as a result of them and that promote mutual support.

Putman explains social capital as confidence, cooperation and long term relations under a framework of set rules to achieve a common objective. Putman states than social capital is the key to civic commitment and represents a social measurement of community health. As people are more connected with each another, the trust between them will be greater and the individual and collective benefits will increase.

There is a close relationship between social capital and community development because of the potentiality that social capital has to generate changes. Putman sustains that social capital can be divided into bonding social capital that forms among homogeneous groups of people and bridging social capital that generates through the interaction of heterogeneous groups that seek to achieve common objectives. Temkin and Rohe (1997) discuss the existence of communitarian socio-cultural capital and have determined that this is the bonding capital that allows individuals to identify themselves with their community, to act and commit with it and therefore work in the restoration of its natural resources.

Gittell and Vidal (1998), examine other forms of social capital such as influence capital, which is related to material support that could not be individually obtained; and social support, which is related to the psychological support to overcome challenges. On the other hand, Granovetter (1973, 1983) emphasizes the importance of the weak bonds to disperse ideas and opportunities between people of different social groups, which involve a greater mobility. These bonds can also enable the bonding between people that are external to the communities or surrounding neighborhoods and the industrial areas or the inner city. In contrast, when structural gaps are present (Burt 1992), that is to say when there are individuals that do not benefit from the connectivity with others, these gaps are usually exploited by companies or politicians, who are a hindrance to development by seeking to fulfill their own interests.

It is clear that without the presence of internal social networks and external bridges it is very difficult to reach community development because through them, individuals can achieve greater goals than the ones that they would have reached individually (Gittell and Vidal 1998; Ballón et al. 2009), whereas communities with structural gaps that do not allow them fully interact tend to stay under-developed. It is important to stress that networks that are not committed to their community, but with their personal interests can also exist and prevent the community development. Nevertheless, Ostrom (1999) adverts that social capital is constructed and abrupt changes in population, technology, economy etc. can negatively affect the community's institutionalism, which adapts to slow changes but not to quick ones. Ideally each level of social capital (individual, community and inter-community) is interconnected and although the creation of individual social capital does not guarantee the formation of community or inter-community capital, individual social capital is a precursor of community social capital, and this community social capital enables the accumulation of individual capital. The understanding of network formation and the way in which collective norms are established, will promote community development and therefore, the management, conservation and restoration of natural resources (Haverlook and Rist 2004). We cannot conceive a society or community that values its natural capital but does not work in favor of the construction of social capital.

10.6 TRADITIONAL RESOURCE MANAGEMENT PRACTICES

Many ecosystems have suffered the consequences of population growth and external pressures and therefore, they need to be recovered. The restoration of these ecosystems usually includes recovering the ecosystem *per se* and the recovery of traditional ecological management practices. This process also entails recuperation of indigenous language and knowledge, because it is usually passed through oral traditions and remains within those that speak the language (Ramos-Prado et al. 2004; SER 2004). Furthermore, in many developing countries people continue applying sustainable traditional cultural methods. These practices look for an interaction within the environment reinforcing ecosystem's health and sustainability (Maffi 2005). Moreover, these traditional methods foster the construction of social capital (Whitelaw and McCarthy 2008). In this sense, ethnoecological restoration promotes new sustainable and culturally appropriate land use practices, under contemporary limitations of rural people. However, ethnoecological restoration depends on people's involvement to succeed.

It is relevant to mention that in order to achieve restoration in Mexico, where the landscape is a highly fragmented mosaic, the unit of work is the forest patch. Ecological systems are like a set of mosaics organized in a discontinuous and nested hierarchy system (Vega-Peña 2005). The traditional ecological units of study, such as individual, population, community and ecosystem are not longer functional in highly fragmented landscapes. The forest patch has an explicit space component and its characteristics (form and dynamics) depend, partly, on their evaluation scale (Vega-Peña 2005). These concepts lead us to consider the successional patch as the ecological restoration unit and that a large scale restoration project would have to include several restoration patches at different successional stages.

Clearly secondary succession patches are management units, used by many Meso-American groups that can promote restoration. In secondary succession, restoration patches are clusters of native species with different ecological strategies and regeneration stages. These species ensembles improve general growth and survival probabilities for the whole system. These groups of individuals positively modify the micro-environment while growing, since they are food sources or settlements for different animals, such as seeds scatters which disperse seeds and then abundance and diversity of plant species increases. Moreover, the production of leaf debris increases the amount of organic matter and soil fertility. In comparison to exotic species, native species facilitate the establishment of tolerant shade seedlings and accelerate succession stages, whereas exotic species can turn aside or stop the successional process (Vázques-Yanes and Batis 1996).

10.7 ETHNOECOLOGICAL RISK ASSESSMENT

The above sections presented an ethnoecological approach for achieving sustainability practices, placing emphasis on biological and cultural factors to focus on ecosystem and agrosystem restoration. We will now present a case study from the Totonacapan area at Veracruz State to illustrate the assessment of ethnoecological risk. The Totonaca ethnic group mainly inhabits the Totonacapan area. This culture, as well as, the ecosystems of the region is at high risk. The area has a high rate of poverty and environmental degradation (Medellín 1988). Then, how can we motivate conservation and restoration in a highly degraded and poverty area? What elements should a project consider? And how can we apply an ethnoecological approach using existing Totonaca knowledge of resource management? The answers are very complex and have change, evolved and adapted while facing different challenges.

The Totonacas are one of the most numerous ethnic groups of Mexico, (Masferrer-Kan 2004). The total population is nearly 375,000 inhabitants. The speakers of Totonaca language are around of 240,000 (INEGI 2000). They used to inhabit a vast territory, from the northern half of the Gulf of Mexico, to the top of the Sierra Madre Oriental. The warm and humid climate of this region supported tropical rain forests and a diverse and buoyant agriculture (Kelly and Palerm 1952). Nowadays, they have lost territory, natural resources and were pushed to the highlands of the Sierra Madre Oriental. Nevertheless, after hundreds of years of domination they still conserve their identity and essential parts of their belief system (Masferrer-Kan 2004), as well as their land use systems and natural resource management strategies (Medellín 1988, Del Ángel-Pérez and Mendoza 2004).

The first stage of the ethnoecological approach is the evaluation and diagnosis of the ecological and cultural systems, with participative actions, establishing main lines of applied research that we have already in progress:

1. A participative land use plan for the Municipality of Papantla, which is the main county in the Totonacapan area, using conservation, production, management and restoration actions.
2. Community-based micro-enterprises of vanilla and certified seeds and seedlings of native species.
3. Agroforestry systems in secondary forest patches.
4. Demonstrative productive restoration parcels.
5. A model for environmental services and payments to small landowners.
6. A training and technology transference program.

The participative land use plan is the starting point for achieving a better management. At the municipal level, categorization of the different types of land use allows to assess the impacts over them and to determine the causes of degradation. To conduct the preliminary diagnosis we used cartography at 1:50,000 and 1:20,000 levels, satellite images and relevant literature (PLADEYRA 2002) to generate different thematic maps. Once, the main types of land use were determined, we established the degree of conservation and degradation and the level of impact on the natural resources. For each category of land use, planning and management activities were proposed considering quality and ecological fragility, based on the degree of conservation and degradation of the different categories of land use (Table 10.1). Finally, ethnoecological strategies were designed and suggested in collaboration with community members (Del Amo-Rodríguez, et al. 2008 a,b).

10.8 PARTICIPATIVE PRODUCTIVE PROJECTS TO DIMINISH RISK

A second stage of the ethnoecological approach is to design and implement participative projects that will consider people's needs and diminish risk. Problems need to be defined and described in terms of the interactions and flows of matter, energy and disturbances of the system, their origins, the degrees and levels of degradation. Furthermore, social, economic and political factors have to be contemplated to achieve viable restoration actions. This is task that requires the

Table 10.1 Planning and management strategies in terms of their quality and ecological fragility.

Ecological fragility: topography -relief and slope- soil, precipitation, resistance and resilience.	Ecological quality: vegetation cover degree, structure and biodiversity
	Environmental policies for land use: a) conservation, b) protection, c) restoration, d) use, e) technological development, f) value added, g) commercialization
	Environmental management units: a) conservation-protection, b) conservation-use, c) conservation-restoration, d) restoration-use, e) compensation
	Planning and management systems: a) rural reserves, b) extractive systems, c) agroforestry units, d) diversified farming , e) rural micro-enterprises

integration of an interdisciplinary research work group, and the adaptation of the methodology to the problem scale. In the following paragraphs, we list some of the strategies that might be proposed for conservation and restoration actions based on our experience (Del Amo-Rodríguez 2002; Vergara-Tenorio and Cervantes-Vázquez 2009; Del Amo-Rodríguez et al. 2007; Del Amo et al. 2010).

a) Conservation

- The establishment of protected areas that include threatened vegetation communities and that consider an integral management for its conservation. These protected areas could be created at a municipal level or take the form of Community of Private Reserves.
- The creation of local germplasm banks associated to micro-enterprises and managed by local community members.
- The designing of ecological corridors and the protection of forest fragments and patches of vegetation in areas that have been used for agricultural, ranching and forestry purposes.
- The implementation of payments for environmental services or other types of financial support strategies.

b) Restoration

- Restoration requires the intentional reintroduction of native species through pioneer tree species, and the elimination or control, as much as possible, of invading and harmful exotic species.
- The enrichment of secondary vegetation is an option and can achieved by clearing and planting seedling in corridors, and thinning the canopy in order to modify the competing conditions and the species composition. Establishment of agroforestry systems with native valuable tree species.

c) Rural community-based micro-enterprises as a productive conservation strategy

It is necessary to design strategies to help communities valuing their cultural practices, and allows them to increase their well being without sacrificing traditional conservation and environmental management practices (Martinez-Ramos and García-Orth 2007). For instance, the creation and consolidation of small rural enterprises may provide an alternative income for producers in the Totonaca, which can be employed for managing, restoring and eventually conserving natural resources, as depicted by Hamilton (2008). In our project we applied two models for the processing and commercialization of vanilla, and for timber and non-timber native forest seeds. Our research not only focused on the implementation of this innovative technology, but in the strengthening the sense of identity between locals settlers and their endemic natural resources, encouraging the proximity of community members to their place of origin, and understanding the operation of the micro-enterprise as an integral system, approaches consistent with Macqueen et al. (2006).

Community participation, the formation of societal networks and the strengthening of social capital constitute essential elements for the successful operation of rural micro-enterprises (Moyano-Estrada 2009). Participation entails to actively work on problem solving instead of allowing other people to take individual decisions that would affect the whole group (Guajardo et al. 2004). The reinforcement of social capital might result in: a) benefits when economic damages generally take place; b) the generation of socio-emotional assets that contribute to the socioeconomic well-being; and c) empowerment to influence others (Atria et al. 2003).

Vanilla (*Vanilla planifolia Andrews*) is considered the most beneficial crop in the humid tropic, but paradoxically, it is an overexploited and underused resource (Soto-Arenas 2006). Although native from Mexico where it was first cultivated (Rebolledo 2007), Mexico only accounts for the 3.43% of the world-wide production for human consumption of this orchid, being Madagascar the most important producer and exporter worldwide (Pascale 2004). The Totonacas developed the traditional cultivation of vanilla in secondary forests (Soto-Arenas 2006). In the mid-19th century vanilla was introduced to the Indian Ocean islands and found the appropriate conditions for its large scale production in Madagascar (Romeu 1995). Nowadays the tendency of small and medium rural enterprises is to imitate the production methods of larger companies, without considering the particular cultural, socio-economic and ecological conditions of their region. The tropic's great diversity of biotic resources represents an opportunity and an advantage in the creation of micro-enterprises based on local raw materials. These are currently being commercialized without an added value and through appropriate marketing strategies might occupy special market niches (Chiriboga 2007).

10.9 CONCLUSION

The scientific, social and educational challenges arisen from our research, present inquires on conciliating theory with actions. Scientists, philosophers, environmental managers or educators should lead us to a coherent quest for socially, equitable and sustainable societies. Without doubt, to achieve sustainability we need to work on interdisciplinary actions involving ecology, economy, social and cultural elements, as well as emphasizing on the processes and not only the products. Current technological processes are jeopardizing the development of societies and ecological balance. Coinciding with other authors and philosophers, we will not preserve and restore ecological and cultural, diversity by abusing the ecosystems carrying capacity; at the same time we will not reach a global balance if we generate or accept regional or local inequities; and we will never attain social justice if we lack solidarity.

We recognize that many rural communities are at risk of losing their traditional natural resource's knowledge, and are struggling adapting to new ways of life and values, pressured by globalization and modernization. Therefore, it is important to recover and enhance traditional identities and preserve traditional practices and resource management systems. In Mexico ethnic groups have been negatively impacted by exotic cultures since colonial times, destroying and risking their natural and cultural heritage. Traditional resource management has been replaced by modern management schemes, based on scientific/industrial approaches, which has inflicted natural resources and augmented poverty and marginalization of the local dweller, particularly ethnic groups. Therefore the current proposal deals with the risk of biocultural resources erosion, by implementing integral and participatory management strategies, which go from conservation, restoration and productive activities, to the recovery of social and cultural capitals, throughout the application of the ethnoecological management model.

ACKNOWLEDGMENTS

We are grateful to Esli Suárez Zurita for assisting us in the formatting and translation of the manuscript.

REFERENCES

Attali J. 1982. Los tres mundos: Para una teoría de la post-crisis. Madrid (Spain): Ediciones Cátedra. 148 p.

Atria R, Siles M, Arriagada I, Robinson LJ, Whiteford S. 2003. Capital social y reducción de la pobreza en América Latina y el Caribe: En busca de un nuevo paradigma. Santiago de Chile: Comisión Económica para América Latina y el Caribe (CEPAL) and University of Michigan. 590 p.

Ballón E, Rodríguez J, Zeballos M. 2009. Fortalecimiento de capacidades para el DTR: Innovaciones institucionales en gobernanza territorial. Documento de Trabajo N° 53. Programa Dinámicas Territoriales Rurales. Santiago de Chile: Centro Latinoamericano para el Desarrollo Rural. 121 p.

Bessette G. 2004. Involving the community: A guide to participatory development communication. Penang: International Development Research Centre. 162 p.

Boulding K. 1966. The Economics of the coming spaceship Earth. In: Jarrett H, editor. Environmental quality in a growing economy: Essays from the sixth RFF forum. Baltimore (MD): John Hopkins Press. p 3–19.

Bourdieu P. 1986. The forms of capital. In: Richardson JG, editor. Handbook of theory and research for the sociology of education. New York (NY): Greenwood Press. p 241–258.

Burt RS. 1992. Structural holes: The social structure of competition. Cambridge (MA): Harvard University Press. 313 p.

Callicott JB. 1998. En busca de una ética ambiental. In: Kwiatkowska T, Issa J, editors. Los caminos de la ética ambiental: Una antología de textos contemporáneos. México: Consejo Nacional de Ciencia y Tecnología (CONACYT) and Plaza y Valdés. p 85–160.

Capra F. 1996. The web of life: A new scientific understanding of living systems. New York (NY): Anchor. 347 p.

Carlsson L, Berkes F. 2005. Co-management: Concepts and methodological implications. Journal of Environment Management, 75(1): 65–73.

Chazdon RL. 2008. Chance and determinism in tropical forest succession. In: Carson W, Schnitzer S, editors. Tropical forest community ecology. Hoboken (NJ): Blackwell Publishing. P 384–408.

Chiriboga M. 2007. Comercialización y pequeños productores. Boletín Intercambios [Internet]. [cited 2010 Mar 4]; 85: 1–45. Available from: http://www.rimisp.org/boletines/bol85/doc1.zip

Choi YD, Temperton VM, Allen EB, Grootjans AP, Halassy M, Hobbs RJ, Naeth MA, Torok K. 2008. Ecological restoration for future sustainability in a changing environment. Ecoscience 15(1): 53–64.

Coleman JS. 1988. Social capital in the creation of human capital. American Journal of Sociology 74: S95–S120.

Coleman JS. 1990. Foundations of social theory. Cambridge (MA): Harvard University Press. p 300–321.

Constanza R, D'Arge R, De Groot R, Farber S, Grasso M, Hannon B, Limburg K, Naeem S, O'Neill RV, Paruelo J, Raskin RG, Sutton P, Van den Belt M. 1997. The value of the world's ecosystem services and natural capital. Nature 387: 253–260.

Covington W, Niering WA, Starkey E, Walker J. 1998. Ecosystem restoration and management: Scientific principles and concepts [Internet]. Asheville, USA. [cited 2005 Feb 25]. Available from http://www.srs.fs.usda.gov/pubs/misc/misc_covingtion.pdf

Daly HE. 1992. From adjustment to sustainable development: The obstacle of free trade. In: Nader R, editor. The case against "free trade": GATT, NAFTA, and the globalization of corporate power. Berkeley (CA): Earth Island Press. p 121–132.

Daly, HE. 1997. De la economía de un mundo vacío a la economía del mundo lleno. In: Goodland R, editor. Medio ambiente y desarrollo sostenible: Más allá del informe Brundtland. Madrid (Spain): Trotta Editorial. p 37–50.

Del Amo-Rodríguez S, editor. 2002. La leña el energético rural en tres microregiones del sureste de México. México D.F. (Mexico): Editorial Plaza y Valdés, Programa Acción Forestal Tropical y el Consejo Nacional para la Enseñanza de la Biología. 190 p.

Del Amo-Rodríguez S, Ramos-Prado JM. 1993. Use and management of secondary vegetation in a humid-tropical area. Agroforestry Systems (21)1: 27–42.

Del Amo Rodríguez S, Vergara-Tenorio MC, Altamirano-Flores I. 2007. Rescatando y revalorando nuestros frutales nativos por medio de la creación de bancos de germoplasma in situ. LEISA Revista de Agroecología 23(2): 30–33.

Del Amo-Rodríguez S, Vergara-Tenorio MC, Ramos-Prado JM, Jiménez-Valdés L, Ellis EA. 2008a. Plan de ordenamiento ecológico de participación comunitaria del Municipio Zozocolco, Veracruz. Editorial de la Universidad Veracruzana. 131 p.

Del Amo-Rodríguez S, Vergara-Tenorio MC, Ramos-Prado JM, Jiménez-Valdés L, Ellis EA. 2008b. Plan de ordenamiento ecológico de participación comunitaria del Municipio Espinal, Veracruz. Editorial de la Universidad Veracruzana. 114 p.

Del Amo-Rodríguez S, Vergara-Tenorio MC, Ramos-Prado JM, Porter-Bolland L. (2010). Community landscape planning for rural areas: A model for biocultural resource management. Journal of Society & Natural Resources 23: 436–450.

Del Angel-Pérez AL, Mendoza MA. 2004. Totonac homegardens and natural resources in Veracruz, Mexico. Agriculture and Human Values 21(4): 329–346.

Dunn RR. 2004. Recovery of faunal communities during forest tropical regeneration. Conservation Biology 18(2): 302–309.

[FAO] Organización de las Naciones Unidas para la Agricultura y la Alimentación (Roma). 2007. Situación de los bosques del mundo [Internet]. Roma: FAO. [cited 2010 Jan 10]; Available from ftp://ftp.fao.org/docrep/fao/009/a0773s/a0773s.zip

Freire P. 1970. Pedagogy of the oppresed. New York (NY): Herder and Herder. 186 p.

Freire P. 1990. La naturaleza de la educación: Cultura, poder y liberación. Barcelona (Spain): Ediciones Paidós. 204 p.

Geilfus F. 2008. 80 tools for participatory development: Appraisal, planning, follow-up and evaluation. San José: Inter-American Institute for Cooperation on Agriculture (IICA). 208 p.

Gittell R, Vidal A. 1998. Community organizing: building social capital as a development strategy. Thousand Oaks (CA): Sage Publications Inc. p 13–55.

Giampietro M. 2005. Multi-scale integrated analysis of sustainability: A methodological tool to improve the quality of narratives. International Journal of Global Environmental Issues 5(3): 119–141.

Granovetter MS. 1973. The strength of weak ties. American Journal of Sociology. 78(6): 1360–1380.

Granovetter MS. 1983. The strength of the weak tie: A network theory revisited. Sociological Theory. 1: 201–33.

Guajardo L, Espinosa G, Hernández O. 2004. La participación campesina en la formulación de proyectos productivos como una alternativa de estrategia para el desarrollo rural. Santiago de Chile: Proyecto Regional de Cooperación Técnica para la Formación en Economía y Políticas Agrarias y de Desarrollo Rural en América Latina (FODEPAL). 24 p.

Habermas J. 1984. The theory of communicative action Volume 1: Reason and the rationalization of society. Boston (MA): Beacon Press. 465 p.

Hambler C. 2004. Conservation: Studies in biology. Cambridge: Cambridge University Press. p 275–307.

Hamilton ND. 2008. Rural lands and rural livelihoods: Using land and natural resources to revitalize rural America. Drake Journal of Agricultural Law 13(1): 180–204.

Haverkort B, Rist S. 2004. Towards Co-evolution of knowledge and sciences: No short cut in integrating local and global knowledge. In: Bridging Scales and Epidemiologies: Linking local knowledge with global science in multi-scale assessments. Conference Program and Abstracts: March 17–20; Alexandria, Egypt. p. 68–69.

Haverlook B, Vantt Hooff K, Hiemstra W, editors. 2002. Antiguas raíces, nuevos retoños: El desarrollo endógeno en la práctica. La Paz. Bolivia. Plural Editores. 336 p.

[INEGI] Instituto Nacional de Estadística y Geografía (México). 2000. XII Censo general de población y vivienda [Internet]. México D.F. (Mexico): Instituto Nacional de Estadística y Geografía. [cited 2010 Mar 17]; Available from: http://www.inegi.org.mx/est/contenidos/Proyectos/ccpv/cpv2000/default.aspx

Kelly IT, Palerm A. 1952. The Tajín Totonac. Washington (DC): Smithsonian Institution. 369 p.

[PLADEYRA] Planeación, Desarrollo y Recuperación Ambiental (Mexico). 2002. Potencial de recarga de acuíferos y estabilización de ciclos hídricos en areas forestadas [Internet]. México D.F. (Mexico): Instituto Nacional de Ecología and Secretaría de Medio Ambiente y Recursos Naturales. [cited 2010 Mar 17]; Available from: www.ine.gob.mx/descargas/dgipea/recarga_acuiferos_est.pdf

Macqueen DJ, Bose S, Bukula S, Kazoora C, Ousman S, Porro N, Weyerhaeuser H. 2006. Working together: Forest-linked small and medium enterprise associations and collective action. Gatekeeper Series No. 125. London: International Institute for Environment and Development. 24 p.

Maffi L. 2005. Biocultural diversity for endogenous development. COMPAS Magazine 9: 18–19.

Márquez-Huitzil R. 2005. Fundamentos teóricos y convenciones para la restauración ecológica: Aplicación de conceptos y teorías a la resolución de problemas en restauración. In: Sánchez O, Peters E, Márquez-Huitzil R, Vega E, Portales G, Valdez M, Azuara D, editors. Temas sobre restauración ecológica. México D.F.: Instituto Nacional de Ecología. p 159–168.

Martínez-Ramos M, García-Orth X. 2007. Sucesión ecológica y restauración de las selvas húmedas. Boletín de la Sociedad Botánica de México (80): 69–84.

Masferrer-Kan E. 2004. Totonacos: Pueblos indígenas del México contemporáneo. México D.F. (Mexico): Comisión Nacional para el Desarrollo de los Pueblos Indígenas and Programa de las Naciones Unidas para el Desarrollo México. 39 p.

Medellín SM. 1988. Arboricultura y silvicultura tradicional en una comunidad totonaca de la costa [dissertation].[Veracruz (Mexico) Instituto Nacional de Investigaciones Sobre Recursos Bióticos. 347p.

Mizrahi AP, Ramos-Prado JM, Jimenez-Osornio J. 1997. Composition, structure and management potential of secondary vegetation in a dry tropical forest. Forest Ecology and Management (96)3: 273–282.

Morin E. 1984. Ciencia con consciencia. Anthropos, Barcelona, España. 376 p.

Moyano-Estrada E. 2009. Capital social, gobernanza y desarrollo en áreas rurales. Ambient@ 88, Septiembre: 112–126.

Myers SC. 1984. The capital structure puzzle. Journal of Finance 39(3): 575–592.

Naveh Z. 2004. Ecological and cultural landscape restoration and the cultural evolution towards a post-industrial symbiosis between human society and nature. Restoration Ecology 6(2): 135–143.

Novo M. 2003. El desarrollo sostenible, sus implicaciones en los procesos de cambio. Polis Revista de la Universidad Bolivariana (1) 5: 1–19.

Ostrom E. 1999. Coping with tragedies of the commons. Annual Review of Political Science 2: 493–535.

Pascale B. 2004. RAPD genetic diversity in cultivated vanilla: *Vanilla planifolia*, and relationships with *V. tahitensis* and *V. pompona*. Plant Science 167(2): 379–385.

Putman RD. 1993. The prosperous community: social capital and public life. The American prospect 4(13): 35–42.

Putman RD. 1995a. Bowling alone: America's declining social capital. Journal of Democracy 6(1): 65–78.

Putman RD. 1995b. Turning in, turning out: the strange disappearance to social capital in America. Political Science and Politics 28(4): 664–683.

Quiroga-Martínez R. 2003. Para forjar sociedades sustentables. Polis Revista de la Universidad Bolivariana 1(5): 1–19.

Ralston SJ. 1997. La civilización inconsciente. Barcelona (Spain): Anagrama. 219 p.

Ramos-Martín J, Giampietro M. 2005. Multi-scale integrated analysis of societal metabolism: learning from trajectories of development and building robust sceneries. International Journal of Global Environmental Issues 5(3): 225–263.

Ramos-Prado JM, Del Amo-Rodríguez S, Gómez-Pompa A, Allen, E. 2004. Trial and error: Three approaches to tropical forest restoration. Proceedings of the Ecological Society of America's; 2004 Jul 30–Aug 7; Portland (O); Available from: http://abstracts.co.allenpress.com/pweb/esa 2004/document/34534

Ramos-Prado JM, Del Amo-Rodríguez S. 1992. Enrichment planting in a secondary forest in Veracruz, Mexico. Forest Ecology and Management. 54(1–4): 289–304.

Rebolledo AJ. 2007. Estudio etnoecológico de la vainilla (*Vanilla planifolia Andrews*) en la zona del Totonacapan [dissertation]. [Veracruz (México)]: Universidad Veracruzana. 74 p.

Rist S. 2002. Si estamos de buen corazón, siempre habrá producción. Caminos de la renovación de formas de producción y vida tradicional y su importancia para el desarrollo sostenible. La Paz, Bolivia: Plural Editores. 505 p.

Romeu E. 1995. La vainilla, de Papantla a Papantla: El regreso de un cultivo. Biodiversitas 1(1): 10–13.

Roth R. 2004. Spatial organization of environmental knowledge: Conservation conflicts in the inhabited forest of northern Thailand. Ecology and Society [Internet]. [cited 2010 Mar 17]; 9(3): 5. Available from: http://www.ecologyandsociety.org/vol9/iss3/art5/

[SER] Society for Ecological Restoration (US). 2004. Principios de SER International sobre Restauración Ecológica [Internet]. Tucson (AZ): SER. [cited 2010 Feb 20]; Available from: http://www.ser.org/pdf/REV_Spanish_Primer.pdf

Soto-Arenas MA. 2006. La vainilla: Retos y perspectivas de su cultivo. Biodiversitas (66):1–9.

Temkin K, Rohe W. 1997. Social capital and neighbourhood stability: An empirical investigation. Housing Policy Debate 9(1): 61–87.

Toledo-Manzur VM. 1992. Utopía y naturaleza: El nuevo movimiento ecológico de los campesinos e indígenas de América Latina. Nueva Sociedad 122: 72–85.

[UNRISD] United Nations Research Institute for Social Development (US). 2002. New information and communication technologies: Social development and cultural change [Internet]. Geneva: UNRISD. [cited 2010 Feb 20]; Available from http://www.unrisd.org/engindex/publ/ list/dp/dp86/dp86-02.htm.

Vázquez-Yanes C, Batis A. 1996. Adopción de árboles nativos valiosos para la restauración ecológica y la reforestación. Boletín de la Sociedad Botánica de México 58: 75–84.

Vázquez-Yanes C, Orozco-Segovia A. 1994. Signals for seed and response to gaps. In: Caldwell MM, Pearcy RW, editors. Exploitation of environmental heterogeneity by plants. San Diego (CA): Academic Press. p 209–236.

Vega-Peña EV. 2005. Algunos conceptos de ecología y sus vínculos con la restauración. In: Sánchez O, Peters E, Márquez-Huitzil R, Vega E, Portales G, Valdez M, Azuara D, editors. Temas sobre restauración ecológica. México D.F.: Instituto Nacional de Ecología. p 147–154.

Vergara-Tenorio MC, Cervantes-Vázquez JR. 2009. Riesgo, ambiente y percepciones en una comunidad rural totonaca. Economía, Sociedad y Territorio 9(29): 145–163.

Whitelaw G, McCarthy D. 2008. Governance, social capital and social learning: Insights from activities in the Long Point World Biosphere Reserve and Oak Ridges Moraine, Ontario, Canada. In: Opermanis O, Whitelaw G, editors. Economic, social and cultural aspects in biodiversity conservation. Rīga: The University of Latvia Press. p 123–130.

Chapter 11

Atmospheric aerosols: A case study from the Caribbean Region

Antonio E. Carro-Anzalotta and Lisbeth San Miguel-Rivera

SUMMARY

Aerosols are small (1 nm–100 μm) and very dynamic particles suspended in the atmosphere. Some of these particles are directly emitted into the atmosphere and others are products of atmospheric synergism. They affect air thermodynamics, climate, ecosystems, agriculture and human health at local, regional and global scale. The net global albedo is increased by atmospheric aerosols counteracting global warming. The climate of cities is strongly determined by these suspended materials. Most aerosols originate from natural sources such as mineral dust, sea spray, volcanoes, natural fires and biogenic processes. Human activities add dust, soot, sulfates, nitrates and volatile organic compounds (VOCs). Although anthropogenic aerosols are a small fraction (10%) of the global total, their toxicity and concentrations in populated areas make them a hazard to human health. In the Caribbean Region, natural aerosols include the Sahara Air Layer, ash from Le Soufriere volcano in the Lesser Antilles, marine and biogenic sources. Manmade aerosols include PM_{10}, $PM_{2.5}$, SO_x, NO_x and VOCs. This chapter will review the assessment of aerosols and their impacts on climate and human health.

11.1 INTRODUCTION

Atmospheric aerosols are important components of Earth atmosphere in terms of thermodynamics, climate, ecosystems and human health. They consist of small particles (1 nm–100 μm) solid, liquid or solid-liquid suspended in a gas. Most aerosols originate from natural sources but a great deal of environmental and health problems are caused by anthropogenic sources. Aerosol size and chemical composition vary with location and time. Besides, aerosols are ubiquitous in air and are often observable as dust, smoke, and haze. Both natural and human processes contribute to aerosol concentrations (Remer et al. 2009).

Aerosols have a significant effect on the climate system and in human health. Atmospheric aerosols particles and trace gases affect the quality of our life in many different ways. In polluted urban environments, they influence human health and deteriorate visibility (Kumala et al. 2008). The human effects include premature mortality, aggravation of respiratory and cardiovascular diseases (Voorhees and Uchiyama 2008; Brunekreef and Holgate 2002).

On the global scale aerosols affect climate by altering the atmospheric heat budget. These affect the Earth's energy budget by scattering and absorbing radiation (direct effect) and modifying the microphysical and radiative properties of clouds (indirect effect) and the environment in which the clouds develop. Such effects can change precipitation patterns as well as cloud extent and optical properties (Remer et al. 2009), or the planetary albedo range (Kiehl and Trenberth 1997).

Aerosol studies involve a wide range of scientific knowledge: chemistry, physics, biology, geology and meteorology in settings like laboratory, open field, ships, airplanes, satellites and numerical models. Interactions among atmospheric suspended matter make aerosols synergy a complex and largely unknown scientific area. A common problem is lack of integration of data produced by new studies. Aerosol assessment strategies require research, monitoring, data base creation and integration of information for decision and policy making (Fuzzi et al. 2006).

11.2 SOURCES OF AEROSOLS

Mineral dust, sea salt, volcanic ash and forest fires are the primary (95) sources of natural aerosols in the atmosphere. At a global scale, North Africa, the Middle East and arid regions of India and China are the main sources for mineral dust aerosol in the atmosphere. Saharan dust crosses the Atlantic Ocean adding nutrients to the oceanic water column and affecting the climate all the way to the Caribbean Region and Central America (Figure 11.1). Dust from Asia brings nutrients to Pacific and Indian oceans. Dust storms pose a serious environmental threat to agriculture and ecosystems. Volcanic activity can produce enormous amounts of aerosols over a short time span during cataclysmic eruptions affecting climate on a global scale for periods of years (Pitari and Mancini 2002).

Manmade aerosols made only a 10% fraction of the global total; their chemical nature, mobility and concentration are a serious threat on human health, ecosystems, agriculture and exposed materials. Anthropogenic aerosols are formed predominantly by combustion and gas to particle conversion resulting in average particle sizes in the sub-micron accumulation mode size range (Stier et al. 2007). Common problems associated with anthropogenic aerosols include acid deposition smog and thermal inversions.

11.3 AEROSOL SYNERGY: FORMATION OF NUCLEATION, CONDENSATION, AND COAGULATION SUBSTANCES

Once released into the atmosphere, primary aerosols can interact in different ways with other components. The products of these interactions vary depending on chemical-physical nature of the particle, humidity, temperature, sunlight and other factors, making them varied and difficult to predict (Kulmala et al. 2005). The formation of new particles and their subsequent growth appears to occur almost everywhere in the atmosphere (Kulmala et al. 2004). As described by Yu and Luo (2009), the particles in the troposphere either come from in-situ nucleation (secondary particles) or direct emission (primary particles). Trace gases continually interact with existing aerosol particles. Sulfur, nitrogen, carbon oxides, and volatile organic compounds (VOC_s) are common primary pollutants that can nucleate to form liquid substances that continue to react and grow from nano-to micro-scale (Yu and Luo 2009).

The growth rate of particles is greater at higher temperatures and in cities. Condensation reactions bring gasses into liquids releasing heat. The most common condensation reaction in aerosols is water vapor (g) over salt particles to form drop-lets and clouds (heterogeneous nucleation). Salt particles present in air act like a condensation nuclei which is hygroscopic. Dust, soot, and smoke aerosols are other type of condensation nuclei. Since condensation releases heat, the amount and type of aerosols affect atmospheric thermodynamics (Kulmala et al. 2005).

Particles can increase their size by colliding and adhering to other particles in a process called coagulation. Water droplets within a cloud coagulate to produce rain. Kulmala et al. (2004) describes the sequence in the growth of a particle as:

1. Condensation of nucleating vapors (after homogeneous nucleation).
2. Activation of soluble vapors.
3. Heterogeneous nucleation.
4. Charge-enhanced condensation.
5. Self-coagulation.
6. Multi-phase chemical reactions.

The observations in the study of Kulmala et al. (2004) conclude that strongly that vapors are responsible for the main growth of small clusters or particles are different from those driving the nucleation, at least in the boreal forest conditions.

11.4 AEROSOLS AND THE THERMAL BALANCE OF EARTH

Radiative forcing is the difference between the incoming-outgoing radiations in Earth's atmosphere. Most aerosols have negative radiative forcing values because they tend to scatter and reflect sunlight creating a cooling effect on the atmosphere. This effect has been documented in great volcanic eruption events like the Tambora 1815 and Pinatubo 1991 eruptions (De Angelis et al. 2003; Pitari and Mancini 2002). The 1816 Tambora volcano eruption created climatic anomalies like the "Year without a summer". This event resulted in the worst famine of the 19th century with thousands of deaths. It has been observed that the global-average surface temperature could have declined by 0.2–0.3°C for one to three years following a major eruption (De Angelis et al. 2003).

Volcanic sulphate aerosols have a strong influence on the atmospheric system by changing its chemical composition and altering its radiation balance (Timmreck and Graf 2006). These additional stratospheric particles play an important role in the chemical budget of the middle atmosphere, producing catastrophic ozone losses and add significant radiative forcing on the global climate for a few years (Pitari and Mancini 2002). The injection of SO_2 to the stratosphere to reduce the effects of Global warming has been recently suggested (Prata et al. 2007).

Clouds affect the atmosphere thermodynamics by reflecting and transporting energy. The effects of aerosols on climate can direct if they scatter and absorb incoming solar radiation, or indirect if they act as cloud condensation nuclei thereby altering radiative properties of clouds (Tunved et al. 2003). Atmospheric particles perturb the Earth's energy budget indirectly by acting as cloud condensation nuclei and thus changing cloud properties and influencing precipitation (Yu and Luo 2009). The effects of aerosols on the radiative properties of Earth's cloud cover are referred to as indirect climate forcing.

11.5 THERMAL INVERSIONS AND CITY CLIMATE

Thermal inversions are events that usually occur in cities located on valleys surrounded by mountains. On normal atmospheric conditions, temperature in the troposphere decreases with altitude. Inversions include a layer of cold air or fog overlying a city for days or weeks with stable conditions creating smog and accumulating anthropogenic pollutants and aerosols. These conditions prevent the development of convection and pollution dispersal (Kulmala et al. 2008). A significant number of health problems related to atmospheric aerosols and fog droplets are believed to be due to particles having diameters less than $10\,\mu m$ because these particles can penetrate deep into the respiratory system (Framptom 2001). Pollution generated by thermal inversions requires quantitative assessment in order to establish limits and monitor air quality daily to make decisions based on real time data (Kulmala et al. 2008).

11.6 CLIMATE EFFECTS OF AEROSOLS

Aerosols exert a variety of impacts of the environment. Particular matter (PM), near in the surface; have long been recognized as affecting pulmonary function and other aspects of human health. Nitrates and sulfates aerosols have a major role in acidifying the surface downwind of gaseous sulfur and nitrogen sources. Particles deposited far downwind fertilize iron poor water in far oceans, and Sahara layer (SAL) reaching the Amazon Basin is through to contribute nutrients to the rainforest soil. Aerosols also interact strongly with solar and terrestrial radiation in several ways (Kahn 2009).

During the last century, the Earth's surface temperature increased by $0.6°C$, reaching the highest levels in the last millennium (Kaufman et al. 2002). This temperature change is attributed to a balance between absorption of incoming solar radiation and emission of thermal radiation from the Earth system. Anthropogenic aerosols are intricately linked to the climate system and to the hydrologic cycle. The major effect of aerosols is to cool the climate system by reflecting sunlight. Depending on their composition, aerosols can absorb sunlight in the atmosphere, further cooling the surface but warming the atmosphere in the process. These effects on the temperature profile, along with the role of aerosols as cloud condensation nuclei, impact the hydrologic cycle (Kaufman et al. 2002).

In polluted regions, aerosol particles share the condensed water during cloud formation, downsizing the cloud droplets in 20 to 30%, hence reducing precipitation. Furthermore, they increase the sunlight reflectance in more than 25% (Kaufman et al. 2002) and cooling the Earth's surface. Modifications in rainfall generation change the thermodynamic processes in clouds, and consequently the dynamics of the atmospheric (heat engine) that drives all of weather and climate (Andreae 2005).

11.7 THE CARIBBEAN REGION AND PUERTO RICO

The Caribbean Region includes most of the tropics of the western hemisphere with Venezuela and Colombia on the south, Central America on the west and the Antilles (West Indies) on the north and east. Puerto Rico is the easternmost of the Greater Antilles, with its center at (18°15′ N; 66°30′ W). The main climatic influence for Puerto Rico and most of the Caribbean Region are trade winds from the north-east.

In the Caribbean, a large fraction of natural aerosols includes desert and soil dust, sea salt particles, and volcanic ash (Kahn 2009). Natural aerosols include the SAL and volcanic material from Soufriere Hills Volcano at the Lesser Antilles island of Montserrat along with sea salts and minor amounts of biogenic material. Pett-Ridge (2009) describes the Caribbean Region as the downwind zone from the Earth's largest dust source.

11.8 THE SAHARA AIR LAYER (SAL)

In the tropical Atlantic the concentration and the variability of aerosol from Africa are among the largest, if not the largest, in the world. It is well known that the mineral dust from the West Africa is often transported into tropical Atlantic region north of the intertropical convergence zone (ITCZ) within a layer of the lower troposphere known as the SAL (Huang et al. 2009). Great quantities of African dust are carried over large areas of the Atlantic and to the Caribbean during much of the year (Figure 11.1). African dust can reach the Caribbean most strongly in boreal summer.

Figure 11.1 The Sahara Air Layer drifting eastward toward the Caribbean Region (Prospero and Lamb 2003).

The Sahara and bordering regions mobilize and suspended considerable quantities of aerosols in the atmosphere, about 1000 and 3000 Tg/yr (Li et al. 2004). SAL has a substantial influence in the regional radiative budget (Moulin 1997). Aerosols radiative process can modify the atmospheric and surface conditions to affect precipitation. At the microphysical level, mineral dust may serve as cloud condensation nuclei (CCN) to facilitate warm rain formation. However, precipitation suppression due to desert dust is also possible (Huang et al. 2009).

The Caribbean Region is well known for its hurricane season. Dunion and Velden (2004) found a connection between the SAL and the suppression of the tropical cyclone (TC) activity. The SAL may adversely affect the TC formation, increasing the atmospheric stability, reducing moisture in the middle troposphere and intensifying wind shear in the environment surrounding the tropical cyclone developing.

Saharan dust has also important influences on nutrient dynamics and biogeochemical cycling of both oceanic and terrestrial ecosystems in North Africa and due to frequent long-range transport dust across the Atlantic Ocean and among others. SAL is transported thousands of kilometers and its influence extends as far afield as Amazonian, the southeast USA, the coral reefs of the Caribbean, and northern Europe. Furthermore, SAL loadings may also have considerable climatic significance through a range of possible mechanisms, and the frequency of dust events can change substantially in response to climate changes both in the long-term and short-term (Middleton and Goudie 2001).

Garrison et al. (2003) indicated that African and Asian dust air masses transport chemical and viable microbial contaminants to downwind ecosystems in the Americas, the Caribbean, and the northern Pacific and may be adversely affecting those ecosystems. Atmospheric flux of pathogenic microorganism may be responsible for the widespread distribution of some diseases occurring on coral reef and associated habitats. Human health is also adversely affected, primary by inhalation of known or suspected components in dust events, including nonpathogenic and pathogenic viable microorganism; chemical contaminants such as carcinogens, toxins, endocrine disruptors, and toxic metals; and small particles that may trigger other physiological reactions (Garrison et al. 2003).

11.8 LE SOUFRIERE

Volcanoes are perforations in the Earth's crust through which molten rock and gases escape to the surface (Blanchard-Boehm 2004). The volcanic eruption expels vast quantities of ash particles and emits gases into the atmosphere, to reside in the stratosphere where they affected the radiative balance of the atmosphere and the Earth's climate (Jones et al. 2007; Prata et al. 2007). Moreover, volcanoes are sources of sulfur dioxide (SO_2), which, along the sulfur containing gases produced by ocean biologic and the decomposition of organic mater. Those are examples of gases that can be converted in secondary aerosols by chemical processes in the atmosphere (Kahn et al. 2009). Approximately, 0.1 ± 0.01 Tg was injected into the stratosphere in form of SO_2 (Prata et al. 2007).

The islands of the Lesser Antilles in the eastern Caribbean Sea represent a large volcanic arc extending from Puerto Rico and the Virgin Islands in the north to

Grenada in the south (Martin et al. 2005). Since July 1995 Soufriere Hills Volcano in the Island of Montserrat (16.7° N, 62.2° W, 915 m asl) has erupted intermittently affecting most of the Caribbean area. Its location at the north-east of the Caribbean combined with trade winds E-NE often result in aerosol events that dominate the effects of the SAL in the region (Figure 11.2). Prata et al. (2007) utilized three differ-ent satellite based measurements to track the aerosols of May 2006 Soufriere erup-tion through the Caribbean Region, across Central America into the Pacific Ocean during 23 days and over 18,000 km.

During this event, a highly buoyant eruption column of ash and gases rose to heights of at least 17 km. The eruption cloud also contained copious amounts of SO_2 and some hydrogen chloride. The sulphate aerosol causes climate effect in the Caribbean Region. Ash blankets from the eruptions will favor albedo (Jones et al. 2007). The Soufriere Hills SO_2 cloud would cause <0.01 K of global surface cooling (Jones et al. 2007; Prata et al. 2007).

11.9 EFFECTS OF ACID RAIN

Originally explained in 1852 by Robert Angus Smith in Manchester England, acidic deposition includes rain, fog, clouds, dew and frost. Normal rain tends to have a pH value of 5.6 due to the presence of CO_2. Sulfur, nitrogen and carbon oxides mix with cloud droplets or water vapor to produce precipitation or deposition with a lower than normal pH. Acid rain usually carries other pollutants like metals depend-ing on where it forms (Chehregani and Kavianpour 2007).

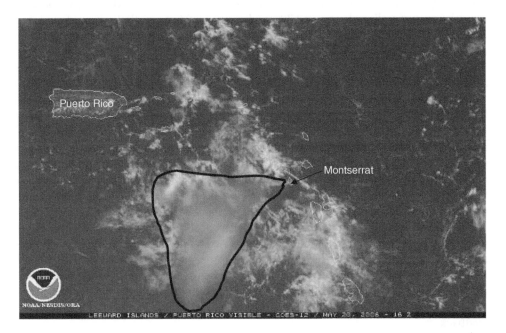

Figure 11.2 Soufriere Hills Volcano aerosols plume dispersion through the Caribbean Region (Prata et al. 2007).

Atmospheric pollution in the form of acid rain compromises the health of forest ecosystems, affects the chemical balance of surface waters and soils and causes erosion of buildings materials. The ecological effects of acid rain are most clearly seen in the aquatic environments (Astel et al. 2008). Acid rain causes a cascade effects that harm or kill individual fish, reduce fish population numbers or completely eliminate fish species from water body, and decrease biodiversity (EPA 2009a; Shea 2008). Moreover, researchers found that acid rain causes slower growth, injury, or death of forest (Chehregani and Kavianpour 2007; Balasubramanian et al. 2007). It has also been implicated in forest and soil degradation (EPA 2009b).

Sulfates and nitrates that form in the atmosphere from sulfur dioxide (SO_2) and nitrogen oxide (NO_x) emissions contribute to visibility impairment. Sulfate particles account for 50 to 70 percent of the visibility reduction in the eastern part of the United States (EPA 2009c). In addition, acid rain and the dry deposition of acidic particles contribute to the corrosion of metals and the deterioration of paint and stone. These effects significantly reduce the societal value of buildings, bridges, and cultural objects (EPA 2009d).

11.10 RISK MANAGEMENT CONCERNS

11.10.1 Health effects overview

On a global scale, respiratory conditions results in a large amount of disability and premature death. Air pollution is a risk factor for a number of respiratory conditions, and is directly associated with over half a million deaths per year (Buckeridge et al. 1998). Pandey and Nathwani (2003) indicated that the mortality risk in most polluted area is 17% higher than in least pollute areas. Estimates of exposure and risk for acute respiratory infections, chronic obstructive pulmonary disease, and lung cancer, in several epidemiological studies, show that over 1.6 million premature deaths, and nearly 3% of the global burden disease, were attributable to indoor air pollution from solid fuels in 2000 (Ezzati 2005).

Overall, epidemiologic evidence suggest that air pollution, especially fine combustion source pollution, common to many urban and industrial environments is an important risk factor for cardiopulmonary disease and mortality. Reviewers often note that evidence of health effects due to acute or short-term exposures is stronger that the evidence for chronic effects due to longer-term exposure (Pope 2000). People with heart or lung diseases, children and older adults are the most likely to be affected by air pollution exposure. However, healthy people may experience temporary symptoms from exposure to elevated levels of air pollutants (EPA 2009e).

Scientists and government officials used risk assessment to estimate the increased risk of health problems in people who are exposed to different amounts of air pollutants. Additionally, with the risk estimates and other factors, the government can set regulatory standards to reduce people's exposures to toxic air pollutants and health problems (EPA 2009f). The suggested assessment activity includes monitoring aerosols, development a data base, create educational programs and government-citizen feedback.

These activities required involvement and commitment of the different sectors of society as government, universities, private sector, community organizations and citizens.

The Caribbean Region needs governments more interested in research and monitoring of air pollutants, giving a major role to the scientific community and raising awareness among citizens about the effects of these on human health and the environment.

11.11 AEROSOLS AND HUMAN HEALTH EFFECTS

The health effects of air pollution have been subject to intense study (Brunekreef and Holgate 2002; Reyes et al. 2000). Air pollution in Latin America and Caribbean countries has been a major environmental health issue for people living in metropolitan areas. These issues arise as a consequence of urban development, population growth, and industrialization (Reyes et al. 2000).

Exposure to pollutants such as airborne particulate matter (PM) and ozone has been associated with increases in mortality and hospital admissions due to respiratory and cardiovascular diseases (Brunekreef and Holgate 2002). Organic compounds, ultrafine particles (UFP$_s$), biologic components, and transition metals are few of the constituents that reportedly exert some type of adverse effect on human health (Reyes et al. 2000).

11.12 PARTICULAR MATTER

Adverse health effects and increased mortality associated with particular air pollution originated from combustion sources are a major concern for population health and quality of life. Fine particulate pollution typically contains a mixture of soot, acid condensates, sulfate, and nitrate particles primary produced by combustion of fossil fuels in transportation, manufacturing, and power generation. The toxic nature of these substances and their ability to penetrate deep into the lungs pose a risk to health. Various pollution exposure studies found statistically significant relationship between increased PM/ozone levels and premature mortality/morbidity (Happo et al. 2008; Voorhees and Uchiyama 2008; Pandey and Nathwani 2003). Also, increased levels of particulate air pollution are associated with asthma exacerbations, increment in respiratory symptoms, decreased lung function, augment of medication use, and hospital admissions (Chalupa et al. 2004).

PM comprises a mixture of several compounds, including carbon centered combustions particle, secondary inorganics, and crustal derived particles. These compounds may contribute with different potential to the PM induced health effects. In practical terms, PM$_{10}$ can penetrate into lower respiratory system and PM$_{2.5}$ can penetrate into the gas-exchange region of the lung (Brunekreef and Holgate 2002). UFP$_s$ $< 0.10\,\mu m$ in diameter are ubiquitous in ambient particulate pollution and dominate ambient particle number and surface area concentrations, both indoor and outdoor for their small size (Frampton 2001). The UFP$_s$ may contribute to health effects of PM because of their high surface area, oxidant capacity ability to evade macrophage phagocytosis, and propensity for inducing pulmonary inflammation (Chalupa et al. 2004).

Numerous scientific studies have linked particle pollution exposure to a variety of problems, including increased respiratory symptoms, such as irritation of the airways, coughing, or difficult breathing (Voorhees and Uchiyama 2008; Schwarze PE et al. 2006; Chalupa et al. 2004). The Environmental Protection Agency (2009a) indicates that people exposure to UFP$_s$ is related with decreased lung function, aggravated asthma, development of chronic bronchitis, irregular heartbeat, nonfatal heart attacks, and premature death in people with heart or lung disease.

11.13 CARBON, SULPHUR AND NITROGEN OXIDES

Atmospheric pollution also causes the appearance and incidence of chronic bronchitis, optic irritation and lung carcinoma among urban populations. The release of carbon dioxide and carbon monoxide in the atmosphere affects the central nervous system even at low concentrations (Alam 2007). Carbon, sulfur and nitrogen oxides are increasingly associated with occurrence of allergies, chest problems and cancer (Astel et al. 2008). When inhaled, sulphur dioxide (SO_2) and nitrogen oxides (NO_x) irritate the respiratory system (Alam 2007).

SO_2 and NO_x interact in the atmosphere to form fine sulfate and nitrate particles that can be transported long distances by winds and inhaled deep into people's lungs. Many scientific studies have identified a relationship between elevated levels of fine particles and increases illness and premature death from heart and lung disorders, such as asthma and bronchitis (Chalupa et al. 2004; Brunekreef and Holgate 2002; Reyes et al. 2000).

11.14 OZONE

The NO_x react with volatile organic compounds (VOC_s) and form ozone (Sousa et al. 2008). Ozone impacts on human health include a number of morbidity and mortality risks associated with lung inflammation, including asthma and emphysema (Babin et al. 2008).

Scientists have been studying the effects of ozone on human health for many years (Babin et al. 2008; Sousa et al. 2008). They found that ozone can cause several types of short-term health effects in the lungs like irritate the respiratory system, reduce lung function, aggravate asthma, inflame and damage the lining of the lung. Ozone damage can also occur without any noticeable signs. Sometimes there are no symptoms, or they are too subtle to notice. In fact, EPA 1999 establishes that ozone continues to cause lung damage even when the symptoms have disappeared.

Gao et al. (2010) indicates that ozone depletion leads to elevated levels of Ultraviolet-B (UVB) at the Earth's surface, resulting in an increase of health risk. The UVB causes nonmelanoma skin cancer and plays a major role in malignant development (Burke and Wei 2009). In addition, UVB has been linked to cataracts (EPA 2009g).

11.15 BIOMASS FUELS AND COAL

Worldwide, almost 3 billion people use biomass (wood, charcoal, crop residues, and animal dung) and coal as their main source of energy for cooking, heating and other household needs (Edelstein et al. 2008; Padhi and Padhy 2008;

Ezzati 2005). Exposure to biomass smoke is therefore a major public health issue (Edelstien et al. 2008).

Burning biomass emits smoke that contains high levels of pollutants like CO, SO_2, NO_x, formaldehyde, benzo[a]pyrene and benzene, which are hazardous for human health (Padhi and Padhy 2008). Inhaling these pollutants result in respiratory morbidity and mortality in children and women (Edelstien et al. 2008). Moreover, high exposures to air pollutants in biomass smoke have been associated with acute respiratory infections, chronic obstructive pulmonary disease, lung cancer, asthma, nasopharyngeal and laryngeal cancers, tuberculosis and diseases of the eye (Ezzati 2005; Mishra 2003).

11.16 CONCLUSION

Aerosols should be considered as one of the most important components of the atmosphere any place in the world. Their effects on materials, ecosystems and human health have been well studied but there is a need for more data and to improve the integration of existing information.

The Caribbean Region is severely impacted with natural and anthropogenic aerosols. This is mainly due to the Sahara Air Layer, the volcanic ash from the Soufriere Hills Volcano and recently industrial and population growth that converts this Region into a high risk area for human health and the environment. Corrective actions should be taken to reduce the risk on ecosystems and human populations.

Risk assessment strategies for this region should include constant monitoring of natural and anthropogenic sources and the integration of aerosol produced conditions into the daily climate forecasts. New technology like satellites, probes, analytic instruments and computer modeling have provided an enormous amount of information to understand aerosols synergy. Risk management should receive the same attention as research and data base creation for every particular location. The Caribbean Region needs more commitment from government and citizens regarding risk assessment of atmospheric aerosols for both health and environmental effects.

REFERENCES

Alam SM. 2007. Environmental Pollution: a silent killer. Economic Review 8: 65–6.

Andreae MO, Jones CD, Cox PM. 2005. Strong present-day aerosol cooling implies a hot future. Nature 435(7046): 1187–90.

Astel AM, Walna B, Simeonov V, Kurzyca I. 2008. Multivariate statistics as means of tracking atmospheric pollution trends in Western Poland. Journal of Environmental Science and Health Part A 43(3): 313–28.

Babin S, Burkom H, Holtry R, Tabernero N, Davis-Cole J, Stokes L, DeHaan K, Lee D. 2008. Medicaid patient asthma-related acute care visit and their associations with ozone and particles in Washington, DC, from 1994-2005. International Journal of Environmental Health Research 18(3): 209–21.

Balasubramanian G, Udayasoorian C, Prabu PC. 2007. Effects of short-term exposure of simulated acid rain on the growth of acacia nilotica. Journal of Tropical Forest Science 19(4): 198–206.

Blanchard-Boehm RD. 2004. Natural hazard in Latin America: tectonic forces and storm fury. Social Studies 95(3): 93–105.

Buckeridge D, Gozdyra P, Ferguson K, Schrenk M, Skinner J, Tam T, Amrhein C. 1998. A study of the relationship between vehicle emissions and respiratory health in an urban area. Geographical & Environmental Modelling 2(1): 23–42.

Brunekreef B, Holgate ST. 2002. Air pollution and health. Lancet 360: 1233–42.

Burke KE, Wei H. 2009. Synergistic damage by UVA radiation and pollutants. Toxicology and Industrial Health 25: 219–24.

Chalupa DC, Morrow PE, Oberdörster G, Utell MJ, Frampton MW. 2004. Ultrafine particle deposition in subjects with asthma. Environmental Health Perspectives 112: 879–882.

Chehregani A, Kavianpour F. 2007. Effects of acid rain in the developmental stages of ovules and seed proteins in bean plants (Phaseolus vulgaris L.). American Journal of Plant Physiology 2(6): 367–72.

De Angelis M, Simões J, Bonnaveira H, Taupin J, Delmas R. 2003. Volcanic eruptions record in the Illimani ice core (Bolivia): 1918-1998 and Tambora periods. Atmospheric Chemistry and Physics 3: 1725–41.

Dunion JP, Velden CS. 2004. The impact of the Saharan air layer on the Atlantic tropical cyclone activity. Bulletin of the American Meteorological Society 85(3): 353–65.

Edelstein M, Pitchforth E, Asres G, Silverman M, Kulkarni N. 2008. Awareness of health effects of cooking smoke women in the Gondar Region of Ethiopia: a pilot survey. BMC International Health and Human Rights 8(10): 1–7.

[EPA] Environmental Protection Agency (USA). 1999. Smog-who does it hurt? What you need to know about ozone and your health. Environmental Protection Agency Technical Report EPA-452/K-99-001. 8 pp.

[EPA] Environmental Protection Agency (USA). 2009a. Effects of acid rain- surface waters and aquatic animals [Internet]. Washington, DC. Available from: http://www.epa.gov/acid-rain/effects/surface_water.html

[EPA] Environmental Protection Agency (USA). 2009b. Effects of acid rain- forests [Internet]. Washington, DC. Available from: http://www.epa.gov/acidrain/effects/forests.html

[EPA] Environmental Protection Agency (USA). 2009c. Effects of acid rain- visibility [Internet]. Washington, DC. Available from: http://www.epa.gov/acidrain/effects/visibility.html

[EPA] Environmental Protection Agency (USA). 2009d. Effects of acid rain- materials [Internet]. Washington, DC. Available from: http://www.epa.gov/acidrain/effects/materials.html

[EPA] Environmental Protection Agency (USA). 2009e. Health and environment [Internet]. Washington, DC. Available from: http://www.epa.gov/air/particlepollution/health.html

[EPA] Environmental Protection Agency (USA). 2009f. Risk assessment for toxic air pollutants: a citizen's guide [Internet]. Washington, DC. Available from: http://www.epa.gov/ttn/atw/3_90_024.html

[EPA] Environmental Protection Agency (USA). 2009g. Health and environmental effects of ozone layer depletion [Internet]. Washington, DC. Available from: http://www.epa.gov/Ozone/science/effects/index.html

Ezzati M. 2005. Indoor air pollution and health in development countries. Lancet 366: 104–6.

Framptom MW. 2001. Systemic and cardiovascular effects of airway injury and inflammation: ultrafine particle exposure in humans. Environmental Health Perspectives 109(4): 529–32.

Fuzzi S, Andreae M, Huebert B, Kulmala M, Bond T, Boy M, Doherty S, Guenther M, Kanakidou M, Kawamura K, Kerminen V, Lohmann U, Russell L, Pöschl U. 2006. Critical assessment of the current state of scientific knowledge, terminology, and research needs concerning the role of organic aerosols in the atmosphere, climate, and global change. Atmospheric Chemistry and Physics 6: 2017–38.

Gao Z, Gao W, Chang N. 2010. Detection of multidecadal changes in UVB and total ozone concentrations over the continental US with NASA TOMS data and USDA ground-based measurements. Remote Sensing 2(1): 262–77.

Garrison VH, Shinn EA, Foreman WT, Griffin DW, Holmes CW, Kellogg CA, Majewski MS, Richardson LL, Ritchie KB, Smith GW. 2003. African and Asian dust: from desert soils to coral reefs. Bioscience 53(5): 469–80.

Happo MS, Hirvonen MR, Hälinen AI, Jalava PI, Pennanen AS, Sillanpää M, Hillamo R, Salonen R. 2008. Chemical composition responsible for inflammation and tissue damage in the mouse lung by coarse and fine particle samples from contrasting air pollution in Europe. Inhalation Toxicology 20: 1215–31.

Huang J, Zhang C, Prospero JM. 2009. Aerosol-induced large-scale variability in precipitation over the tropical Atlantic. Journal of Climate 22: 4970–88.

Jones MT, Sparks RS, Valdes PJ. 2007. The climate impact of supervolcanic ash blankets. Climate Dynamics 29: 553–64.

Kahn RA, Yu H, Schwartz SE, Chin M, Feingold G, Remer LA, Rind D, Halthore R, DeCola P. 2009. Chapter 1 Introduction. In: Chin M, Kahn RA, Schwartz SE, editors. Atmospheric aerosol properties and climate impacts, A report by the US Climate Change Science Program and the Subcommittee on Global Change Research. Washington DC (USA): National Aeronautics and Space Administration. p 9–20.

Kaufman YJ, Tanré D, Boucher O. 2002. A satellite view of aerosols in the climate system. Nature 419(6903): 215–23.

Kiehl JT, Trenberth KE. 1997. Earth's annual global mean energy budget. Bulletin of the American Meteorology Society 78(2): 197–208.

Kulmala M, Laakso L, Lehtinen K, Dal Maso M, Anttila T, Kerminen V, Hõrrak U, Vana M, Tammer H. 2004. Initial steps of aerosol growth. Atmospheric Chemistry and Physics 4: 2553–60.

Kulmala M, Petäjä T, Mönkkönen P, Koponen IK, Dal Maso M, Aalto PP, Lehtinen K, Kerminen VM. 2005. On the growth of nucleation mode particles: source rates of condensable vapor in polluted and clean environments. Atmospheric Chemistry and Physics 5: 409–16.

Kulmala M, Kerminen V, Laaksinen A, Riipinen I, Sipila M, Ruuskanen TM, Sogacheva L, Hari P, Bäck J, Lehtinen KEJ, Viisanen Y, Bilde M, Svenningsson B, Lazaridis M, Tørseth K, Tunaved P, Nilsson ED, Pryor S, Sørensen L, Hõrrak U, Winkler PM, Swietlicki E, Riekkola M, Krejci R, Hoyle C, Hov Ø, Mthre G, Hansson H. 2008. Overview of the biosphere-aerosol-cloud-climate interactions (BACCI) studies. Tellus 60B(3): 300–17.

Li F, Vogelmann AM, Ramanathan V. 2004. Saharan dust aerosol radiative forcing measured from space. Journal of Climate 17: 2558–70.

Martin JW, Wishner K, Graff JR. 2005. Caridean and sergestid shrimp from the Kink'em Jenny submarine volcano, southeastern Caribbean Sea. Crustaceana 78(2): 215–21.

Middleton NJ, Goudie AS. 2001. Sahara dust: sources and trajectories. Royal Geographical Society 26: 165–81.

Mishra V. 2003. Effect of indoor air pollution from biomass combustion on prevalence of asthma in the elderly. Environmental Health Perspectives 111: 71–7.

Moulin C, Lambert CE, Dulac F, Dayan U. 1997. Control of atmospheric export of dust from North Africa by the North Atlantic Oscillation. Nature 387(6634): 691–94.

Padhi BK, Padhy PK. 2008. Domestic fuels, indoor air pollution, and children's health. Annals of the New York Academy of Sciences 1140: 209–17.

Pandey MD, Nathwani JS. 2003. Canada wide standard for particulate matter and ozone: cost-benefit analysis using a life quality index. Society for Risk Analysis 23(1): 55–67.

Pett-Ridge JC. 2009. Contributions of dust to phosphorus cycling in tropical forests of the Luquillo Mountains, Puerto Rico. Biogeochemistry 94: 63–80.

Pitari G, Mancini E. 2002. Short-term climatic impact of the 1991 volcanic eruption of Mt. Pinatubo and effects on atmospheric tracers. Natural Hazards and Earth Systems Sciences 2: 91–108.

Pope CA. 2000. Review: epidemiological basis for particulate air pollution health standards. Aerosol Science and Technology 32: 4–14.

Prata A, Carn S, Stohl A, Kerkmann J. 2007. Long range transport and fate of a stratospheric volcanic cloud from Soufriere Hills volcano, Montserrat. Atmospheric Chemistry and Physics 7: 5093–5103.

Prospero JM, Lamb PJ. 2003. African droughts and dust transport to the Caribbean: climate change implications. Science 302(5647): 1024–27.

Qian W, Quan L, Shi S. 2002. Variations of the dust storm in China and its climate control. Journal of Climate 15: 1216–30.

Remer LA, Chin M, DeCola P, Feingold G, Halthore R, Kahn RA, Quinn PK, Rind D, Schwartz SE, Streets D, Yu H. 2009. Executive summary. In: Chin M, Kahn RA, Schwartz SE, editors. Atmospheric aerosol properties and climate impacts, A report by the US Climate Change Science Program and the Subcommittee on Global Change Research. Washington DC (USA): National Aeronautics and Space Administration. p 1–8.

Reyes DR, Rosario O, Rodríguez JF, Jiménez BD. 2000. Toxic evaluation of organics extracts from airborne particulate matter in Puerto Rico. Environmental Health Perspectives 108(7): 635–40.

Schwarze PE, Øvrevik J, Låg M, Refsnes M, Nafstad P, Hetland RB, Dybing E. 2006. Particulate matter properties and health effects: consistency of epidemiological and toxicological studies. Human & Experimental Toxicology 25: 559–79.

Shea SB. 2008. Acid rain, rain go away: some Adirondack lakes are showing promising sings of recovery. New York State Conservationist 62(5): 12–7.

Sousa SIV, Pereira MMC, Martins FG, Alvim-Ferraz CM. 2008. Identification of regions with high ozone concentrations aiming the impact assessment on childhood asthma. Human and Ecological Risk Assessment 14(3): 610–22.

Stier P, Seinfeld J, Kinne S, Boucher O. 2007. Aerosol absorption and radiative forcing. Atmospheric Chemistry and Physics 7: 5237–61.

Timmreck C, Graf H. 2006. The initial dispersal and radiative forcing of a Northern Hemisphere mid-latitude super volcano: a model study. Atmospheric Chemistry and Physics 6: 35–39.

Tunved P, Hansson H, Kulmala M, Aalto P, Viisanen Y, Karlsson H, Kristensson A, Swietlick E, Dal Maso M, Ström J, Komppula M. 2003. One year boundary layer aerosol size distribution data from five Nordic background stations. Atmospheric Chemistry and Physics 3: 2183–2205.

Voorhees AS, Uchiyama I. 2008. Particulate matter air pollution control programs in Japan—an analysis of health risks in the absence of future remediation. Journal of Risk Research 11(3): 409–21.

Yu F, Luo G. 2009. Simulation of particle size distribution with a global aerosol model: contribution of nucleation to aerosol and CCN number concentrations. Atmospheric Chemistry and Physics 9: 7691–7710.

Chapter 12

The potential of *Thillandsia* sp. to monitor PAHs from automobile sources in Puerto Rico

Yolanda Ramos-Jusino and Eddie N. Laboy-Nieves

SUMMARY

Transportation includes different mobile sources of air quality providing services and comfort, but causing serious health problems of which most people are unaware. Poly Aromatic Hydrocarbons (PAHs) containing matter are one of the precursors of such problems; it may enter the human body via the respiratory and digestive track, and the skin. In cities, mobile sources, including working machinery contribute to the main atmospheric emissions. Quantitative risk estimates of PAHs as air pollutants are very uncertain because of the lack of useful and reliable data. In Puerto Rico (PR) the PAHs impacts on marine environment are evidence of potential public risks. The lack of an environmental health network shows need of more research and awareness. To form a basis for action against PAHs in the environment, frequent monitoring of selected species under various conditions is essential. This chapter overviews the needs for PAHs monitoring and review how bromeliad *Tillandsia sp* can be employed in PR for this work where 15 species are found, the most abundant *Tillandsia recurvata*. Innovation and appropriate experimental design is required to attain this goal. These plants enhance environmental indoor and outdoor data acquisition and consequently foster environmental risk assessment studies and sustainable development.

12.1 INTRODUCTION

Air quality plays an important role in human development and health, and the general environmental quality for terrestrial ecosystems (Godish 2004). As humans evolved, the atmosphere began to be inflicted by their activities, which decreased its natural state by the dumping of gaseous and aerosols contaminants. Air pollution remains a major problem and poses continuing risks to health, particularly by the increase in lung cancer risk (Briggs 2003; Paolo V and Husgafvel-Pursiainen 2005). The United States Environmental Protection Agency (herein EPA) has committed to substantially reduce average inhalation cancer risks and potential noncancerous health risks exposure from mobile source air toxics, given that inhalation of air toxics continue to be a major public health concern (Touma et al. 2007). Consequently, there is an immediate need to improve the monitoring and emission inventory capabilities to formulate air pollution control and management strategies (Sharma et al. 2004).

12.1.1 Challenges to monitor and model air quality

Because of resource limitations and practical considerations, it is necessary to utilize approaches other than air quality monitoring to determine the impact of pollution (Godish 2004). The air pollution emission factors are representative values that attempt to relate the quantity of a pollutant released to the ambient air with an activity associated with the release of that pollutant. In most cases, these factors are simply averages of all available data of acceptable quality, and are generally assumed to be representative of long-term averages (EPA 2008, 2010).

Mobile source emissions must be inventoried on the spatial and temporal scales over which pollution problems are observed (Sawyer et al. 2000). Emissions estimates can be achieved using a transportation model that can predict dynamic vehicle operating characteristics such as acceleration and deceleration, and combining these with modal emission data. Isakov and Venkatran (2006) found that a comprehensive analysis of on-road emission estimates is not feasible. They focused on three factors thought to have a substantial influence on emissions uncertainty: number of vehicles, age distribution of vehicles, and emission factors. The models cannot be used to identify "hot spots" (areas in immediate to major roads), where the air concentrations exposure, and/or risk might be significantly higher (Strum et al. 2007).

Several line source models mostly Gaussian-based, have been suggested to predict pollutant concentrations near highways/roads (Sharma et al. 2004). Gaussian dispersion modeling assumes: (1) steady-state conditions; (2) wind speed, diffusion and direction characteristics of the plume are constant; (3) no chemical transformation take place; and (4) wind speeds are greater than or equal to 1 m/sec Godish (2004). Gaussian models, in addition to their user friendly nature and simplicity are appealing conceptually as they are consistent with the random nature of the turbulence of the atmosphere. However, comparison of experimental results with these models predictions indicated many deficiencies and limitations (Sharma et al. 2004; Rodier 2007).

12.2.2 Poly Aromatic Hydrocarbons (PAHs)

PAHs are persistent organic pollutants which remain solid at ambient atmospheric conditions. They are produced in the combustion chemistry of organic materials and fuels, and are commonly found in the elemental carbon fraction of atmospheric aerosols (Godish 2004). Two and three rings PAHs are mainly in the gas phase, four ringed PAHs are in both gas phase and particulate phase, and five and six ringed PAHs are mainly attached to particles. Particulate phase is observed predominantly in fine fractions with a diameter ranging between 0.01 and 0.5 μm (Bostrom et al. 2002). The best characterized individual source of PAHs is vehicle emissions (Bostrom et al. 2002), but the full impact of traffic on air quality and health is not known well understood (Marchand et al. 2004).

Commercial aircrafts are responsible for 13% of transportation-related fossil fuel consumption and approximately 2% of anthropogenic carbon dioxide emissions. The assessment of aviation's global emissions revealed that fuel burn and nitrogen oxide emissions increased to over 12% and 15.5% from 2000 to 2005

(Kim et al. 2007). It is predicted that air travel will grow from 2 to 5% per year for the next 20 years (Malwitz et al. 2009), thus aggravating the PAH emission scenario if contingency measures are not taken to correct the quality and efficiency of fossil fuels combustion. For instance Malwitz et al. (2009) recommended that reducing vertical separation minimum between aircrafts from 2000 to 1000 feet for cruise altitudes between 29,000 and 41,000 feet, allows more efficient flight trajectories, and reduced fuel burn and related emissions by 1.5% to 3%.

Most PAHs are potential human carcinogens. They may enter the human body via the respiratory and digestive track, and the skin. The sequential or simultaneous exposure of humans to these compounds initiates DNA mutations, increasing the risks of cancer in direct exposed individuals and their conceived children (MDH 2000; Bostrom et al. 2002). Experimental evidence suggests that parental exposure PAH, which occurs primarily through tobacco smoke, occupational exposure, and air pollution, could increase the risk of cancer during childhood (Cordler et al. 2004).

The most carcinogenic PAH are to a high extent associated with airborne particles because of the lipophilic properties of PAHs fractions of the compounds are likely to be retained in the lung tissue and attain high local concentration. At higher exposure levels the capacity of the airway epithelium to retain the compounds becomes saturated, and nonlinearities between exposure and airway target dose can be expected. Therefore risk assessment based on animals exposed to high levels of PAH may underestimate cancer risk in humans after decades of environmental exposure (Bostrom et al. 2002).

The vehicle exhaust system is the larger contributor to PAHs emissions in central parts of large cities, according to Bostron et al. (2002). These emissions are related to fuel type, driving conditions, ambient temperature, gases exhausted after treatment, and engine mechanical conditions. While three-phased catalytic converters reduce PAHs emissions from 80% to 90%, diesel fuel vehicles produce more PAHs than their gasoline engine counterpart. Sawyer et al. (2000) alerted that future emissions will depend primarily upon the deterioration and maintenance of used vehicles and the effectiveness of inspection and repair programs, than by new car emission standards. Bostrom et al. (2002) also reported that road dust from threat wear of tires and asphalt may also contribute to the PAHs levels in ambient air.

12.3 AIR QUALITY MONITORING IN THE USA AND PUERTO RICO

Models to evaluate regulation compliance and to forecast air quality are as important as the monitoring stage to determine exposition risks and limits for emissions. Continuous research to enhance the transportation subset of atmospheric models and their application in conjunction with quality monitoring is needed for an effective air quality control of PAHs emissions (EPA 2008). However, studies of PAHs emission factors in the USA seems to be poorly reliable because of insufficient samples and unrecognized sampling and analyses processes (EPA 2010). For EPA, an excellent emission factor quality rating means that multiple tests were performed on the same source using sound methodology and reported in enough detail for adequate validation. These tests are not necessarily EPA reference methods, although preferred by the Agency.

Under the 1970 Clean Air Act Amendments, the USA Congress authorized EPA to regulate hazardous pollutants. As a result, toxic substances are regulated under the National Emission Standards for Hazardous Air Pollutants (NESHAP). Many state air pollution control agencies identified and designated more toxic air pollutants than the ones listed. Since hazardous air pollutants are carcinogenic, EPA interpreted the statutory language as requiring an emission standard of zero, thus it became reluctant to regulate emissions of economically important substances (Godish 2004). Despite this relevant effort, emissions data are often lacking or uncertain for many airborne contaminants. Currently available ambient-air emission inventories of PAHs either fail to account for population-based activities (such as residential wood combustion and motor vehicle activity) and/or report 'total PAH' or particulate organic matter emissions instead of individual compounds (Lobscheid et al. 2004).

The EPA AP42 Compilation of Air Pollutant Emission Factors was published in 1968, and has been submitted to several revisions. Volume I includes stationary point and area emission factors, and Volume II includes mobile source emission factors. Volume I is currently in its fifth edition, Volume II is no longer maintained as such. Instead roadway air dispersion models for estimating emissions from on road and from non-road vehicles and mobile equipments are the analytical focus (EPA 2010). AP42 contains emission factors for different pollutants (except from indoor sources), assumes uniform outdoor concentration, and does not reflect elevated concentrations and exposure near roadways (Strum et al. 2007).

12.4 PAHS RISK ASSESSMENT SITUATION

In USA, more than 35 million people live within 100 m of a major roadway; the growth of vehicular miles traveled and traffic congestion in urban areas may exacerbate adverse health effects (EPA 2008). Risk estimates from PAHs exposure in the USA are primarily based on peer reviewed laboratory studies of animals exposed to specific toxics, but human are exposed to a combination of pollutants including more than five hundred PAHs (Bostrom et al. 2002). The USA Center for Disease Control (CDC) provides a national network of information about hazards monitoring and exposure and health surveillance to improve the health of communities. Since 2004, the CDC has allocated funds for the establishment and operation of the Coordinated Environmental Health Network. Puerto Rico, a USA territory located in the Greater Antilles archipelago in the Caribbean Sea, has not solicited such funds, yet. This inaction in an island where about 40% of the typical family income is spent on the ownership and maintenance of private vehicles (UMET 2009); the disease statistics seems to be in conflict with the reports on medical papers (Ortiz et al. 2009); and where cancer is the second cause of deaths in both sexes (DH 2010); reveals poor research and public interest in environmental and health risk assessments. Therefore, research is essential to understand priorities and needs in order to develop strategies to control and manage health risks, particularly related to cancer incidence (Ortiz et al. 2009).

Elevated levels of PAHs are found in estuaries and coastal marine waters near heavily populated areas. Studies of marine pollution in Puerto Rico had evaluated PAHs concentrations and impacts on the biota (Mayol-Bracero et al. 2001; Rodríguez et al. 2007; Pait et al. 2008). Chemical characterization of submicron atmospheric aerosols carried by tropical trade winds entering Puerto Rico from the northeast suggest a that the organic component (mostly volatile) is from natural origin and did not show the presence of PAHs (Mayol-Bracero et al. 2001). The Guayanilla Bay in the south coast of Puerto Rico, has shown a long term exposure to petrogenic hydrocarbons with a characteristically high PAHs content and a high concentration of PAHs degrading bacteria (Rodríguez et al. 2007). In this Bay, the load and mutation rate in mangroves, together with the known mutagenicity of PAHs, strongly suggest a cause-effect relationship (Rodriguez et al. 2007). The chemical analysis of sediments samples from southwest Puerto Rico indicated that automobile emissions may be an important source of PAHs in that coastal environment (Pait et al. 2008).

Ultrafine fraction of particulate matter and high levels of benzene of engine exhaust has been associated to elevated levels of DNA strands breaks in taxi drivers, residents living near intense traffic roads, and suburban and rural dwellers (Avogbe et al. 2005). It has been demonstrated that the uptake of airborne ultrafine particles by target cells is not necessary for transfering toxicants from the particles to the cells. Epidemiologic evidence supports an association between inhalation of fine and ultrafine ($\leq 2.5\,\mu m$) ambient particle matter and increases in cardiovascular/respiratory morbidity and mortality (Penn et al. 2005; Kocbach et al. 2006; Armstrong et al. 2004). Men at high levels of occupational PAHs have increased the risk of prostate cancer (Ribick et al. 2006). Trucking industry workers have an elevated risk of lung cancer with increasing years of work (Garshick et al. 2008).

Penn et al. (2005) and Bostrom (2002) reported that about 500 PAHs have been detected in air, but often the measurements only include benzene (a) pyrene (B[a]P) as representative of the whole group. According to these authors, although B[a]P has been used as an indicator of carcinogenic PAHs and PAH derivates, its suitability as an indicator has been questioned, mainly because the most-cited quantitative risk of lung cancer is based on an increased risk of lung cancer among coke-oven workers. The PAHs profile of relevant emissions today probably differ from those of coke-oven emissions with regard to the relative contribution of B[a]P. An evaluation of B[a]P alone may underestimate the carcinogenic potential of airborne PAH mixtures, because other co-occurring substances are carcinogenic as well. PAHs are absorbed onto particles which may also play a role in their carcinogenicity, a fact that affirm that individual PAHs models underestimate the risk (Bostrom et al. 2002; Penn et al. 2005).

Lobscheid and McKone (2004) estimated the relative magnitude of emissions from four major sources of PAHs to ambient air: on-road motor vehicles, including light-duty gasoline vehicles and diesel-powered buses and medium and heavy duty trucks; residential wood combustion; and power generation from external combustion boilers. Ambient air is dominated by smaller PAHs molecules like phenantherene, pyrene and fluoranthene, which are important to assess human exposure (Pereira et al. 2007).

PAHs may exert both mutagenic (genotoxic) and epigenetic actions as complete carcinogens, like the B[a]P. Fluoranthene and dioxins only exert epigenic actions. Cancer incidence is related to the probability of initiation times of promoter substances, which lead to a raised cancer risk above some threshold dose. A mutagenic compound not acting as a promoter may, on the basis of a negative animal cancer test be classified as non-carcinogen. Such a compound is expected to interact with inherited and acquired conditions, increasing the incidence of tumors. One example of a PAHs a previously classified as non-carcinogen is floranthene, which occur at relatively high concentrations in vehicle emissions (Bostrom et al. 2002). Their small size (24 ± 6 nm) compared to wood smoke (31 ± 7 nm) represents a great risk for public health (Kockbach et al. 2006).

Ohura et al. (2004) assessed PAHs risk assessment in indoor environments near heavy traffic roads. Indoor and outdoor gaseous PAHs compositions differed significantly, mainly because indoor gases were affected by insect repellents and heating sources. Particulate PAHs concentrations indoors were significantly affected by cigarette smoking, the age and type (wood) of the house, and outdoor PAHs concentrations. The existence of a cancer risk from exposure to PAHs is beyond reasonable doubt, despite the uncertainty of the exposure–response relationship (Armstrong et al. 2004).

Refined air quality modeling approaches are necessary for monitoring, evaluating, and assessing PAHs in ethnical or economically stratified subpopulations in an urban area. For neighborhood assessment incorporating site specific data can lead to improvement in modeled concentrations estimates (Isakov and Venkatram 2006). Also, it has been demonstrated that monitoring PAHs with the bromeliad *Tillandsia* can be an innovative strategy to study this pollutant in different countries where these plants are endemic (Brighigna et al. 2002; Pereira et al. 2007; Segala-Avilés et al. 2008; Zambrano-García et al. 2009). These plants may be important cost effective tools for data acquisition enhancement about emission factors of PAHs, trace metals and other toxics.

12.5 *TILLANDSIA*: A POTENTIAL NATURAL PAHS BIOMONITOR FOR PUERTO RICO

Monitoring airborne metals, PAHs and other bioaccumulative pollutants (biomonitoring) is a technique applied to explore contaminant occurrence and dispersion trends, identify emission sources, estimate atmospheric deposition, and relate biological/ecological changes to air pollution (Zambrano-García et al. 2009). Since bioaccumulative air pollutants usually travel in particles, biomonitoring is done preferentially with organisms that rely on the atmosphere as primary source of moisture and nutrients, such as lichens, mosses and some vascular plants like bromeliads. Porwollik (2007) reported that about 50% of the bromeliad family lives as epiphytes growing on other plants in the axels of tree branches or in similar ecological niche. One of the most cosmopolitan bromeliad is represented by the genus *Tillandsia*. Cedeño-Maldonado (2005) described this genus as caulescent or acaulescent herbs-like bromeliads with variable habit and size, represented in the Western Hemisphere by nearly 540 species ranging from the southeastern United States to South America and the West Indies. Their leaves are spirally distributed along the stem and tend to look like spikes (Figure 12.1). Flowers are bisexual and characterized by short-pedicellates, symmetric sepals and petals free. Their fruits are encapsulated and have many fusiform or plumose seeds. They live mainly as epiphytes in tree branches, cliffs and rocks.

Tillandsia has been reported as a good monitor for bioaccumulation of air pollutants (Brighigna et al. 2002; Pereira et al. 2007; Zambrano–García et al. 2009).

Figure 12.1 Common *Tillandsia* species from the West Indies: *T. recurvata* (left) and *T. usneoides* (right) @Eddie N. Laboy-Nieves.

It may be useful as a common natural receptor in comparative studies within and among Latin American countries and the southern USA, thus reducing the potential variability typical from using different species (Zambrano-García et al. 2009). These plants are perennial and have a very slow growth. Their trichomes protect from damaging UV radiation, reflect most of the incoming light, and collect moisture from fog and dew. They trap dust, the only source of nutrients for the plant's growth and shield the stomata from excess loss of water. Trichomes increase the rate of water absorption and reduce the rate of water loss due to evaporation, while providing a defense against insects. Successful transplantation of *Tillandsia* from rural into urban areas was reported by Brighina et al. (2002) and Zambrano-García et al. (2009). The above facts give *Tillandsia* a high suitability to contribute for monitoring air pollutants.

The age-related variability in pollutant bioaccumulation and quality controls assurance methodology for sampling are relevant factors to consider in a pollution research that use *Tillandsia*. Research is currently under way to reduce the uncertainty associated with measurement data and models parameters and structures in biomonitoring (Angerer et al. 2006). Pereira et al. (2007) identified endocrine disrupters substances (industrial chemicals and synthetic/ natural hormones) while using *Tillandsia* to monitor air pollutants at Volta Redonda (VR), a highly industrialized city in Rio de Janeiro, Brazil. These included polychlorinated biphenyls, polychlorinated dibenzo-p-dioxins, dibenzofurans, and organochlorine pesticides (chlordane, lindane, hexachlorobenzene), which are recognized to inflict humans and wildlife, mainly by their widespread use, emission patterns, persistence in the environment and bioaccumulative properties. This study measured total deposition rates using funnels covered with polyurethane (PU foams). Simultaneously to the sampling of total deposition, the use of *Tillandsia usneoides* for ambient air monitoring of persistent organic pollutants was investigated. This species has a relatively low growth rate (3–4 cm/year) and is well adapted for hot and dry environments. Moreover *Tillandsia usneoides* revealed itself to be an excellent biomonitor for mercury in indoor/outdoor contamination near gold shops in the Amazon basin. Four deposition and biomonitoring systems were installed in VR around metallurgical industrial facilities. Two sampling systems were set up in the Itatiaia National Park (PNI), a nature reserve about 65 km west from VR, used as a control site.

In the VR very low deposition rate was found at the most contaminated sites, in contrast high levels in the corresponding biomonitor sample. This was certainly due to losses since the PU foams were partially to totally destroy by acid rain. The mass fragmentograms of the deposition and corresponding biomonitor sample were very similar. The homologue profiles found in the samples from PNI indicated that the main contamination pathway is the long range transport of these pollutants. These results not only confirmed the benefits of *Tillandsia* as biomonitors, but also that PAHs can be transported through long distances and persist in the environment as an aerosol.

In Florence, Italy, Brighina et al. (2002) employed *Tillandsia caput-medusae* and *T. bulbosa*, for monitoring PAHs, showing an increasing trend in time of PAHs bioaccumulation. PAHs data were obtained using gas chromatography-mass spectrometry (GC/MS) analysis of plant extracts. Results were compared with instrumentally recorded rain and PM_{10}, indicating that trichome-operated physical capture of aerial particles was prominent in PAHs bioaccumulation on tillands. Scanning electron microscope observations confirmed the role of the trichomes to capture particles (Brighina et al. 2002).

Mezquital Valley (MV), a Mexican wastewater-based agricultural and industrial region, is a "hot spot" of regulated air pollutants emissions. Using *Tillandsia recurvata*, Zambrano-García et al. (2009) detected spatial patterns and potential sources of 20 airborne elements and 15 PAHs. Results indicated a high deposition of bioaccumulative air pollutants at MV. Since *Tillandsia recurvata* reflected the regional differences in exposition, these authors recommended the species as a biomonitor to compare results from southern USA to Argentina, where it is native. Segala-Avilés et al. (2008) reported that pollution affected the structure of the leaves of *Tillandsia usneoides* causing alterations in the scales, density of stomata, and epidermis thickness. The percentage of anomalous scales may potentially be used as an alternative bioindicator parameter or air quality.

There are 15 species of the genus *Tillandsia* inhabiting Puerto Rico (Table 12.1), where the most common is *Tillandsia recurvata*, popularly called ball moss or "*nido de gungulén*" (Cedeño-Maldonado 2005; USDA 2009). Given that this species grows as a small sphere on power or phone lines along roads and on trees in the coastal plains, it offers the highest potential as biomonitor of PAHs from transportation sources in Puerto Rico.

12.6 RISK MANAGEMENT

Modern air pollutants associated with road traffic and the use of chemicals for domestic, food and water treatment and pest control, are rarely present in excessive large concentrations, so effects on health are usually far from immediate or obvious (Briggs 2003). However, currently, no systems exist at the state or national level to track many of the exposures and health effects that may be related to environmental

Table 12.1 List of *Tillandsia* species inhabiting Puerto Rico: E = Endemic, WI = West Indies,
P = Pantropical America, C = Common, R = Rare.

Species	Distribution	Abundance in Puerto Rico
ariza-juliae	WI	R
borinquensis	E	C
bulbosa	P	C
fasciculata	P	C
festucoides	P	R
flexuosa	P	R
lineatispica	E	R
polystachia	P	C
pruinosa	P	R
recurvata	P	C
setacea	P	C
tenuifolia	P	C
usneoides	P	C
utriculata	P	C
variabilis	P	C

hazards. Existing environmental hazard, exposure, and disease tracking systems are not linked together, thus it is difficult to study and monitor relationships among hazards, exposures, and health effects (Angerer et al. 2006).

Exposure to ambient particulate matter from transportation exhaust emission has been associated with a number of adverse human health effects (Bostrom et al. 2002; Angerer et al. 2006; EPA 2008). Aerosols size, surface area, and chemistry seem to influence the negative effects in human health (Kockbach et al. 2006). Also, PAHs aerosols from asphalt and tires shall be considered for the assessment of exposition and risk because they may also contribute to the PAHs levels in ambient air (Bostrom et al. 2002).

Biomonitoring has the potential to revolutionize environmental health research and practice (Angerer et al. 2006). Integrating biomonitoring and toxicological data into a risk assessment process will allow for a scientifically sound decision making process in many institutions, like EPA. Biomonitoring air contaminants not only will provide data to determine levels of exposition or source specificity, but will also enhance modeling of the environmental state (EPA 2010). Ambient air quality indicators derived from the use of *Tillandsia* for biomonitoring, toxicological and bioremediation research can provide information about emission factors of PAHs, trace metals and other toxics (Sabzali et al. 2009). Evaluations specific to a regional scenario are needed to adequately quantify population risks.

Bioremediation of atmospheric PAHs has been demonstrated to be a successful technique (Maier et al. 2000; Kanaly and Harayama 2000; Makkar and Rockne 2003; Angerer et al. 2006). Interests in the biodegradation mechanisms and environmental fate of PAHs is prompted by their ubiquitous distribution and their potentially deleterious effects on human health (Kanaly et al. 2000). However the epigenic activity of PAHs derivates in the carcinogenic process are almost unknown (Bostron et al. 2002).

The impacts of air pollution on forests and marine waters may be substantial; PAHs can be transported long distances so dispersion and transformation in the ambient is an area in which formal research is required (Mayol-Bracero et al. 2001; Bostrom et al. 2002; Rodríguez et al. 2007; Pait et al. 2008; ICP 2009; Zambrano-García et al. 2009). Dispersion models are important for this analysis and Puerto Rico needs to get involve in both areas. It is also necessary to quantify to what extend transportation is degrading indoor air quality. People are most of the time in an indoor ambient at work, home, shopping or recreation (EPA 2010; Pait et al. 2008; Ortiz 2009).

A typical sequence of an environmental investigation considers the sampling design, laboratory analysis, interpretation of data, and the application of the analysis of that data for the decision-making process for the betterment and management of that environment. These decisions may require remedial, preventive or punitive measures which could have substantial consequences in terms of human health as well as economics. Therefore, it is of utmost importance that these investigations ensure that decisions are based on the best possible, suitable and reliable information. That should be the case for biomonitoring researches (EPA 1988).

As stated in the previous section, the potential to employ *Tillandsia* to reduce pollutants, monitor toxics and to improve indoor air quality in high traffic and dense residential zones have been demonstrated. Appealing only to the instrumental value of *Tillandsia* for biomonitoring can be an effective policy strategy to manage health risk from atmospheric pollution hazards. However, this species of plant like all living things, have an intrinsic value that commands our environmental ethics upon its employment for biomonitoring. Since *Tillandsia* is part of ecosystems, any attempt to take advantages of the characteristics of these plants for environmental and health activities shall guarantee their protection. Future research ought to consider this ethical issue, which was not explicitly mentioned in the aforementioned studies, but are relevant to diminish environmental disputes (DesJardins 2006).

Finally, it is important to educate people about the impacts of transportation in the environmental and public health quality of Puerto Rico. The fact that this ~9,000 km^2 and 4 million inhabitants island has one of the largest paved road densities in the world (3.4 km/km^2) and where there are 574 vehicles/1000 inhabitants (López-Marrero and Villanueva-Colón 2006), implies that the majority of its population and the peripheral natural environment is highly vulnerable to the effects of traffic-related air pollutants. Awareness about how roadways and vehicles may impact human health and natural resources is necessary to develop a strong commitment on public mass transportation systems and rigorous air quality control in the construction and maintenance of roadways.

12.7 CONCLUSION

In Puerto Rico the majority of the population and the peripheral natural environment are highly vulnerable to the effects of traffics-related emissions, the best characterized source of polycyclic aromatic hydrocarbons (PAHs). These compounds represent an unknown relevant risk to health and environmental quality. Their most adverse effect is cancer, the second cause of death in Puerto Rico. The south west of Puerto Rico show high PAHs concentration in marine environments and vehicles emissions seem to be the relevant pollution source.

The Environmental Protection Agency regulations will reduce emissions from vehicles but the adverse effects to health will continue to be an important environmental problem. The PAHs emissions factors from AP42 have many limitations to quantify the potential risks in Puerto Rico. Local and regional data is needed, and monitoring with *Tillandsia* can be a cost effective alternative for this purpose. The resultant local data will provide more reliable information to improve traffic models for environmental quality evaluations. Of the 15 species of *Tillandsia* inhabiting Puerto Rico, the most abundant and cosmopolitan is *Tillandsia recurvata*, which has a potential for biomonitoring PAH's and other atmospheric pollutants. This will require the development of a coordinated protocol between the EPA and research institutions. Consequently, using *Tillandsia* for risk assessment studies of PAHs from transportation sources will be more scientifically reliable and will allow comparison from different scenarios.

REFERENCES

Angerer J, Bird MG, Burke TA, Doerrer NG, Needham L, Robinson SH, Sheldon L, Zenick H. 2006. Strategic biomonitoring initiatives: moving the science forward. Toxicological Sciences 93(1): 3–10.

Armstrong B, Htchinson E, Unwin J, Fletcher T. 2004. Lung cancer risk after exposure to polycyclic aromatic hydrocarbons: A review and meta-analysis. Environmental Health Perspectives [Internet]. [cited 2010 Jan 26]; 112 (9): 970–978. Available from: http://www.ncbi.nlm.nih.gov/pmc/articles/PMC1247189/

Avogbe PH, Ayi-Fanou L, Autrup H, Loft S, Fayomi B, Sanni A, Vinzents P, Moller P. 2005. Ultrafine particulate matter and high-level benzene urban air pollution in relation to oxidative DNA damage. Carcinogenesis 26(3): 613–620.

Bostrom CL, Gerde P, Hanberg A, Kyrklund T, Rannug A, Tornqvist M, Victorin K, Westerholm R. 2002. Cancer risk assessment, indicators, and guidelines for Polycyclic Aromatic Hydrocarbons in the ambient air. Environmental Health Perspectives 10(3): 451–488.

Brighigna L, Papini A, Mosti S, Cornia A, Bocchini P, Galletti G. 2002. The use of tropical bromeliads (*Tillandsia spp*) for monitoring armospheric pollution in the town of Florence, Italy 50(2): 577–585.

Briggs D. 2003. Environmental pollution and global burden of desease. British Medical Bulletin [Internet].[cited 2010 Jan 13]; 68: 1–24. Available from: http://bmb.oxfordjournals.org/cgi/reprint/68/1/1.pdf

Cedeño-Maldonado J.A. 2005. Bromeliaceae. In: Acevedo-Rodríguez P, Strong MT, editors. Monocotyledons and Gymnosperms of Puerto Rico and the Virgin Islands. Washington DC. Smithsonian Institution. p 199–229.

Cook R, Strum M, Touma JS, Palma T, Thurman J, Ensley D, Smith R. 2007. Inhalation exposure and risk from mobile source air toxics in future years. Journal of Exposure Science and Environmental Epidemiology 17: 95–105.

Cordler S, Monfort C, Fillipini G, Preston-Martin S, Lubin F, Holly EA, Peris-Bonet R, McCredie M, Choi W, Little J, Arsian A. 2004. Parental exposure to polycyclic aromatic hydrocarbons and the risk of childhood brain tumors. American Journal of Epidemiology [Internet]. [cited 2009 Nov 4];159(12): 1109–1116. Available from http://aje.oxfordjournals.org/cgi/reprint/159/12/1109

Tietenberg T, Lewis L. 2009. Environmental & natural resource economics. Addison-Wesley. 660p.

DesHardins JR. 2006. Environmental ethics an introduction to environmental philosophy. Wadsworth. 286p.

[DH] Department of Health (PR). 2010. Datos de cáncer [Internet]. Estado Libre Asociado de Puerto Rico. [cited 2010 Jan 22]. Available from: www.salud.gov/pr/RCancer/Reports/Pages/default.aspx

[EPA] Environmental Protection Agency (USA). 1988. Spatial autocorrelation: Implications for sampling and estimation [Internet]. [cited 2010 Jan 18]; 10 p. Available from: http://www.epa.gov/esd/cmb/research/papers/ee107.pdf

[EPA] Environmental Protection Agency (USA). 2008. Emission and air quality modeling tools for near-roadway applications [Internet].[cited 2010 Jan 18]; 40 p. Available from: http://ntis,gov/search/product.aspx?ABBR=PB2009/03941

[EPA] Environmental Protection Agency (USA). 2010. Emission factors & AP 42 [Internet]. Technology Transfer Network Clearinghouse for Inventories & Emission Factors. [cited 2009 Oct 15]. Available from: http://www.epa.gov/ttn/chief/ap42/

Garshick E, Laden F, Hart JE, Rosner B, Davis ME, Eisen EA, Smith TJ. 2008. Lung cancer and vehicle exhaust in trucking industry workers. Environmental Health Perspectives 116(10): 1327–1332.

Godish T. 2004. Air quality. Lewis Publishers. 460p.

Institute of Medicine. 1988. The future of public health [Internet].[cited 2009 Oct 15]; 20p. Available from: http://books.nap.edu/catalog/1091.html

[ICP] The International Cooperative Programme. 2009. Air pollution and vegetation [Internet]. [cited 2010 Oct 5]; 50 p. Available from: HTTP://icpvegetation.ceh.ac.uk/publications/annual.html

Isakov V, Venkatram A. 2006. Resolving neighborhood scale in air toxics modeling: A case study in Wilmington, CA. Journal of the Air & Waste Management [Internet]. [cited 2010 Jan 17]; 56: 559–569. Available from: http://www.epa.gov/asmdnerl/peer/products/148683_Isakov_Air.pdf

Kanaly RA, Harayama S. 2000. Biodegradation of high-molecular-weight polycyclic aromatic hydrocarbons by bacteria. Journal of Bacteriology [Internet].[cited 2010 Jan 28]; 182 (8): 2059–2067. Available from: http://jb.asm.org/cgi/content/full/182/8/2059

Kim BY, Fleming GG, Lee JJ, Waitz IA, Clarke JP, Balasubramanian S, Malwitz A, Kima K, Locke M, Holsclaw CA, Maurice LQ, Gupta ML. 2007. System for assessing aviation's global emissions (SAGE), part i: model description and inventory results. Journal of Transportation Research. D (12): 325–346.

Kocbach A, Li Y, Yttri KY, Casee FR, Schwarzem PE, Namork E. 2006. Physicochemical characterisation of combustion particles from vehicle exhaust and residential wood smoke. Particle and Fiber Toxicology 3(1): 1–10.

Lobscheid AB, McKone TE. 2004. Constraining uncertainties about the sources and magnitude of ambient air exposures to polycyclic aromatic hydrocarbons (PAHs): The state of Minnesota as a case study. Lawrence Berkeley National Laboratory. Lawrence Berkeley National Laboratory. [cited 2009 Nov 10]; 22p. Available from: http://escholarship.org/uc/item/09f8m320

López-Marrero T, Villanueva-Colón N. 2006. Atlas Ambiental de Puerto Rico. Editorial Universidad de Puerto Rico. 160 pp.

Makkar RS, Rockne KJ. 2003. Comparison of systhetic surfactants and biosurfactants in enhancing biodegradation of polycyclic aromatic hydrocarbons. Environmental Toxicology and Chemistry [Internet]. [cited 2010 Jan 28]; 22(10): 2280–2292. Available from: http://tigger.uic.edu/~krockne/paper12.pdf

Malwitz A, Balasubramanian S, Fleming G, Yoden T, Waitz I. 2009. Impact of the reduced vertical separation minimum on the domestic United States. Journal of Aircraft 16(1): 148–156.

Marchand N, Besombes JL, Chevron N, Masclet P, Aymoz G, Jaffrezo JL. 2004. Polycyclic aromatic hydrocarbons (PAHs) in the atmospheres of two French alpine valleys: sources and temporal patterns. Atmospheric Chemistry and Physics Discussions 4:1167–1181.

Mayol-Bracero OL, Rosario O, Corrigan CE, Morales R, Torres I, Pérez V. 2001. Chemical characterization of submicron organic aerosols in the tropical trade winds of the Caribbean using gas chromatography/mass spectrometry. Atmospheric Environment 53: 1735–1745.

[MDH] Minnesota Department of Health (USA). 2000. Children's Health Risks from Chemical Exposures [Internet]. [cited 2010 Jan 20]. Available from: http://www.health.state.mn.us/divs/eh/children/healthrisks.html

Ohura T, Amagal T, Fusaya M, Matsushita H. 2004. Polycyclic aromatic hydrocarbons in indoor and outdoor environments and factors affecting their concentrations. Environmental Science Technology 38(1):77–83.

Ortiz A, Calo W, Suárez-Balseiro C, Maura-Sardo M, Suárez E. 2009. Evaluación bibliométrica de la investigación sobre cáncer en Puerto Rico, 1903–2005. Revista Panamericana de Salud Pública [Internet]. [cited 2009 Oct 22]; 25(4): 353–361. Available from: http://www.scielosp.org/pdf/rpsp/v25n4/10.pdf

Pait AS, Thitall DR, Jeffrey CFG, Caldow C, Manson AL, Lauenstein GG, Christensen JD. 2008. Chemical contamination in southwest Puerto Rico: An assessment of organic contaminants in nearshore sediments. Marine Pollution Bulletin 56: 580–606.

Paolo V, Husgafvel-Pursiainen K. 2005. Air pollution and cancer: biomarker studies in human populations. Carcinogenesis [Internet]. [cited Jan 26, 2009]; 26(11): 1846–1855. Available from: http://carcin.oxfordjournals.org/cgi/reprint/26/11/1846.pdf

Penn A, Murphy G, Barker S, Henk W, Penn L. 2005. Combustion-derived ultrafine particles transport organic toxicants to target respiratory cells. Environmental Health Perspectives 113(8): 956–963.

Pereira MS, Waller U, Reifenha W, Torres JPM, Malm O, Korner W. 2007. Persistent organic pollutants in atmospheric deposition and biomonitoring with *Tillandsia usneoides* (L.) in an industrialized area in Rio de Janeiro State, south east Brazil– Part I: PCDD and PCDF. Chemosphere 67:1728–1735.

Porwollik V. 2007. Conservation strategy for the bromeliad society international [Internet]. University of Applied Sciences Eberswade (Germany): Forestry Department. [cited 2009 Nov 1]; 82p. Available from: http://www.bsi.org/bsi_info/conservation/BSI_Conservation

Ribick BA, Neslund-Dudas C, Nock NL, Schultz LR, Eklund L, Rosbolt J, Bock CH, Monaghan KG. 2006. Prostate cancer risk from occupational exposure to polycyclic aromatic hydrocarbons interacting with the GSTP1 lle105Val polymorphism. Cancer Detection and Prevention 30(5): 412–422.

Rodier CJ. 2007. Beyond uncertainty: modeling transportation, land use, and air quality in planning. Mineta Transportation Institute (CA): San José State University. [cited 2010 Jan 17]; 74p. Available from: http://www.dot.ca.gov/newtech/researchreports/reports/2007/mpdelinglanduse-mtireport.pdf

Rodríguez NJ, Massol A, Imam SH, Zaidi BR. 2007. Microbial utilization of toxic chemicals in surface waters of Guayanilla Bay, Puerto Rico: impact of seasonal variation. Caribbean Journal of Science 43 (2): 172–180.

Sabzali A, Gholami M, Sadati MA. 2009. Enhancement of benzene biodegradation by variation of culture medium constitutents. African Journal of Microbiology Research 3(2): 77–81.

Sawyer RF, Harley RA, cadle SH, Norbeck JM, Slott R, Bravo HA. 2000. Mobile sources critical review: 1998 NARSTO assessment. Atmospheric Environment 24(2000): 2161–2181.

Segala-Avilés D, Baesso Moura B, Domingos M. 2008. Structural Analysis of Tillandsia usneoides L. Exposed to Air Pollutants in Sao Paulo City-Brazil. Water and Air Soil Pollution (189):61–68.

Sharma N, Chaudhry KK, Chalapati-Rao CV. 2004. Vehicular pollution prediction modeling: a review of highway dispersion models 24(4): 409–435.

Stone B, Mednick AC, Holloway T, Spak SN. 2009. Mobile source CO_2 mitigation through smart growth development and vehicle fleet hybridization [Internet]. [cited 2010 Jan 23]; 43: 1704–1710. Available from: http://pubs.acs.org/doi/pdf/10.1021/es8021655

Tietenberg T, Lewis L. 2009. Environmental & natural resource economics. Addison-Wesley. 660p.

Zambrano-García A, Medina-Cayotzin C, Rojas-Amaro A, López-Veneroni D, Chang-Martínez L, Sosa-Iglesias G. 2009. Distribution and sources of bioaccumulative air pollution at Mezquital Valley, Mexico, as reflected by the atmospheric plant *Tillandsia recurvata* L. Atmospheric Chemistry and Physics 9:5809–5852.

[UMET] Universidad Metropolitana (Puerto Rico). 2009. Centro de Estudios de Desarrollo Sostenible. Sustainability of Land Use in Puerto Rico [Internet].[cited 2009 Dec 1]; 171p. Available from: www.proyectosambientales.info

[USDA] United States Department of Agriculture. 2009. Guide to the Ecological Systems of Puerto Rico [Internet]. [cited 2010 March 20]; 444p. Available from: http://www.tropicalforestry.net/

Chapter 13

Degradation of PCBs and spilled oil by bioenhancing agents

Eileen C. Villafañe-Deyack and Eddie N. Laboy-Nieves

SUMMARY

Oil spills and PCBs contamination have significantly impacted our natural and social environment, mainly from pollution vectors associated to transportation and leakage from waste disposal and industrial facilities. The large volumes of hydrocarbons produced, used, and disposed constitutes the basis to affirm that a large portion of the Earth surface and subsurface is been inflicted by these petrochemicals. Furthermore, the inadequate management of equipment containing PCBs represents a potential source of toxic discharges to oceanic, terrestrial and limnetic ecosystems. Remediation technologies have been developed to mitigate the hazards PCBs represent to the environment and human health, but they tend to be expensive, particularly for developing countries. In this scenario, bioremediation is gaining public acceptance. This technology has been successfully studied and applied to situations such as Valdez oil spill on Alaska and Morris Berman on San Juan beaches. It has been reported to be an eco-efficient alternative to mitigate contamination from oil spill and similar chemicals. Surfactants compounds have been extensively proven to degrade hazardous wastes. The use of surfactants on in-situ remediation is a suitable alternative for PCBs final disposal due to high costs of transportation, incineration and other approved remediation procedures. Their use also eliminates long-term responsibilities, do not generate secondary hazardous compounds and is cost effective.

13.1 INTRODUCTION

Modern society is surrounded by an ample spectrum of products derived from crude oil, especially fuels. Shipping by tankers is the most effective means for moving oil from its source to energy-demanding countries. The shipping traffic is directly proportional to the consumption of petroleum products, which presently shows a declining tendency, from 76.7 to 45.4 million barrels per day between 2000 and 2009, respectively (DOE 2010). Despite this demand diminution, the sea remains congested of tankers in the six major trade routes (Figure 13.1), commonly referred to as *choke points* (Ornitz 1996). In February 2010 the movement of crude oil by

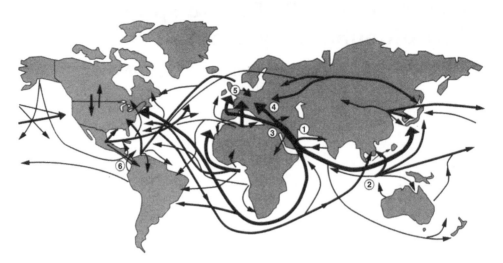

Figure 13.1 Shipping lanes and crude oil flow [million barrels per day]:
(1) Strait of Hormuz [14], (2) Strait of Malacca [7],
(3) Suez Canal [1], (4) Bosporus [2], (5) Rotterdam [1], (6) Panama Canal [1]
(adopted from Ornitz 1996)

pipelines, tankers and barges in the United States, was 36.3 millions barrels (EIA 2010). The recent explosion and sinking of the floating oil rig Deepwater Horizon into the Gulf of Mexico, spilled hundreds of thousands gallons of diesel fuel and an unknown amount of crude oil, becoming one of the worst offshore drilling disasters in recent U.S. history (Kaufman and Robertson 2010). Associated to these spills are large volumes of hydrocarbons which are produced, used, and disposed, constituting a large portion of the surface and subsurface contamination throughout the world (Sheyla et al. 2010). Therefore, oil spills and contamination in the trade routes, exploration site, and leakage from waste disposal or industrial facilities, will continue to significantly impact the natural environment and human health.

This chapter evaluates the environmental threat of oil spills, the chemical nature of polychlorinated aromatic hydrocarbons (PCBs), and the potential for mitigating hydrocarbon-related disasters through bioenhancing agents. Two case studies are presented to demonstrate that oils spills and the inadequate management of equipment containing PCBs, represent a hazard to soil and to surface and groundwater.

13.2 CHEMISTRY BEHIND CRUDE OIL

13.2.1 Hydrocarbons

Hydrocarbons are chemicals composed only of hydrogen (H) and carbon (C), and HC derivatives that contain oxygen, nitrogen, sulfur and halogens. They exist at ambient conditions as gases, volatile liquids, semi-volatile substances, and solids (Godish 2004). Crude petroleum and its distillates contain hydrocarbons and derivatives which are the most used fraction for petrochemical production. The ultimate fate of hydrocarbons as pollutants is determined by their chemical and environmental characteristics (Camilli

2009). Naturally occurring deposits as well as unintentional spills can be selectively volatilized into the atmosphere. Heavier oil fractions may adsorb to detritus, while wind, waves or other mixing actions in aquatic system may result in the formation of oil emulsions (Skimmer et al. 2010). Microorganisms can also metabolize and cometabolize some of the hydrocarbons especially the lighter compounds.

Accidental shoreline spills have been well documented, for instance the 1989 Exxon Valdez in Alaska (Short et al. 2007), and in Puerto Rico, the 1968 Ocean Eagle, 1973 Zoe Colocotronis, 1975 Z-102, 1978 Peck Slip, and the 1993 Morris J. Berman (Ornitz 1996; Kurtz 2008). Some spills are classified as intentional, like those reported in the Persian Gulf during the 1991 Desert Storm War (Obuekwe et al. 2009). Following the Exxon Valdez, oil transporters took bolder steps to control over the shipping of petroleum products along the increasingly sea lanes and vulnerable points where oil travels (Ornitz 1996). Despite the Valdez lessons, it can be inferred from Kaufman and Robertson (2010) that oil spills continues endangering natural ecosystems and human health, and it is unknown how much more can these systems take to naturally degrade hydrocarbons.

13.3 POLYCHLORINATED BIPHENYLS (PCBS)

PCBs are chlorinated aromatic hydrocarbons known as askarels, which are resistant to fire, do not conduct electricity, and have low volatility at normal temperatures (Adebusoye et al. 2008). PCBs are extremely resistant to natural chemical and biological degradation, hence their high toxicity to animals and humans (Wang et al. 2009). Since 1972, the US Environmental Protection Agency (EPA) prohibits PCBs use on open spaces. EPA regulates PCBs final disposal through the Toxic Substances Control Act, using nine remediation technologies described in Table 13.1. Regulations also provides for the use of alternative methods or innovative technologies equivalent to incineration (Varanasi et al. 2007; Parelo 2010).

Application of surfactants and bioenhancing agents

The remediation of contaminated sites by microorganisms is dependent on the bioavailability of the contaminants. According to Strong-Gunderson and Palumbo (1995), many pollutants in soil and subsurface environments are not soluble in water and persist in nonaqueous-phase liquids (NAPLs) or are partitioned into soil organic matter. Any substance existing as a NAPL or sorbed to the soil matrix is not available for microbial degradation since biodegradation takes place in the liquid phase. Therefore, rates of degradation are limited by mass transfer problems even when other parameters are optimized. Adding surface-active agents, such as surfactants, helps to catalyze solubilization and to transform contaminants into forms suitable for biodegradation

Several methods exist for the remediation of petroleum contaminants: mechanical removal and disposal in a hazardous waste landfill, incineration of hazardous wastes, physico-chemical processes, and bioremediation, being the first two the most common procedure. Physico-chemical processes include: 1) air stripping for remediating groundwater contaminated with volatile organic compounds (VOCs); 2) soil vapor extraction to remove VOCs *in situ* or from stockpiled excavated soils; 3) carbon adsorption to remove the soluble contaminants (adsorbent) from water; and

Table 13.1 Summary of PCBs remediation technologies.

Remediation Technology	Method Description
Incineration	Standard method used to destroy PCBs. Application of high temperature, over 1,000°F in the presence of oxygen in order to destroy PCBs.
Thermal Desorption	Indirect application of heat to remove PCBs from the contaminated matrix.
Solidification–Stabilization	To reduce PCBs mobilization. Solidification method encapsulate the contaminated material in cement or asphalt. Stabilization method uses reagents to stabilize contaminated compounds.
Soil washing	Water wash to remove PCBs not adsorbed on clay. Wash is treated as wastewater.
Vitrification	Use of heat to melt contaminated materials until a vitreous material is formed. The product is treated through pyrolysis or oxidation process.
Chemical dehalogenations	Chorine removal through alkaline reactions. Commonly used with Thermal Desorption
Solvent extractions	Use of solvent mixtures to extract contaminated material from its matrix in order to reduce the material to be treated.
Bioremediation	Use of microorganisms to biodegrade contaminated compounds.
Sanitary Landfill disposal	Contaminated material below regulated risk concentrations is disposed on sanitary landfills following the necessary precautions.

4) steam stripping to remove volatile and semivolatile compounds from groundwater or waste water (Kulkarni et al. 2008). With respect to remediation, most of the current available technologies are insufficient (especially for low-level contaminant concentrations) and are costly (Kulkarni et al. 2008).

Bioremediation is an environmentally sound method, since it destroys organic contaminants and, in most cases, does not generate secondary waste products. Bioremediation has gained increasing acceptance and publicity since EPA and Exxon Company, demonstrated its effectiveness on Alaskan beaches inflicted by the Exxon Valdez oil spill (Hailong and Boufadel 2010). It has been used successfully in the *ex-situ* removal of crude oil on sandy beach after the Morris J. Berman spill (Kurtz 2008). The effectiveness of bioremediation is a function of the availability and concentration of the pollutant, the activity of the microbial population, and the factors that control the reaction rates (Banat et al. 2000; Badour et al. 2003). Other key aspects affecting the degradation of synthetic chemicals in soil are related to the substance concentration, depth of contaminant, soil types, soil microorganisms and acclimatation, physical environments (including pH, temperature, oxygen availability, redox potential and moisture content of the soil), and external carbon sources (Robles-González et al. 2008).

Bioremediation can occur either via aerobic or anaerobic pathways, by indigenous heterotrophic bacteria and by engineered microorganisms, through cometabolic processes. Biorestoration is useful for hydrocarbons, especially water insoluble compounds like the carcinogenic benzene, which are difficult to remove by other means (Singh and Lin 2008). Oil contaminants are only degraded by aerobic organisms (Robles-González et al. 2008). Biodegradation has been used to treat aquifers contaminated with petroleum hydrocarbons. However, low aqueous solubility and high affinity to sorb onto soil particles, reduce the availability of hydrocarbons to microorganisms (Atlas 1995).

Bioremediation has several characteristics that make it a versatile and cost effective option. This technique can treat, transform and mineralize low concentration contaminants, avoid harsh chemical and physical treatment, minimize site disruption, eliminate long-term liability, and it can be coupled to other treatment technologies (Banzhaf 2010). Optimization is achieved by adding nutrients, surfactants, bacteria, or a combination of all three.

The mechanism for the enhancement and degradation of organic compounds in the presence of surfactants is not well understood (Franzetti et al. 2009; Salihu et al. 2009). Plate et al. (2008) emphasized that subsurface contaminants are difficult to remove because they adsorb to soil matrix surfaces and can be retained by capillary action as an immiscible, separate phase. Hence, bioavailability is achieved by partitioning contaminants into the aqueous phase. Surfactants have the ability to increase aqueous concentrations of poorly soluble compounds and interfacial areas between immiscible fluids, thus potentially improving the accessibility of biodegradation of microorganisms (Plate et al. 2008). Surfactants mobilize chemicals in soils or sediments by lowering the capillary forces and by raising the solubility of the matrix, making it available for biodegradation. Surfactants have been proposed for the bioremediation of hydrocarbon-contaminated soils and PCBs removal (Parelo 2010).

Surfactants are amphipathic molecules that tend to migrate to surfaces and interfaces or create new molecular surfaces by forming aggregates called micelles. This critical micelle concentration (CMC) is unique to each surfactant. Above the CMC a constant monomers concentration is maintained in equilibrium with micelles; below the CMC, only surfactants monomers are in solution (Rahman and Gakpe 2008). Supra-CMC levels resultes in virtually complete inhibition of biodegradation due to interference with substrate transport into the cell or to reversible physical-chemical interferences with the activity of enzymes and other membrane protein involved in hydrocarbon degradation (Rahman and Gakpe 2008). Various explanations have been proposed for the inhibitory effect of surfactants: the partitioning of hydrophobic substances into surfactant micelles, lowering the available substrate concentration in aqueous solution to microorganisms due to micellization; toxicity by the surfactant or high concentration of solubilized hydrocarbons at high concentrations; and preferential metabolisms for the surfactant (Plate et al. 2008). Finally, surfactants are emulsifying agents that drop the interfacial tension between two immiscible liquids (Brown 2007).

Surfactant studies have assessed their ability to enhance the removal of PCBs by soil washing methods (Ehsan et al. 2006). As the aqueous surfactant concentration increases, a critical concentration is reached at which this molecule become micelles. Adsorption of the hydrophobic PCBs into the hydrophobic interior of the micelles increases the apparent solubility of the PCBs and thereby enhances their transport (Vasilyeva and Strijakova 2007). The addition of surfactants associated with chemical oxidation was proposed to overcome three limitations of hydrophobic compounds biodegradation: adsorption to surfaces, weak solubility, and availability to microorganisms (Vasilyeva and Strijakova 2007). Nonionic surfactants affect the

PCBs degradation efficiency when chemical and biological treatments are integrated (Patria 2008).

Fertilizers have also been found to increase remediation rates by stimulating bacterial growth. The enhancement of biodegradation by fertilizers relies on the fact that essentials nutrients, such as nitrogen, phosphorous and iron, are limiting to oil-degrading microbes (Strong-Gurderson and Palumbo 1995). For instance, the oleophilic fertilizer Inipol EAP22 is designed to adhere to oil and was used in the bioremediation of the Exxon Valdez oil spill (Short et al. 2007). The addition of surfactant-like products increases the effluent pollution level, but at concentrations higher than CMC, these products interfered or slowed down microbial degradation (Strong-Gunderson et al. 1994).

13.4 CASE STUDY I. BIOAUGMENTATION EXPERIMENTS ON WEATHERED CRUDE OIL CONTAMINATED SOIL FROM MEXICO

This case study presents the use of two commercially available nutrient-surfactant fertilizer products for their effect on enhanced biodegradation rates of crude oil contaminated soils in southern Mexico. Energy generation is Mexico's main source of foreign currency since 1950. However, poor management and processing of crude oil had led to oil spills, which have contaminated vast extensions of land and water close to refineries or oil exploration sites. The national oil company "Petróleos Mexicanos (PEMEX)" primarily contaminates these sites with weathered hydrocarbons as a consequence of extracting millions of barrels of oil (Ferrera-Cerrato et al. 2006).

PEMEX and the Mexican Petroleum Institute (IMP) have taken the leadership in correcting environmental problems and are very concerned about restoring several contaminated sites. Remediation outcomes represent a very important scientific and environmental challenge, because they can be applied globally, especially by developing countries. To address these issues and evaluate appropriate technologies, an international research team was established among IMP, the US. Department of Energy Oak Ridge National Laboratory (DOE), Universidad Nacional Autónoma de Mexico, and Universidad Metropolitana of Puerto Rico. The team studied the ability of surfactants BioTreat[2] and Inipol EAP22 to enhance the biodegradation rates of weathered crude oil contaminated soil. IMP provided samples of bacteria isolated from a hazardous waste spill in Mexico, which were analyzed in collaboration with DOE.

The soil used in these experiments were contaminated with hydrocarbons during the last 60 years and had indigenous microbial population adapted to the site. This was the primary reason to use bacteria isolated from Mexico versus from the United States, since specially adapted bacteria are difficult to distribute and often cannot compete with indigenous population due that laboratory growing conditions differ significantly from the site they were applied (Rodrigues et al. 2009).

The Biolog System (Strong-Gunderson and Palumbo 1994) was used to characterize the physiology of each bacterium from the IMP bacteria sample. The bacterial community was not individually characterized, however, it was maintained as facultative anaerobes. Only three (3) Gram-negative bacteria were identified through the Biolog System, and their characteristics are described in Table 13.2.

Table 13.2 Characterization of the IMP microbial community.

Bacteria	Characteristics	Degradation Potential
Citrobacter freundii	Suspected to cause diarrhea and possibly extra intestinal infections.	Can biosynthesized aromatic compounds (Choi et al. 2007)
Pseudomonas citronellolis	Small clinical interest.	Allow the biodegradation of recalcitrant branched hydrocarbons through a mechanism of enzyme recruitment. Efficient in the degradation of isoprenoid compounds (Kishore and Mukherjee 2007)
Klebsiella pneumonia A.	Pathogenicity attributed to the production of a heat-stable enterotoxin. Cause pneumonia and urinary tract infections.	Catabolize aromatic compounds through the biodegradation pathway of a meta-cleavage converting methyl benzoates to pyruvates (Rodrigues et al. 2009)

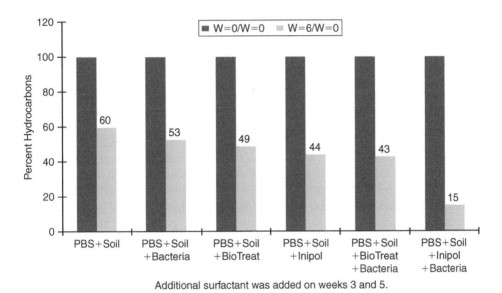

Figure 13.2 Hydrocarbon concentration of crude oil contaminated soil samples, weeks 0 and 6 (Villafañe-Deyack 1999).

In order to screen surfactants used in spilled oil bioremediation Gautam and Tyagi (2006) suggested that they have to meet three criteria: be biodegradable; must not serve as a preferred substrate in the presence of oil and; must not be toxic to indigenous bacteria. It had been demonstrated that the addition of surfactant-like products increases the solubilization of the contaminant concentration at the aqueous phase (Strong-Gunderson et al. 1994). Figure 13.1 shows that after six weeks, the use of Inipol with soil, native bacteria, and the IMP bacteria sample degraded 86% of the hydrocarbons. The application of BioTreat and Inipol demonstrated that bacteria were influenced by the availability of essentials nutrients than with

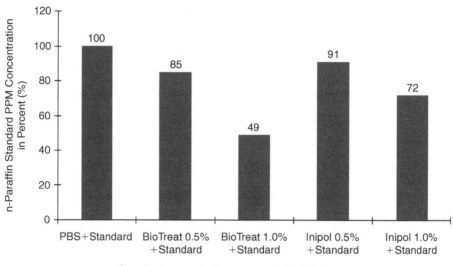

Standard concentration extracted with 0.0, 0.5 and 1.0%.

Figure 13.3 Percent of n-Paraffin Standard extracted with different surfactant concentrations (Villafañe-Deyack 1999).

the concentration of hydrocarbon in the aqueous phase, a finding that is consistent with similar treatments reported by Gillespie and Strong-Gunderson (1997) and Ayotamuno et al. (2007).

Concentrations higher than critical micelle concentration (CMC) tend to interfere with soil extractions or to slow microbial degradation (Brown 2007; Laha et al. 2009). The potential interference of surfactants (0.0/0.5/ 1.0 and 10%) with the soil contaminant extraction assay was compared with the extraction efficiency of the n-Paraffins Standard mixture. Results suggested that Inipol 0.5% and BioTreat 0.5% are the best surfactant concentration to extract crude oil contaminants from soil slurries and hydrocarbon standards (Figure 13.2).

The contaminant heterogeneity of the aqueous phase from vials of the same size with different soil volumes and mixing area was analyzed. The difference in vials size showed to be more significant than in soil volumes with both surfactants, as reported by Menéndez-Vega et al. (2007) and Sunarso and Ismadji (2009). This demonstrated the importance of an adequate surface area for interaction between contaminated soil samples and the liquid phase to make hydrocarbons contaminants available for microbial degradation at the aqueous phase. The results of from this case study demonstrated the effect BioTreat and Inipol, as bioenhancing agents and to be non-toxic to indigenous or introduced bacteria. They are environmentally safe and can be used without threaten to human health in bioremediation efforts (Atlas 2007; Delille et al. 2007).

13.5 CASE STUDY II. APPLICATION OF SURFACTANTS ON PCBS DEGRADATION IN PANAMA

This investigation describes and analyses the use of a nutrient-surfactant and its effect on PCBs degradation. It was conducted in three phases: bench scale analysis,

Table 13.3 Pilot project results for PCBs degradation.

Sample	Time (hours)	PCB 1260 (ppm)
Control	0	650
Wastewater	24	7.6
Wastewater	72	0.5

Table 13.4 Field project results for PCB degradation.

Sample	Time (days)	PCB 1260 (ppm)
Control	0	141
Wastewater	5	16.4
Wastewater	13	5.0

two pilot projects performed in Carrasquilla and Cerro Patacón, and a field project in Río Hato. The bench scale analyses were performed in Puerto Rico, while the pilot and field projects were done in Panamá.

The bench scale study pursued determining an adequate PCBs degradation treatment. Results demonstrated that high surfactant concentrations interfered with standard extraction procedures, as reported by Brown (2007). Solid-Phase Extraction (EPA Method 608 ATP 3 M 0222) was used as an alternative method under this condition

The pilot project emphasized the degradation of PCB extracted from transformers at the Carrasquilla Power Plant in Panamá City, which is operated by the Unión Fenosa Company. Samples were analyzed at LAISA Laboratory (Panama) following EPA-8082A analytical methods (EPA 2000). Results of PCB degradation are shown in Table 13.3.

The field project was performed with 135 transformers stored in the Río Hato Power Plant, a facility constructed following EPA regulations requirements. Transformers were sampled to determine PCBs concentration and characteristics. The wastewater resulted from this treatment was disposed at the local sanitary landfill. All samplings were performed by Unión Fenosa personnel with the supervision of government representatives from the Health and Environment Ministries. PCB analysis was performed at LAISA Laboratory, which results are presented in Table 13.4.

The Applied and Experimental Microbiology Laboratory of the University of Panamá was also required to perform a bench scale study to corroborate PCBs degradation of the field project. Their results showed a reduction in PCBs concentration from 450 ppm to 125 ppm after 13 treatment days, an outcome consistent with Hernández (2005). Panama government agencies sent samples to the certified International Analytical Group Laboratory (IAG) in Florida (USA), to analyze the oil and aqueous phase for PCBs and hazardous compounds included under the Resource Conservation and Recovery Act (RCRA) regulation. No hazardous compounds were detected. The final pH of the wastewater from the PCB degradation treatment was nearly neutral (7.6), which is not harmful to the environment. Similar results have been reported by (Agarwal et al. 2007). A second pilot project was performed with the presence of representatives from Panama's Health and Environment Ministries at the Cerro Patacón Sanitary Landfill, to analyze possible effects of wastewater to be disposed. Analyses were performed at University of Panama, as requested by the Government. Results are summarized in Table 13.5.

This study evidences the effectiveness of a surfactant degrading PCBs from standards solutions and dielectric contaminated oils. The bench scale study demonstrated PCBs' degradation with initial contact and after a long-term treatment, a finding which was confirmed with the pilot and field project. It was found that the targeted pollutants became homogeneous and water-soluble after contact with chlorine ions present in the water matrix, a result similar to the one reported by Robles-González (2008). PCB degradation neutralized the alkaline ions resulting from the treatment, as demonstrated by Brown (2007). Results confirmed that there were no hazardous compound in wastewaters, thus it was plausible to be disposed at the Cerro Patacón Sanitary Landfill. The pilot project at Cerro Patacón brought a collateral benefit: wastewaters help accelerating the degradation of lixiviates and organic wastes, eliminating pestilent odors and decreasing total suspended solids (refer to Table 13.5). The process demonstrated to be environmentally sound and safe, therefore it is applicable for the management of similar facilities in developed and developing countries.

The enhancing role of bacteria for bioremediation

The driving force behind the hydrocarbon bioremediation presented in the two case studies is bacteria, like the species identified in Table 13.2. They can metabolize and

Table 13.5 Cerro Patacón Sanitary Landfill wastewater treatment analysis.

Sample	Parameter	Initial Result	Final Result
Wastewater	pH	7.6	7.6
	Nitrates	<0.1 ppm	<0.1 ppm
	Nitrites	<0.1 ppm	<0.1 ppm
	Hydrocarbons	591 ppm	591 ppm
	Oil and Grease	700 ppm	700 ppm
Landfill Lixiviate	Odor	Pestilent	Pestilent
(Lixiviate)	pH	7.6	7.6
	Total Suspended Solids	44 ppm	44 ppm
	Benzene	<0.01 ppm	<0.01 ppm
	1,4-Dichlorobenzene	<0.01 ppm	<0.01 ppm
Wastewater + Fecal Waste	pH	8.3	8
	Total Suspended Solids	465 ppm	283 ppm
	Odor	Pestilent	Non-detectable
Wastewater + Lixiviate	Odor	Pestilent	Non-Detectable
	pH	7.6	7.6
	Benzene	<0.01 ppm	<0.01 ppm
	1,4-Dichlorobenzene	<0.01 ppm	<0.01 ppm
Wastewater +	Odor	Pestilent	Non-Detectable
Lixiviate + Fecal Waste	pH	8.3	8
	Total Suspended Solids	465 ppm	283 ppm

use these compounds as a source of energy to develop and maintain their cell mass (Plate et al. 2008). These bacteria are heterotrophic, which means that they depend on external sources of carbon to satisfy their energy and cellular growth needs as many other organisms, including humans (Laha et al. 2009). A successful microbial degradation requires a set of conditions: (1) compound(s) must be biodegradable; (2) they must be of a form that produces adequate energy and carbon for the organisms to use and assimilate and; (3) the environmental setting must be hospitable for the organisms (Ojo 2007). Crucial environmental factors include pH, temperature, moisture, terminal electron acceptor content, and micronutrients such as nitrogen and phosphorus (Hosokawa et al. 2009).

As long as the physical and chemical characteristics do not change, the bacteria population dynamics will not be altered (Andreoni and Gianfreda 2007), and biodegradation will flow. The bacteria dominance is a function of their capacity to use hydrocarbons as a primary source of carbon and energy in their metabolism and their ability to better compete in the new environmental setting (Short et al. 2008). Although the metabolic pathway can be aerobic, when oxygen in the system is depleted or limited, denitrification processes dominate (Bowden et al. 1995; Hosokawa et al. 2009), inflicting the bacteria capacity to biodegrade hydrocarbons.

Scientists believe that it is unnecessary to introduce non-indigenous organisms to biologically address hydrocarbon in the environment (Oboh et al. 2007). Specially adapted bacteria are difficult to distribute and often cannot compete with the indigenous population since their laboratory growing conditions differs significantly from the site they were applied; as a result, they soon become a minor element of the microbial community (Singh and Lin 2008). Where indigenous bacteria are insufficient in number or capability to degrade the existing compounds and bioaugmentation, it is required to enhance bioremediation (Salinas-Martínez et al. 2008).

13.4 RISK ASSESSMENT

The persistence of petroleum pollution depends on the quantity and quality of hydrocarbon mixture and on the properties of the affected ecosystem. In some environments, petroleum hydrocarbon persists indefinitely whereas under another set of conditions the same hydrocarbons may be completely biodegraded within a few hours or days (Atlas 1995; Ornitz 1996; Kaufman and Robertson 2010). Remediation technologies have the greatest potential for risk reduction if they are effective in removing the more carcinogenic, high molecular weight compounds. Bioremediation techniques can be considered a promising alternative to clean oil spills using microbial processes to reduce the concentration and/or the toxicity of pollutants (Rosa and Triguis 2007).

The April 2010 oil spill in the Gulf of Mexico, like other recorded accidental and intentional crude leakage onto natural environments, evidence actual scenarios of petroleum hydrocarbons and PCBs serious environmental and health risks. Although physical, biological and chemical technologies can be applied for many sites inflicted by these contaminants, many variables will dictate just how devastating this slick will ultimately be to the ecosystem, including whether it takes days or months to seal the leaking oil and whether winds keep blowing the oil ashore (Kaufman and Robertson

2010). Technologies based on natural biological processes are often received with mistrust and lack of understanding, although these are most promising, as demonstrated with the two case studies herein. The ability to isolate high numbers of certain oil-degrading bacteria from oil-polluted environment is commonly taken as evidence that these microorganisms are the active degraders of that environment (Okerentugba and Ezeronye 2003; Khashayar and Mahsa 2010). Although, hydrocarbon degraders may be expected to be readily isolated from an oil-associated environment, the same degree of expectation may be anticipated for microorganisms isolated from a total unrelated environment such as domestic wastewater (Ojo 2006).

Ecological and health risk assessment can help identify environmental problems, establish priorities, and provide a scientific basis for regulatory actions. There are standard frameworks and guidance documents in the United States and several other countries to apply risk assessment for the regulation of chemicals and the remediation of contaminated sites (Sutter II 2007). Professional judgment and the mandates of a particular statute will be the driving forces in making decisions. Bioremediation works well for remediating soils and aqueous phases contaminated with petroleum hydrocarbons and PCBs, but its implications for risk assessment are uncertain. For petroleum hydrocarbons in soil and PCBs, international regulatory guidance on the management of risks from contaminated sites is now emerging (Sutter II 2007). There is also growing support for the move toward compound-specific risk-based approaches for the assessment of hydrocarbon-contaminated land (Rosa and Triguis 2007; Khashayar and Mahsa 2010).

Bioremediation projects ought to meet each country's specific environmental statutory framework applicable for the site and contaminants treatment (Laha et al. 2009). Decision-making regarding appropriate methods for dealing with PCBs and petroleum contaminated sites shall be based on protection of human health and the environment, both present and future, to minimize risk. Local dwellers ought to be aware that they have to transform (or even revolutionize) their line of thoughts about the best practices to protect their environment, so that they trigger changes in decision makers for the betterment of nature (Laboy-Nieves 2008).

13.3 CONCLUSIONS

Crude oil contains thousands of hydrocarbons with different structures. The widespread of petroleum products parallels the number of oil contaminated sites. Bioenhancing PCBs and spilled oil has been demonstrated as an eco-efficient method for the natural degradation of these contaminants and the restoration of polluted sited. Bioremediation is emerging as one of several alternative technologies for removing pollutants from the environment, restoring contaminated sites, and preventing further pollution. This environmental restoration is needed because certain chemicals can accumulate in the environment to levels that threaten human health and environmental quality. To strengthen this process, characteristics that affect pollutant migration and the ability of microorganisms to grow, requires site-specific feasibility studies, like the one herein described and analyzed for weathered hydrocarbons and PCBs at re1ative high concentration sites in Mexico and Panama. Developing countries ought to consider bioremediation, because it is cost effective technique that has been proven to mitigate the hazards of hydrocarbon toxicity, but it also provides the databank to establish a risk assessment scenario particular for each country. Research is needed to decrease environmental and health risk of hazardous waste treatments and to provide fundamental understanding of the critical factors that determine the outcome of bioremediation.

REFERENCES

Adebusoye S, Ilori M, Picardal F, Amund O. 2008. Cometabolic degradation of polychlorin-ated biphenyls (PCBs) by axenic cultures of *Ralstonia* sp. strain SA-5 and *Pseudomonas* sp. strain SA-6 obtained from Nigerian contaminated soils. World Journal of Microbiology and Biotechnology 24(1): 61–68.

Agarwal S, Al-Abed S, Dionysiou D. 2007. In situ technologies for reclamation of PCB-contaminated sediments: Current challenges and research thrust areas. Journal of Environmental Engineering 133(12): 1075–1078.

Andreoni A, Gianfreda L. 2007. Bioremediation and monitoring of aromatic-polluted habi-tats. Applied Microbiology and Biotechnology 76(2): 287–308.

Atlas R. 2007. Microbial hydrocarbon degradation – bioremediation of oil spills. Journal of Chemical Technology & Biotechnology 52(2):149–156.

Atlas, RM. 1995. Bioremediation. Chemical and Engineering 3:32–42.

Ayotamuno M, Okparanma R, Nweneka E, Ogaji S, Probert S. 2007. Bio-remediation of a sludge containing hydrocarbons. Applied Energy 84(9): 936–943.

Banat I, Makkar S, Cameotra S. 2000. Potential commercial application of microbial sur-factants. Applied Microbiology Biotechnology 53: 495–508.

Banzhaf H. 2010. Economics at the fringe: Non-market valuation studies. Journal of Environmental Management 91(3): 592–602.

Bodour A, Drees K, Maier R. 2003. Distribution of biosurfactant-producing bacteria in undis-turbed and contaminated arid southwestern soils. Applied Environmental Microbiology 69: 3280–3287.

Bowden S, Strong-Gunderson J, Palumbo A. 1995. Effects of low-surfactant concentration on the bioavailability and enhanced biodegradation of volatile organic compounds. In: Book of abstracts for the special symposium of industrial and engineering chemistry division American Chemical Society. Atlanta, Georgia.

Brown D. 2007. Relationship between micellar and hemi-micellar processes and the bioavail-ability of surfactant-solubilized hydrophobic organic compounds. Environmental Science and Technology 41(4): 1194–1199.

Camilli R, Bingham B, Reddy C, Nelson R, Duryea E. 2009. Method for rapid localization of seafloor petroleum contamination using concurrent mass spectrometry and acoustic posi-tioning Marine Pollution Bulletin 58(10): 1505–1513.

Choi S, Lee J, Park S, Kim N, Choo E, Kwak Y, Jeong Y, Woo J, Kim N, Kim Y. 2007. Prevalence, microbiology, and clinical characteristics of extended-spectrum β–lactamase-producing *Enterobacter* spp., *Serratia marcescens, Citrobacter freundii*, and *Morganella morganii* in Korea. European Journal of Clinical Microbiology and Infectious Diseases 26(8): 557–561.

Delille D, Coulon F, Pelletier P. 2007. Long-term changes of bacterial abundance, hydrocar-bon concentration and toxicity during a biostimulation treatment of oil-amended organic and mineral sub-Antarctic soils. Polar Biology 30(7):925–933.

Dercová K, Cicmanová J, Lovecká P, Demnerová K, Macková M. 2007. Ecotoxicity, bio-degradation, and remediation of PCB-contaminated sediments. Journal of Biotechnology 131(2): 242–253.

[DOE] Department of Energy (USA). 2010. International Energy Statistics. [Internet]. [cited 2010 May 02]. Available from: http://tonto.eia.doe.gov/cfapps/ipdbproject/iedindex3. cfm?tid=5&pid=54&aid=2&cid=&syid=2000&eyid=2009&unit=TBPD.

[EIA] Energy Information Administration. 2010. Petroleum Supply Monthly. USA: Department of Energy. 147 p. {cited .. }. Available from: http://www.eia.doe.gov/pub/oil_ gas/petroleum/data_publications/petroleum_supply_monthly/current/pdf/psmall.pdf

Ehsan S, Prasher S, Marshall W. 2006. A washing procedure to mobilize mixed contaminants from soil: I. Polychlorinated biphenyl compounds. Journal of Environmental Quality 35:2146–2153.

[EPA] Environmental Protection Agency (USA). 2000. Resource Conservation and Recovery Act Solid Waste 846 (RCRA SW-846) Polychlorinated biphenyl's (PCBs) by gas chromatography. Revision 1 November 2000.

Ferrera-Cerrato R, Rojas-Avelizapa N, Poggi-Varaldo H, Alarcón A, Cañizares-Villanueva R. 2006. Procesos de biorremediación de suelo y agua contaminados por hidrocarburos del petróleo y otros compuestos orgánicos. Revista Latinoamericana de Microbiología 48 (2): 179–187.

Franzetti A, Bestetti G, Caredda P, La Colla P, Tamburini E (2008) Surface-active compounds and their role in the access to hydrocarbons in Gordonia strains. FEMS Microbiology and Ecology 63: 283–248.

Franzetti A, Caredda P, Ruggeri C, La Colla P, Tamburini E, Papacchini M, Bestetti G. 2009. Potential applications of surface active compounds by *Gordonia* sp. strain BS29 in soil remediation technologies. Chemosphere 75 (6): 801–807.

Gautam K, Tyagi V. 2006. Microbial surfactant: A review. Journal of Oleo Science 55(4): 155–166.

Gillespie M, Strong-Gunderson JM. 1997. Effects of nutrient-surfactant compound on solubilization rates of TCE. Applied Biochemistry and Biotechnology 63–65(1): 835–843.

Godish T. 2004. Atmospheric pollution and pollutants. In: Godish T, editor. Air Quality. Florida (USA). Lewis Publishers. p. 23–69.Hailong L, Boufadel M. 2010. Long-term persistence of oil from the Exxon Valdez spill in two-layer beaches. Nature Geoscience 3: 96–99.

Hernández B. 2005. Determinación de PCBs en aceites, aguas residuales y proyecto piloto utilizando PCBs y surfactante Envi-TS². Laboratorio de Microbiología Experimental y Aplicada, Universidad de Panamá, LAMEXA.

Hosokawa R, Nagai M, Morikawa M, Okuyama H. 2009. Autochthonous bioaugmentation and its possible application to oil spills. Journal of Microbiology and Biotechnology 25(9): 1519–1528.

Kishore D, Mukherjee A. 2007. Crude petroleum-oil biodegradation efficiency of *Bacillus subtilis* and *Pseudomonas aeruginosa* strains isolated from a petroleum oil contaminated soil from North-East India. Bioresource Technology 98(7): 1339–1345.

Khashayar T, Mahsa T. 2010. Biodegradation potential of petroleum hydrocarbons by bacterial diversity in soil. World Applied Science Journal 8(6): 750–755.

Kaufman L, Robertson C. 2010 May. In Gulf Oil Spill, Fragile Marshes Face New Threat. New York Times; 1–2. <http://www.nytimes.com/2010/05/02/us/02spill. html?partner=rss&emc=rss>

Kulkarni P, Crespo J, Afonso C. 2008. Dioxins sources and current remediation technologies. Environmental International 34(1):139–153.

Kurtz R. 2008. Coastal oil spill preparedness and response: The Morris J. Berman incident. Review of Policy Research 25(5): 473–486.

Laboy-Nieves EN. 2008. Ética y sustentabilidad ambiental en Puerto Rico. Actas del Foro Internacional de Recursos Hídricos. [Dominican Republic] INDRHI, Santo Domingo. p 63–74.

Laha S, Tansel B, Ussawarujikulchai A. 2009. Surfactant-soil interactions during surfactant-amended remediation of contaminated soils by hydrophobic organic compounds: a review. Journal of Environmental Management 90(1): 95–100.

Menéndez-Vega D, Gallego J, Peláez A, Fernández de Córdoba G, Moreno J, Muñoz D, Sánchez J. 2007. Engineered *in situ* bioremediation of soil and groundwater polluted with weathered hydrocarbons. European Journal of Soil Biology 43(5–6):310–321.

Nikolopoulou M, Pasadakis N, Kalogerakis N. 2007. Enhanced bioremediation of crude oil utilizing lipophilic fertilizers. Desalination 211(1–3): 286–295.

Oboh B, Ilori M, Akinyemi J, Adebusoye J. 2006. Hydrocarbon degrading potentials of bacteria isolated from a Nigerian bitumen (tarsand) deposit. Nature and Science 4(3): 51–57.

Obuekwe C, Al-Jadi Z, Al-Saleh E. 2009. Hydrocarbon degradation in relation to cell-surface hydrophobicity among bacterial hydrocarbon degraders from petroleum contaminated Kuwait desert environment. International Biodeterioration and Biodegradation 63(3): 273–279.

Ojo O. 2006. Petroleum-hydrocarbon utilization by native bacterial population from a wastewater canal Southwest Nigeria. African Journal of Biotechnology 5(4): 333–337.

Ojo O. 2007. Molecular strategies of microbial adaptation to xenobiotics in natural environment. Biotechnology and Molecular Biology Review 2 (1):1–013.

Okerentugba P, Ezeronye O. 2003. Petroleum degrading potentials of single and mixed microbial cultures isolated from rivers and refinery effluents in Nigeria. African Journal of Biotechnology 2(9): 288–292.

Ornitz BE. 1996. Oil Crisis in Our Oceans. Colorado (USA): Tageh Press. 340 p. Paria S. 2008. Surfactant-enhanced remediation of organic contaminated soil and water. Advances in Colloid and Interface Science 138(1): 24–58.

Perelo L. 2010. Review: In situ and bioremediation of organic pollutants in aquatic sediments Journal of Hazardous Materials 177(1–3): 81–89.

Plante C, Coe K, Plante R. 2008. Isolation of surfactant-resistant bacteria from natural, surfactant-rich marine habitats. Applied and Environmental Microbiology 47(16): 5093–5099.

Rahman P, Gakpe E. 2008. Production, characterization and application of biosurfactants-review. Biotechnology 7(2): 360–370.

Robles-González I, Fava F, Poggi-Varaldo H. 2008. A review on slurry bioreactors for bioremediation of soils and sediments. Microbial Cell Factories 7(5): 1–16.

Robles-González I, Fava F, Poggi-Varaldo H. 2008. A review on slurry bioreactors for bioremediation of soils and sediments Microbial Cell Factories 7:5–25.

Rodrigues D, Sakata K, Comasseto J, Bicego M, Pellizari V. 2009. Diversity of hydrocarbon-degrading *Klebsiella* strains isolated from hydrocarbon-contaminated estuaries. Journal of Applied Microbiology 106(4): 1304–1314.

Rosa AP, Triguis JA. 2007. Bioremediation process on Brazil shoreline. Environmental Science and Pollution Research 14(7): 470–476.

Salihu A, Abdulkadir I, Almustapha M. 2009. An investigation for potential development on biosurfactants. Biotechnology and Molecular Biology Reviews 3 (5): 111–117.

Salinas-Martínez A, Santos-Córdova M, Soto-Cruz M, Delgado E, Pérez-Andrade H, Háuad-Marroquín L, Medrano-Roldán H. 2008. Development of a bioremediation process by biostimulation of native microbial consortium through the heap leaching technique. Journal of Environmental Management 88(1):115–119.

Sheyla R, Couceiro M, Hamada N, Forsberg B, Padovesi-Fonseca C. 2010. Effects of anthropogenic silt on aquatic macroinvertebrates and abiotic variables in streams in the Brazilian Amazon Journal of Soils and sediments 10(1): 89–103.

Short J, Irvine G, Mann D, Maselko J, Pella J, Lindeberg M, Driskell W, Rice S. 2007. Slightly weathered Exxon Valdez oil persists in Gulf of Alaska beach sediments after 16 years. Environmental Science and Technology 41: 1245–1250.

Singh C, Lin J. 2008. Isolation and characterization of diesel oil degrading indigenous micro-organisms in Kwazulu-Natal, South Africa. Journal of Biotechnology 7(12): 1927–1932.

Skinner K, Cuiffetti L, Hyman M. 2009. Metabolism and cometabolism of cyclic ethers by a filamentous fungus, a *Graphium* sp. Applied and Environmental Microbiology 75(17): 5514–5522.

Strong-Gunderson J, Palumbo A, Davidson T, Bergman S. 1994. Effects of commercially available bioenhancers/"surfactants" on the degradation and mobilization of various hydrocarbons. In Book of abstracts for the special symposium of industrial and engineering chemistry division, American Chemical Society. Atlanta, Georgia.

Strong-Gunderson J, Palumbo A. 1995. Bioavailability enhancement by addition of surfactant and surfactant-like compounds. In: Hinchee editor. Microbial process for bioremediation. Columbus, Richland: Battelle Press. p 33–40.

Sunarso J, Ismadji S. 2009. Decontamination of hazardous substances from solid matrices and liquids using supercritical fluids extraction: A review. Journal of Hazardous Materials 161(1):1–20.

Sutter II GW. 2007. Ecological Risk Assessment. Boca Raton, Florida (USA): CRC Press. 538 p.

Varanasi P, Fullana A, Sidhu S. 2007. Remediation of PCB contaminated soils using iron nano-particles. Chemosphere 66(6): 1031–1038.

Vasilyeva G, Strijakova E. 2007. Bioremediation of soils and sediments contaminated by poly-chlorinated biphenyls. Journal of Microbiology 76(6): 639–653.

Villafañe-Deyack E. 1999. Effect of nutrients and bioenhancing agents on the degradation of weathered crude oil contaminants: Bioremediation in the south of Mexico. [Thesis] [Puerto Rico]; Universidad Metropolitana, San Juan.

Wang, I, Wu Y, Lin L, Chang-Chien G. 2009. Human dietary exposure to polychlorinated dibenzo-*p*-dioxins and polychlorinated dibenzofurans in Taiwan. Journal of Hazardous Materials 164(2–3): 621–626.

Chapter 14

Cryptosporidium oocyst transmission in the aquatic environment of Haiti

Ketty Balthazard-Accou, Evens Emmanuel, Patrice Agnamey, Philippe Brasseur, Anne Totet and Christian Raccurt

SUMMARY

Cryptosporidium are intestinal parasites in humans and domestic animals. These opportunistic agents are the major cause of digestive pathology in HIV-infected individuals in developing countries and of acute diarrhea in children under five. In Haiti, *Cryptosporidium* is responsible for 17% of acute diarrheas observed in infants under 2 years of age and 30% of chronic diarrheas in patients infected by HIV. The transmission of *Cryptosporidium* oocysts in young children, HIV-infected individuals, and people living in poor socioeconomic conditions is probably due to the consumption of water or contaminated food. *Cryptosporidium* oocysts have been detected in the surface water, groundwater and in public water supplies in two large cities in Haiti, Port-au-Prince and Cayes. High concentrations from 4 to 1274 oocysts per 100 liters of water were determined in drinking water. Exposure to such concentrations of *Cryptosporidium* in drinking water can generate major biological risks for human health. The aim of this chapter is to review and analyze the different risk factors associated with the presence of *Cryptosporidium* oocysts in the aquatic environments of Haiti.

14.1 INTRODUCTION

Cities and towns produce liquid effluents containing organic and inorganic pollutants, and biological contaminants, in particular pathogenic micro-organisms. Termed 'neighborhood pollution' (Académie des Sciences 1998), urban water generally groups rainwater and urban wastewater. Studies have shown that pollution due to an initial rainfall can be high, as the pollution it contains is as great as and very often greater than that of urban effluent (Chocat et al. 1993). Thus these effluents are a non negligible source of potential contamination for host environments, such as the soil and groundwater (Mikkelsen et al. 1996). Indeed, groundwater is increasingly subject to intensive voluntary discharges of highly polluted effluents, wastewater and runoff water in urban areas (Pitt et al. 1999). Moreover, it becomes patent that the use of groundwater will be restricted in future not because of its shortage, but due to its poor quality (Guillemin and Roux 1994).

In developing countries, there is a strong likelihood of a marked correlation between the groundwater contamination in urban areas and the way in which public services operate. One characteristic of towns and cities in developing countries is the coverage ensured by basic public services, such as drinking water supply, sewage collection and treatment, rainwater drainage and the collection of solid wastes (Emmanuel and Lindskog 2002). Due to reasons relating to the economic problems of these countries, when urban effluents are actually collected, they are most usually discharged into open drainage canals, or into septic tanks equipped with infiltration shafts (Emmanuel 2004). Regarding biological pollution, investigations have been conducted in Haiti. Indeed, oocyts of *Cryptosporidium* has been detected in surface water used as drinking water in Port-au-Prince and in the water supplied by the public water service (Brasseur et al. 2002, 2009). Furthermore, other studies have reported that in Haiti, cryptosporidia are responsible for 17.5% of acute diarrheas observed in infants under 2 years of age (Pape et al. 1987) and 30% of chronic diarrheas in patients infected by HIV (Pape et al. 1983). In Haiti, cryptosporidiosis in children is linked to malnutrition and mannose deficiency related to lectin (Kirkpatrick et al. 2006). A study carried out at Port-au-Prince showed that cryptosporidia oocysts were present in 158 people (i.e. a prevalence of 10.3%). Of this number, 56 adults out of 57 (98%) and seven infants out of 36 (19%) had HIV+. Genotyping led to the identification of three species: 59% *Cryptosporidium hominis*, 38% *C. parvum* and 3% *C. felis* (Raccurt et al. 2006).

Investigations performed on water resources of the town of Cayes (Haiti) have highlighted the presence of *Cryptosporidium* oocysts in surface and groundwater used by the population for domestic purposes (Balthazard-Accou et al. 2009). These results show that the surface water and groundwater of Cayes are contaminated by pollution of fecal origin and are a source of potential biological risk for the health of the population exposed. The objective of this work is to identify and analyze the main determinants participating in the transport (transfer) of *Cryptosporidium* oocysts in the groundwater of Haitian towns, and which increase the probability of occurrence of cryptosporidiosis in consumers of this water. This chapter presents the biology of *Cryptosporidium*, its detection in aquatic ecosystems, the analysis of groundwater microbiological contamination and the risk of drinking water in Haiti.

14.2 BIOLOGY OF *CRYPTOSPORIDIUM*

Cryptosporidium represents the genus of a variety of intracellular parasites that infect vertebrates, including humans, worldwide. *Cryptosporidium* belongs to Apicomplexa (synonym Sporozoa), an obligate parasitic group of eukaryotes (Table 14.1). Species of *Cryptosporidium* parasitize specific host tissues, namely, the intestine, stomach, or trachea. Cryptosporidiosis is a self-limiting disease in healthy hosts but represents a life-threatening problem in immunocompromised individuals for which there is no effective treatment.

The life cycle of this parasite is complex, showing both sexual and asexual stages (Figure 14.1). Oocysts assumes a latent stage from months to several years. Return to favorable conditions rapidly leads to the contrary phenomenon (Bonnard 2001). The persistence of oocysts in environmental compartments can be influenced by their age, i.e. generally the oldest oocysts are more likely to be eliminated by environmental

Table 14.1 Taxonomic classification of *Cryptosporidium* (O'Donoghue 1995).

Taxonomic position	Name	Biological characteristics
Kingdom	*Protists*	Unicellular eucaryotes (mainly characterized by cells with a core having a nuclear envelope)
Phylum	*Apicomplexa/Sporozoa*	– Intracellular parasite – Presence of sporozoits in the cycle – Typical apical complex.
Class	*Coccidia*	– Oocysts containing sporozoits – Asexual and sexual reproduction
Sub-class	*Coccidiasina*	Development cycle comprising generally merogony, gametogony and sporogony
Order	*Eucoccidiorida*	Presence of merogony infecting vertebrates and invertebrates
Sub-order	*Eimeriorina*	Independent development of microgamy and macrogamy
Family	*Cryptosporidiidae*	– Monoxenous biological cycle – Oocysts containing four sporozoite nuclei – Development under the brush border of epithelial cells of colonized intestinal mucous membranes
Genus	*Cryptosporidium*	Only genus of the family of *Cryptosporidiidae*

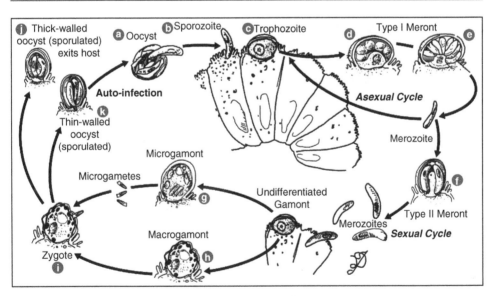

Figure 14.1 Life cycle of *Cryptosporidium*.
Source: *http://www.stanford.edu/group/parasites/ParaSites2005/Cryptosporidiosis/index.html.*

changes and certain chemical disinfectants (Carey et al. 2004). Under natural conditions, fecal matter shelters oocysts from desiccation and increases the impermeability of the wall to small molecules, thereby reducing their exposure to lethal environmental factors (Robertson et al. 1992). The resistance of oocysts in a solid matrix such as soil has become a crucial parameter in understanding their transfer to lower layers. Oocysts persist longer in soil than in water at the same temperature, with a preference for moist silty soils rather than moist clayey or sandy ones (Jenkins et al. 2002). Oocysts can remain viable and infectious in water for several months at temperatures ranging from 0 to 30°C (Fayer et al. 1998b). Other tests have shown that boiling water can kill *Cryptosporidium* oocysts in less than a minute (Fayer et al. 1996). The exposure to sunlight had no effect on the viability of *Cryptosporidium* oocysts, but UV at 265 nm and black light at 365 nm lead to a reduction in the number of viable oocysts (Chauret et al. 1995).

14.3 DETECTING *CRYPTOSPORIDIUM* OOCYSTS IN AQUATIC ECOSYSTEMS

Tests have been carried out on several filtration methods, with varying yields: membrane filtration (Aldom and Chagla 1995), cartridge (Shepherd and Wyn-Jones 1996), and chemical flocculation (Vesey 1993). Cartridge filtration is the choice method for detecting oocysts in environmental samples. Several authors have evaluated different types of cartridge: woven acrylic and polyethersulfone, that permit filtering large volumes (10 to 1000 liters) of water (AFNOR 2001; McCuin and Clancy 2003). Their porosity of 1 μm permits capturing *Cryptosporidium* oocysts and other protozoa with a theoretical sensitivity of 1 to 10 oocysts/liter of distilled water for 100 L of filtered water.

The elution of oocysts is an important step and its efficiency depends on the turbidity of the sample, the type of filter and the method chosen (Musial et al. 1987; Inoue et al. 2003). The oocysts trapped by filter are eluted with detergent solutions that break the hydrophobic interactions between the oocysts, sediment particles and filter fibers (Musial et al. 1987). Sonication of the deposit following centrifugation of the eluate appears to improve the separation of *Cryptosporidium* oocysts (Musial et al. 1987). Standardization of the process has led to the use of saline type PBS solutions supplemented with detergents such as laureth-12 and Tween-80 (AFNOR 2001). Despite their dispersive nature, the relatively high ionic force of PBS (164 mM) tends to leads to the aggregation of oocysts with the debris (Davies et al. 2003). The use of solutions containing dispersive agents (EDTA, sodium pyrophosphate) significantly increases the average percentage of *Cryptosporidium* oocyst recovery in comparison to conventional solutions (Inoue et al. 2003).

Clays, humic and fulvic acids, organic compounds, salts and heavy metals are frequently found in water concentrates (Tsai and Rochelle 2001). Consequently, subjecting samples to a purification phase makes detecting and identifying oocysts easier. The two techniques most frequently used to separate oocysts from environmental debris are density gradient flotation and immunomagnetic separation.

14.4 MICROBIOLOGICAL RISKS ASSESSMENT IN THE DRINKING WATER IN HAITI

14.4.1 Generalities

NRC (1983) defines risk evaluation of the toxic characteristics of a chemical product and the conditions of human exposure to this product. The global methodology of full health risk evaluation relies on an epidemiological approach and it currently appears to be the best adapted tool for quantifying health risks. The general approach of evaluating health risks is based on four steps: identifying the danger, studying the dose-response relationship, estimating exposure, characterizing the risk (NRC 1983).

The evaluation of chemical risks refer to the application of models or toxicological and physicochemical databases, but this analytical approach seems not to be advanced (Gofti et al. 1999). The data used to evaluate risk linked to environmental exposure is often lacking. In addition, biological risk has a large number of characteristics that prevent the simple transposition of the method employed in chemistry to biology. The difference between the estimation of a chemical risk and a microbiological or infection risk, resides in the identification of dose-response functions, and especially in the choice of dose-response model (Zmirou-Navier et al. 2006). Another particularity is represented by the existence of human, animal and environmental reservoirs that are difficult to control (Hartemann 1997).

14.5 THE EPIDEMIOLOGICAL CHAIN

Preventing the transmission of infectious diseases due to the exposure of human beings to polluted food, water, soils and air has always been a major concern in public health and environmental sciences (Haas et al. 1999). The transmission of an infectious agent requires the co-existence of a source of pathogenic or opportunistic agent to contact vulnerable subjects; mode of transmission and; a receptive subject. Regarding the source, increasing numbers of bacteria, viruses, yeasts and other fungi, have been identified and characterized. Likewise, the reservoir can be infected by humans and animals, healthy carriers and sometimes the environment (e.g. *Legionella*). In fact, for most pathogenic agents the environment only acts as an accidental or transient reservoir, as the passage of the microbe in the environment depends on the nature of the infection of the carrier – excretion in the case of eliminated infection, transfer by material or insect in the case of infection (Bonnard, 2001).

Direct and indirect transmission occurs via sometimes novel and complex channels made possible by technological progress (for example, aerosolization), although more is known today about how microorganisms survive in the environment. The factors permitting this survival are multifold (temperature, nutriments, pH, U.V, other organisms) and their presence or absence influence the fate of infectious agents, assessed by T90 (Hartemann 1997).

Receptive subjects have evolved considerably thanks to medical progress and increased longevity, leading to the emergence of vulnerable and highly immunodepressed populations. Experiments involve getting laboratory animals, or volunteers in good health, to ingest different doses of pathogens like *Cryptosporidium* cysts conserved in food, counting the number of cysts excreted in feces and monitoring the appearance of clinical signs (DuPont et al. 1995). These parameters characterize the capacity of the microorganism to cause clinical disturbances in the infected subject, on which even fewer data are available. In the enterovirus family, virulence ranges from 1 to 97%, leading to the recommendation of taking 50% as the average estimation when specific data are lacking (Hartemann 1997). *Cryptosporidium parvum* virulence had been estimated at 100% in the evaluation of the impact on health of the accident of Milwaukee. This impact was proven to be close to the actually observed, thereby validating the hypotheses (Haas 1996).

14.6 CASE STUDY OF PORT-AU-PRINCE

During 2000 and 2001, water distribution points used by the population for water supplies in Port-au-Prince, were analyzed only once each but at different periods by sampling at least 100 liters. Samples were taken from eight reservoirs, seven points of the city drinking water network and three public fountains. The choice of sampling site was decided as a function of the residential districts of HIV+ treated patients. The degree of *Cryptosporidium* pathogenicity varies as a function of age (children under 5) and the immune status of HIV+ subjects (Bras et al. 2007). The exposed population was divided into four categories: (1) immunocompetent subjects aged 5 and over, for contamination risks leading to a low risk of infection; (2) immunodepressed subjects aged 5 and over with severe illness linked to HIV; (3) immunocompetent under 5 years old, for whom the contamination risk lead to a high risk of illness; (4) immunodepressed infants under 5 years old, for whom the contamination risk leads to high risk of illness.

To quantify the microbiological risks due to *Cryptosporidium*, the Bras et al. (2007) method was employed, which is based upon a contamination module and the analysis of oocysts found in 100 liters of filtered water. For microbiological dangers, the models most frequently used were the "exponential" and "Beta-Poisson" models. The following equation was used to determine the probability of infection:

$$Pr = 1 - \exp(-rD) \tag{14.1}$$

Where:
    P = probability of infection of an individual exposed to dose D of oocysts;
    r = the probability of survival of oocysts in drinking water ingested by the host
    D = ingested dose.

The emission was evaluated with the data obtained by analyzing the water used by the population from 18 sites of greater Port-au-Prince between 2000 and 2002 (Brasseur et al. 2002) and on the basis of demographic data available for the different boroughs of the City. The number of oocysts observed during the analysis did not correspond to the actual number of oocysts (AFSSA 2002). The study took into account the daily consumption of water by infants under 5: 0.75 L/d; and children aged 5 and over and adults: 2 L/d (Fawell and Young, 1999).

For immunocompetent subjects, the risk was estimated by using the information in the literature on the infective dose of *Cryptosporidium* (DuPont et al. 1995). For children under 5, the risk seems to be relatively constant (1×10^2) in the different areas studied. Furthermore, the number of persons contaminated by HIV in Haiti is estimated to be about 400,000, i.e. 5% of the total population (IHSI 2003). By accepting that 5% are uniformly distributed over the Country, the risk of infection calculated varies from 1 to 97×10^2, depending on the district and the age of the persons exposed (Bras et al. 2007).

14.7 CASE STUDY OF THE CITY OF CAYES

Balthazard-Accou et al. (2009) highlighted the presence of *Cryptosporidium* oocysts in the surface and ground water of the coastal city of Cayes (population ~150,000), which has a well watered coastal plain (more than 2000 mm/yr). The watershed is divided by three distinct types of aquifer: alluvial unconfined groundwater, karstic, and cracked and segmented carbonate (PNUD 1991a).

Water samples were taken from 15 sites in September, November and December 2007, i.e. at the end of the main rainy season and at the beginning of the main dry season. The sampling sites were chosen as a function of the state of insalubrities of certain districts of the City and information given by dwellers. Six sites were found to be positive for *Cryptosporidium* oocyst, which ranged from 5 to 100 per 100 liters. The results obtained were determined according to the standard method based on the filtration, elution and concentration of oocysts and cysts by immunomagnetic separation and by detection and counting under epifluorescent microscopy (EPA 2001, 2005). This technique reduces the number of false positives and increases microscopic results (Connell et al. 2000).

Nine oocysts per 100 L were obtained from water samples from the public water supply service of Cayes. Detecting *Cryptosporidium* in these waters can be more worrisome for vulnerable populations (elderly, children, immunosuppressed) and can increase mortality rate in the City because of the lack of an effective treatment for an infection with this parasite (MacKenzie et al. 1994; Rose et al. 1991).

The different positive results obtained for *Cryptosporidium* confirmed that the population of Cayes is exposed to health problems linked to water-borne infections. Indeed, a median infectious dose of 132 oocysts, as determined in healthy adult volunteers, provokes human infection in 50% of cases (DuPont et al. 1995). However, a mathematical model based on data from the Milwaukee outbreak suggested that some individuals developed cryptosporidiosis following the ingestion of only one oocyst (Haas and Rose 1994).

14.8 RISK FACTORS ASSESSMENT FOR GROUNDWATER CONTAMINATION

Risk factor refers to a measurable characteristic in a group of individuals or their life context that makes it possible to predict a negative outcome on a specific criterion (Wright and Masten 2005). For Fougeyrollas et al. (1998), a risk factor is an element belonging to an individual or originating from an environment liable to cause an illness, trauma or another negative impact on the development of the human organism. This latter definition is of particular interest as it not only highlights the inherent or hereditary characteristics of an individual, but also (and especially in the case of a population) exposure to environmental risks. Last (2004) underlined that the term risk factor is quite inaccurate, because it may describe an exposure that: increases the probability of a specific result; increases the probability of occurrence of an infection or other outcome (a determinant) and; a determinant that can be modified by an action, leading to a reduction of the probability of occurrence of results.

Regarding drinking water supply, the socioeconomic conditions of Haiti have led to the use of technologies that privilege the use of groundwater rather than surface water (Emmanuel et al. 2004). The aquifer of the Plaine du Cul-de-Sac at Port-au-Prince, and that of the Plaine des Cayes have the highest potential groundwater reserves for their respective regions. They provide a large proportion of the drinking water supplies to the populations of the two conurbations in question, but reserves are contaminated by *Cryptosporidium* oocysts. The exposure of consumers to this contaminated water is a major health risk factor. To address these risk factors it is necessary to identify not only the determinants that are facilitating pollution, but also the actions to ensure the drinkability of the water from these aquifers.

In Haiti, there is no control for the quality of the water distributed by public services (OMS 1998). Urban areas lack basic services for the collection and treatment of wastewater and solid wastes, and the removal of excreta. Latrines and septic tanks feed fecal contamination into alluvial and karstic aquifers. Chlorination is the only method used to treat raw water intended for human consumption (Emmanuel and Lindskog 2002), but it is ineffective in inactivating *Cryptosporidium* oocysts (Korick et al. 1990).

Analysis of the risk factors responsible for groundwater contamination by *Cryptosporidium* oocysts first requires hydrogeological studies of the aquifers concerned, followed by studies of the mechanisms involved in the transport of oocysts from the non saturated to the saturated zone, finally resulting in the adsorption of this colloid on porous materials. Karstic aquifers distinguish the hydrogeology of Haiti, characterized by irregular pores, cracks, fractures and conduits (PNUD 1991a). Following a shower, the rapid and turbulent replenishment of groundwater occurs via the drainage of considerable volumes of non filtered water through large conduits (Denić-Jukić and Jukić 2003).

The higher ground surrounding the plain (the Plaine du Cul-de-Sac) covers a large area and reaches high altitudes in the south (up to 2000 m) while it is narrow and low in the north (altitude of about 1000 m). The largest rivers (Rivière Grise and Fond Parisien) flow from the Massif of Selle and infiltrate the plain (PNUD 1991b); Freshwater infiltrations generally occur through faults between the alluvial formations of the plain and limestone hills. The groundwater of the Plaine du Cul-de-Sac makes up an aquifer system partially open to the sea. Thus its hydrological balance is impacted by circulating underground water from its sources to its outlets, a hydrogeological mechanism of communicating vessels (Simonot 1982). Water supply depends on the frequency and intensity of the rainfall recorded and land use. The water resources of this aquifer are threatened by saline (Simonot 1982), microbiological (Bras et al. 2007), and heavy metal contamination (Emmanuel et al. 2009). The burial of several thousand cadavers, the victims of the earthquake of 12 January 2010, in the Plaine, without specific geological pretreatment, are liable to greatly increase the pollution indexes of this aquifer.

Overlooked by the Massif de la Hotte, with an altitude of over 2000 m, this region receives abundant rainfall (more than 3000 mm/yr) on the summits, which decreases seaward (1400 mm/yr). The sources to replenish groundwater is rainwater infiltration through the karstic limestone of the Massifs. The Plaine des Cayes covers a surface area of $250\,km^2$ and is dominated by mostly limestone mountains. The extent of surface alluvial deposits practically coincides with that of the plain itself. It constitutes the largest directly exploitable underground water resource for the entire region. The depth of the water bearing stratum exceeds $40\,m$ upstream of the plain (PNUD 1991a).

Xin Dai and Jan Boll (2003) observed that oocyst are released from feces or manure on the surface and moves into runoff water, and its transport rate depends on size and settling velocity. Based on its size, the *Cryptosporidium parvum* oocyst is physically classified as a biological colloid. Surface charges measured by the ξ potential of the oocysts have been found to be neutral to slightly negative in most natural waters. Exact values depend on the analytical methods used (Brush et al. 1998). The transport and filtration of such colloids in porous media occurs by hydrodynamic processes such as advection and interactive processes between colloids and solids surfaces (Harvey 1991). The following equation is a simple one-dimensional transport model in a steady-state flow field:

$$\frac{\partial c}{\partial t} + \frac{\rho_b \partial c}{\theta \partial t} = vd\,\frac{\partial^2 c}{\partial x^2} - v\left(\frac{\partial c}{\partial x} + \lambda c\right) \tag{14.2}$$

where ∂ is the partial differential, c is the concentration of *C. parvum* oocysts in suspension, s is the concentration of *C. parvum* oocysts adsorbed reversibly on solids surfaces, ρ_b is bulk density, θ is porosity, d is the hydrodynamic dispersion coefficient, v is the pore advection velocity, and λ is the colloid filtration coefficient. It is commonly observed that velocity enhancement in Equation 14.2 is higher for colloids than for water (DeMarsily 1986). Note that Equation 14.2 accounts for permanent deposition (filtration) of colloids through the first-order term $v\lambda c$ as well as for reversible deposition (sorption) through the second term on the left-hand side.

Several models have been introduced to describe the permanent removal of colloids by filtration onto the solid phase (Logan et al. 1995). The deposition kinetics of *Cryptosporidium* oocyts onto solid surfaces is directly dependent on it size and surface properties. Like viruses and bacteria (Bitton 1975; van Loosdrecht et al. 1987), oocysts can be considered as colloidal particles. If the particles have sufficiently high repulsion, dispersion will resist flocculation and the colloidal system will be stable. However, if no repulsion mechanism exists then flocculation or coagulation will eventually take place (Kuznar and Elimelech 2004; McBride et al. 2002). Little research has been performed on oocyst adhesion to solid surfaces (Kuznar et al. 2004). The addition of enzymes to remove macromolecules enhances adhesion properties (Kuznar and Elimelech 2006), a process that depends on the ionic strength and pH dominating the interaction force between the surface adhesion of the parasite and the filter media (Hsu et al. 2001). When removing surface macromolecules, oocysts do not have electrosteric repulsive interactions with the quartz surface and behave like model colloidal particles (Kuznar and Elimelech 2004). This method provides 91% efficiency (Kuznar and Elimelech 2006).

The risk assessment of *Cryptosporidium* contamination in drinking water from Haitian aquifers is not a goal but a basis for decision-making. In this context of Haiti, the risk management of oocyst hazards and risks for human health may consist on the establishment of a monitoring and control of the microbiological quality of drinking water where technical procedures including analytical measurements and inspection-based should take place. The application of three main factors and the study of their efficiency seem important: (i) the elaboration or adoption of national drinking water guidelines. In relation to analytical measurements, this factor will facilitate the monitoring of exposure characterization and the prioritisation of potential interventions in the context of environmental exposure to contaminated drinking water. (ii) The protection of the groundwater resource. This factor concerns the sanitary inspection as part of surveillance and control of *Cryptosporidium* contamination in water resources. The exposure of drinking water sources of to liquid and solid wastes justifies largely this factor application. In the case of solid wastes, the lixiviate takes care of organic, mineral and metallic pollutants, by extraction of the soluble compounds and contributes, from the percolation of the water through the ground to the pollution of the karstic and alluvial aquifers. This constraint is particularly important for the location of the discharges and the public spaces used for burial of several cadavers after natural disasters (iii) the treatment of raw water to drinking water. Since the *Cryptosporidium* oocyst is a biological colloid, the technical processes used to remove heavy metals in drinking water, such as adsorption, will upgrade the quality of drinking water and reduce or eliminate risks for human health of *Cryptosporidium* oocysts.

14.9 CONCLUSION

Cryptosporidium oocyts have been detected in untreated surface and groundwater from two large cities in Haiti. The transmission of cryptosporidiosis to humans, and in particular the groups at greatest risk – children under five, HIV patients, and the undernourished – occurs through food and water containing *Cryptosporidium* oocyts. Cryptosporidiosis is a frequent cause of diarrhea in children under 2 years-old and in HIV patients. No effective treatment of cryptosporidiosis is available. *Cryptosporidium* oocysts in natural aquatic environments and drinking water leads to a biological hazard linked to the dangerous characteristics of this organism. Since groundwater is used more than surface water in Haiti, this study focused on the need to upgrade the quality of drinking water by identifying and analyzing the risk factors involved in the contamination of groundwater by *Cryptosporidium* oocysts. Given the inefficiency of certain chemical and physical treatment methods used to trap oocysts in raw water, it will be necessary to identify new actions capable of modifying the behavior oocysts in aquatic environments. In the framework of evaluating the biological risks due to *Cryptosporidium* oocysts, the aim would be focus on scientific objectives intended to improve knowledge and better characterize the exposure phase. Therefore a geological and hydrological study would be most useful. The study of adsorption under static and dynamic conditions may lead to assessing not only the capacity of the soil to trap oocysts, and also provide knowledge regarding the hydrodynamics involved in transporting or transferring oocysts from non saturated areas to saturated ones. In the specific case of urban groundwater in Haitian cities, it would be relevant to study new treatment processes to procure health the betterment of the population.

ACKNOWLEDGEMENTS

The authors would like to acknowledge Accent Europe (http://www.accent-europe.fr/), for the English translation of this manuscript.

REFERENCES

[AS] Académie des Sciences (France). 1998. Contamination of soils by trace elements: risk management. Report N°42. 440 p.

[AFNOR]. Agence Française de Sécurité Sanitaire des Aliments (France). 2001. Qualité de l'eau – Detection and enumeration of *Cryptosporidium* oocysts and Giardia cysts – Method of concentration and counting NF T90-455, La Plaine Saint-Denis. Association Française de Normalisation, France. 22 p.

[AFSSA]. Agence Française de Sécurité Sanitaire des Aliments (France). 2002. Report on the Protozoa infections related to food and water: scientific evaluation of risks associated with Cryptosporidium sp. 185 p.

Aldom JE, Chagla AH. 1995. Recovery of *Cryptosporidium* oocysts from water by a membrane filter dissolution method. Letters in Applied. Microbiology, 20:186–187.

Balthazard-Accou K, Emmanuel E, Agnamey P, Brasseur P, Obicson Lilite, Totet A, Raccurt CP. 2009. Presence of *Cryptosporidium* oocysts and Giardia cysts in the surface water and groundwater in the city of Cayes, Haiti. AQUA-LAC, Journal of the International Hydrological Programme for Latin America and Caribbean 1:63–71.

Bitton G. 1975. Adsorption of viruses onto surfaces in soil and water. Water Resources 9: 473–484.

Bonnard R. 2001. The biological risk and risk assessment method. Paris: Institut National de l'Environnement Industriel et des Risques. INERIS (DRC-01-25419-ERSA-RBn-383). 70 p.

Bras A, Emmanuel E, Obiscon L, Brasseur P, Pape JW, Raccurt CP. 2007. Biological risk assessment of *Cryptosporidium* sp. in drinking water in Port-au-Prince, Haiti. Environnement Risques et Santé 6:355–364.

Brasseur P, Eyma E, Li X, Verdier Ri, Agnamey P, Liautaud B, Dei Cas E, Pape JW, Raccurt CP. 2002. Movement of *Cryptosporidium* oocysts in surface water supply and distribution service to Port-au-Prince, Haiti. Colloque International "Gestion Intégrée de l'Eau en Haïti", actes du colloque:172–175.

Brush C, Walter M, Anguish L, Ghiorse W. 1998. Influence of Pretreatment and Experimental Conditions on Electrophoretic Mobility and Hydrofobicity of *Cryptosporidium parvum* oocyst. Applied an Environmental Microbiology 64: 4439–4445.

Carey CM, Lee H, Trevors JT. 2004. Biology persistence and detection of *Cryptosporidium parvum* and *Cryptosporidium hominis* oocyst. Water Reseach 38: 818–862.

Chauret C, Armstrong N, Fisher J, Sharma R, Springthorpe S, Sattar S. 1995. Correlating *Cryptosporidium* and *Giardia* with microbial indicators. Journal American Water Works Association 87: 76–84.

Chocat B, Thibault S, Seguin D. 1993. Urban water supply and sanitation. Lyon:Institut National des Sciences Appliquées. 142 p.

Connell K, Rodgers CC, Shank-Givens H.L, Scheller J, Pope ML, Miller K. 2000. Building a better protozoa data set. Journal American Water Works Association 92: 30–43.

Current WL, Garcia LS. 1991. Cryptosporidiosis. Clinical Microbiology Reviews 4: 325–58.

Davies CM, Kaucner C, Deere D, Ashbolt NJ. 3003. Recovery and enumeration of *Cryptosporidium parvum* from animal fecal matrices. Applied and Environmental Microbiology 69: 2842–2847.

Denić-Jukić V, Jukić D. 2003. Composite transfer functions for karst aquifers'. Journal of Hydrology 274: 80–94.

DeMarsily G. 1986. Quantitative Hydrogeology 1st ed. San Diego (California): Academic Press. 440 p.

DuPont HL, Chappell CL, Sterling CR, Okhuysen PC, Rose JB, Jakubowski W. 1995. The infectivity of Cryptosporidium parvum in healthy volunteers. New England Journal of Medicine 332: 855–9.

Emmanuel E, Lindskog P. 2002. Views on the current situation of water resources in the Republic of Haiti. In: Emmanuel E. et Vermande P. (ed) Actes du Colloque International Gestion Intégrée de l'Eau en Haïti. Port-au-Prince:Laboratoire de Qualité de l'Eau et de l'Environnement, Université Quisqueya. p 30–52.

Emmanuel E. 2004. Evaluation of Health Risks and Ecotoxicological Linked to Hospital effluents. Thesis. Institut National Des Sciences Appliquées De Lyon. 257 p.

Emmanuel E, Pierre MG, Perrodin Y. 2009. Groundwater contamination by microbiological and chemical substances released from hospital wastewater: Health risk assessment for drinking water consumers. Environment International 35: 718–726.

[EPA] Environmental Protection Agency (USA). 2001. Method 1622: Cryptosporidium in water by Filtration /IMS/ FA. EPA-821-R-01-026. US, Environmental Protection Agency, Office of Water. Washington D.C. 57 p.

[EPA] Environmental Protection Agency (USA). 2005. Method 1623: Cryptosporidium and Giardia in water by Filtration /IMS/FA. EPA 815-R-05-002. US, Environmental Protection Agency, Washington D.C. 76 p.

Fayer R, Trout JM, Nerad T. 1996. Effects of a wide range of temperatures on infectivity of Cryptosporidium parvum oocysts. Journal of Eukaryot Microbiology 43: 64 p.

Fayer R, Trout JM, Jenkins MC. 1998. Infectivity of Cryptosporidium parvum oocysts stored in water at environmental temperatures. Journal for Parasitology 84: 1165–9.

Fawell J, Young W. 1999. Exposure to chemicals through water. In: IEH, editor. Exposure Assessment in the Evaluation of Risk to Human Health. Report of a Workshop Organised by the Risk Assessment and Toxicology Steering Committee. Leicester: Institute for Environment and Health. p. 22–23

Fougeyrollas P, Cloutier R, Bergeron H, Côté J, St Michel G. 1998. Classification of Quebec workflow disability. Quebec: International Network on the Disability Creation Process. 34 p.

Guillemin C, Roux J C. 1994. Groundwater pollution in France. Manuel et méthodes. Orléans: BRGM. 231 p.

Gofti L, Zmirou D, Seigle Murandi F. Hartemann Ph, Poleton JL. 1999. Microbiological risk assessment of waterborne: a state of the art and perspectives. Epidemiol Sante Publique. 49: 411–22.

Haas CN, Rose JB. 1994. Reconciliation of microbial risk models and outbreak epidemiology: the case of the Milwaukee outbreak. In: Proceeding of the 1994. Annual Conference: Water Quality. New York: American Water Works Association. p. 517–23.

Haas CN. 1996. How to average microbial densities to characterize risk. Water Research 30: 1036–1038.

Haas CN, Rose JB, Gerba CP. 1999. Quantitative Microbial Risk Assessment. New York: Wiley. 449 p.

Hartemann P. 1997. Microbiological risks. In: Sociéte Française de Santé Publique Actes du colloque Science et décision en santé environnementale. Les enjeux de l'évaluation et de la gestion des risques. 9–11 décembre 1996. Metz (Vandoeuvre-lès-Nancy): SFSP. p. 114–121.

Harvey RW, Garabedian SP. 1991. Use of colloid filtration theory in modeling movement of bacteria through a contaminated sandy aquifer. Environmental Science and Technology 25: 178–185.

Hsu B, Huang C, Pan J. 2001. Filtration Behaviors of Giardia and *Cryptosporidium*-Ionic Strength and pH Effects. Water Research 35: 3777–3782.

Inoue M, Rai SK, Oda T, Kimura K, Nakanishi M., Hotta H, Uga S. 2003. A new filter-eluting solution that facilitates improved recovery of *Cryptosporidium* oocysts from water. Journal of Microbiological. Methods 55: 679–686.

[IHSI] Institut Haïtien de Statistique et d'Informatique (Haiti). 2003. Survey of Living Conditions in Haiti. Port-au-Prince: IHSI. Volume1. 640 p.

Jenkins MB, Bowman DD, Fogarty EA, Ghiiorse WC. 2002. *Cryptosporidium parvum* oocyst inactivation in three soil types at various temperatures and water potentials. Soil Biology & Biochemistry 34: 1101–1109.

Kirkpatrick BD, Huston CD, Wagner D, Noel F, Rouzier P, Pape JW. 2006. Serum mannose-binding lectin deficiency is associated with cryptosporidiosis in young Haitian children. Clinical Infectious Diseases 43: 289–294.

Korich D, Mead J, Madore M, Sinclair N, Sterling COR. 1990. Effects of ozone, chlorine dioxide, chlorine and monochloramine on *Cryptosporidium parvum* oocyst viability. Applied and Environment Microbiology 56: 1423–8.

Kuznar Z, Elimelech M. 2004. Adhesion kinectics of viable *Cryptosporidium parvum* oocysts to quartz surfaces. Environmental Science Technology 38: 6839–6845.

Kuznar Z, Elimelech M. 2006. *Cryptosporidium* oocysts surface macromolecules. Environmental Science and Technology 40: 1837–1842.

Last JM. 2004. Dictionary of Epidemiology. St-Hyacinthe (Canada): Edisem-Maloine, French Edition. 306 p.

Logan A.J, Stevik TK, Siegrist RL, Ronn RM. 2001. Transport and fate of Cryptosporidium parvum oocysts in intermittent sand filters. Water Research 35: 4359–4369.

McBride MB, Baveye P. 2002. Particle interactions in colloidal systems: Diffuse double-layer models, long-range forces and ordering in clay colloids. Soil Science Society American Journal 66: 1207–1217.

MacKenzie WR, Hoxie N, Proctor M, Gradius M, Blair K, Peterson D, Kazmierczak J, Addiss D, Fox K, Rose JB, Davis J. 1994. A massive outbreak in Milwaukee of *Cryptosporidium* infection transmitted through the public water supply. New England Journal of Medicine 7: 331–161.

McCuin RM, Clancy JL. 2003. Modifications to United States Environmental Protection Agency methods 1622 and 1623 for detection of *Cryptosporidium* oocysts and *Giardia* cysts in water. Applied and Environmental Microbiology 69: 267–274.

Mikkelsen PS, Häfliger M, Ochs M, Tjell JC, Jacobsen P, Boleer M. 1996. Experimental assessment of soil and groundwater contamination from two old infiltration systems for road run-off in Switzerland. The Sciences of the Total Environment 189/190: 341–347.

Musial CE, Arrowood MJ, Sterling CR, Gerba CP. 1987. Detection of *Cryptosporidium* in water by using polypropylene cartridge filters. Applied and Environmental Microbiology 53: 687–692.

[NRC] National Research Council (USA). 1983. Risk Assessment in the Federal Government: Managing the Process. Washington DC: National Academy Press. 191 p.

O'Donoghue PJ. *Cryptosporidium* and cryptosporidiosis in man and animals. 1995. International Journal of Parasitology 25: 139–195.

[OMS]. Ministère de la Santé Publique (Haiti). Analysis of health status. 1998. Port-au-Prince: Imprimerie Henri Deschamps. 250 p.

Pape JW, Liautaud B, Thomas F, Mathurin JR, St Amand MM, Boncy M, Pean V, Pamphile M, Laroche AC, Johnson WD. Jr. 1983. The acquired immunodeficiency syndrome in Haiti. Annals of International Medicine 103: 674–678.

Pape JW, Levine E, Beaulieu ME, Marshall F, Verdier R, Johnson WD. Jr. 1987. Cryptosporidiosis in Haitian children. American Journal of Tropical Medicine and Hygiene 36: 333–7.

Pitt R, Clark S, Field R. 1999. Groundwater contamination potential from stormwater infiltration practices. Urban Water 1: 217–236.

[PNUD] Programme des Nations Unies pour le Développement. 1991a. Development and management of water resources. Haiti: Water availability and adequacy. Volume IV Région Centre-Sud New-York (USA): PNUD. 50 p.

[PNUD] Programme des Nations Unies pour le Développement. 1991b. Development and management of water resources: Water availability and adequacy. Volume VI Région Sud-Ouest. New-York (USA): PNUD. 70 p.

Raccurt CP, Brasseur P, Verdier RI, Li X, Eyma E, Panier Stockman C, Agnamey P, Guyot K, Totet A, Liautaud B, Nevez G, Dei-Case E, Pape JW. 2006. Human cryptosporidiosis and species involved in Haiti. Tropical Medicine and International Health 11: 929–934.

Robertson LJ, Campbell AT, Smith HV. 1992. Survival of *Cryptosporidium parvum* oocysts under various environmental pressures. Applied and Environmental Microbiology 58: 3494–500.

Rose JB, Gerba CP, Jakubowsky W. 1991. Survey of potable water supplies for Cryptosporidium andGiardia. Environmental Science Technology 25: 1993–1400.

Shepherd KM, Wyn-Jones AP. 1996. An evaluation of methods for simultaneous detection of Cryptosporidium oocysts and Giardia cysts from water. Applied and Environmental Microbiology 62: 1317–1322.

Simonot M. 1982. The groundwater resources of the region of Port-au-Prince. Current Status and Recommendation. Port-au-Prince: Programme des Nations Unies pour le Développement. 52 p.

Tsai Y, Rochelle PA. 2001. Extraction of nucleic acids from environmental samples. In: Rochelle P.A. editor. Environmental molecular microbiology: protocols and applications. Wymondham Norfolk (UK): Horizon Scientific Press. pp 15–30.

Van Loosdrecht M, Lyklema J, Norde W, Schaa G, Zehnder A. 1987. Electrophoretic mobility and hydrophobicity as a measure to predict the initial steps of bacterial adhesion. Applied Environmental Microbiology 53: 1898–1901.

Vesey G, Slade JS, Byrne M, Shepherd KM, Fricker CR. 1993. A new method for the concentration of *Cryptosporidium* from water. Journal of Applied Bacteriology 75:82–86.

Wright MO, Masten AS. 2005. Resilience processes in development: fostering positive adaptation in the context of adversity. In: Goldstein S and Brooks R, editors. Handbook of resilience in children. New York (NY): Kluwer Academic/Plenum. p. 17–37

Xin D, Boll J. 2003. Evaluation of attachment of *Cryptosporidium parvum* and *Giardia lamblia* to soil particles. Journal of Environmental Quality 32: 296–304.

Zmirou-Navier D, Gofti-Laroche L, Hartemann P. 2006. Waterborne microbial risk assessment:a population based dose-response function for Giardia spp. (E.M.I.R.A. study). BMC Public Health 6: 122 p.

Risk management and wastewater reuse

Mixed tidewater and street water plume into Guánica Bay, Puerto Rico (@Eddie N. Laboy-Nieves)

Chapter 15

Health-related aspects of wastewater reuse in urban truck farming: A case study of Yaoundé, Cameroon

Guy R. Kouam-Kenmogme, Francis Rosillon, Alexandre Nono and Hernanie Grelle-Mpakam

SUMMARY

The reuse of wastewater in urban agriculture in the city of Yaoundé, Cameroon, is a constant source of the spread of waterborne diseases. The wastewater which comes from households, hospitals, and markets, contains high levels of faecal bacteria (more than 10^6 CFU of coliforms and 100,000 CFU of *Streptococci* per 100 ml water). Parasitological analyses revealed the presence of protozoan cysts (*Entamoeba histolytica*: 2–62/litre water, *Giardia*: 0–7/litre water) and helminth eggs (0–37/litre water). Analyses of stool samples from a cohort of 50 individuals revealed that all categories of people concerned by market gardening (the market gardeners, vendors, and consumers) were affected by intestinal amoebiasis uniformly (about 20% of the individuals in each category were infested). Moreover, some cases of cholera and typhoid fever, as well as dermatitis, were reported in those involved in market gardening directly and consumers alike. This chapter presents the results of a study based on the Ecohealth approach, which revealed some risky behaviours and practices that amplify the deteriorating local health conditions. Focused measures taken by state authorities (such as the building of a wastewater treatment plant), non-governmental organizations (through information and awareness-raising campaigns), and the various parties involved in market gardening (*e.g.*, following good practices and rules of hygiene) could usefully be put into effect to promote safe urban truck farming as a way to meet various environmental, social, and economic concerns to meet the population's needs most effectively.

15.1 INTRODUCTION

The use of raw or partially treated wastewater in urban agriculture is a constant and clearly rising trend in developing countries. According to Raschid-Sally and Jayakody (2008), wastewaters are used in urban agriculture in 80% of cities from developing countries. Nearly 200 million people are engaged in this highly controversial practice, which is usually banned but sometimes tolerated by national legislation around the world (FAO 2007). It has been estimated that around 20 million hectares spread over fifty different countries in the world, are irrigated with raw or partially treated wastewater (Scott et al. 2008; Jimenez and Asano 2008). Stepping up this activity is justified by the contributions that it makes to reduce unemployment rates, meeting food needs, recycling wastes, and using vacant land plots. It also helps to improve the natural environment and living standards of people involved in truck farming and the rest of the population (UNDP 1996; Cissé et al. 2002; Olanrewaju et al. 2004; Broutin et al. 2005; Nguegang 2008; Kouam-Kenmogne et al. 2010).

The use of wastewater in truck farming poses the recurrent underlying question of the health risks that are linked to this practice. Wastewater contains pernicious pathogens that are grounds for challenging the quality of products. More important are the sites selected for gardening in Yaoundé, Cameroon, the geographical enclave object of the present case study. The city is located in marshy bottomlands where solid and liquid wastes generated in the city are directly dumped; water treatment facilities in Yaoundé are lacking and rubbish collection remains ineffectual. The ultimate recipients of this waste are characterized by high concentrations of a poisonous cocktail of chemical and microbiological elements (LESEAU 2002; Kouam-Kenmogne et al. 2010).

Under such conditions, it is logical to ask about the health risks to take into account in urban truck farming, given the quality of the wastewater that is used and the degree of pollution of the farmed plots. What about the quality of cultivated products? What risky behaviours and practices occur there? What types of action can be conducted to reduce these health risks, which can be considered the "background noise" of this market gardening activity? Moreover, these practices can be expected to continue for a long time, given the current situation and future of wastewater treatment and use in market gardening. Nor should we forget that this activity, which has positive fall-out, especially socio-economics, warrants adopting a forward-looking approach with a view to re-using wastewater in a sanitary manner.

This chapter portraits health impacts of using wastewater in the urban markets in the City of Yaoundé. The authors aim to present the City's natural setting, describe the Ecohealth approach employed in the study, and discuss gives the outcomes before wrapping up with a series of recommendations.

15.2 GEOGRAPHIC SETTING OF THE CITY OF YAOUNDÉ

Yaoundé, which is the political capital of Cameroon, is located between 11°05′ and 12°15′ East, and 3°05′ and 4°25′ North (Figure 15.1). Yaoundé lies on a hillside site characterized by a three-tiered topography, i.e., the bottomlands, midslope, and summit. It has a Guinean equatorial climate with four well-marked seasons. The monthly mean temperature and annual precipitation is about 25°C and 1600 mm/yr, respectively (Djeuda Tchapnga et al. 2001). The Mefou and Mfoundi Rivers, along with their catchment areas, are the main watercourses draining the City. Native vegetation is represented by trees and shrubs mainly from the genus Celtis, Ulmaceae, and Sterculariaceae, but it has been greatly changed by human activities over the past two decades (UNDP 2000). The soil structure is characterised by red or yellow ferralitic substrate overlying gneisses that were laid down during the pan-African orogenesis, with hydromorphic soils in the valleys (Yongue-Fouateu 1986;

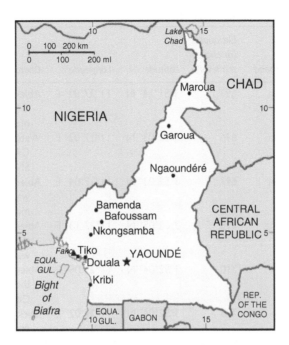

Figure 15.1 Location of Yaoundé, Cameroon.
Source: www.afrique-planete.com.

Segalen 1994). Yaoundé has a heterogeneous and growing population (more than 2 million inhabitants) involved in administrative work, commerce and trade, and agriculture.

15.3 THE ECOHEALTH APPROACH FOR STUDYING URBAN AGRICULTURE

Urban truck farming is confronted with multiple environmental, land tenure, socioeconomic, and sanitary concerns. This complex situation can only be examined validly by developing an approach of integrated and transverse research. The Ecohealth initiative relative to the health of the ecosystems and to the human health is adapted to the problem of urban truck farming. This approach makes reference to three pillars of the sustainable development (environment, economy and society), being health the resultant of these three constituents (COPEH 2010). The Ecohealth approach, although rather recent was already used in some projects of research-action in developing countries (Yonkeu 2005; Houenou 2006; Fayomi 2010).

It is also a question for the researchers, to associate the actors and the populations concerned by this activity in all the stages of the research. This guarantees a better appropriation of the results by the actors and their membership of the effective implementation of measures to improve the situation. Mono-disciplinary sector-based approaches give way to a holistic strategy which leans on diverse ways of acquisition of data through sociological surveys, environmental observations,

Table 15.1 Description of the sampling points.

Sampling point code	Neighbourhood	Elevation (m above sea level)	Latitude	Longitude	Characteristics
KAN₁	Nkolbisson	693	3°51'14" N	11°27'02" E	Abiergué located from 1 to 2 km from the market garden sites
KAN₂	Nkolbisson	696	3°52'03" N	11°27'02" E	Watercourse coming from CAMWATER's (former SNEC) water tank
KAN₃	Nkolbisson	697	3°52'03" N	11°27'04" E	Abiergué, place for sampling water used to water the market garden produce
KAO₄	Oyomabang	698	3°52'05" N	11°27'13" E	Abiergué before it crosses the eutrophic body of water at Oyomabang
KAM₅	Mokolo	710	3°52'04" N	11°29'06" E	Abiergué at Mokolo, before the north entrance of Cité-Verte
KEECV₆	Cité-Verte	723	3°52'04" N	11°29'02" E	Black water from the SIC division of Cité-Verte

campaigns for water monitoring and analysis, and epidemiological studies (Forget and Lebel 2001).

In May 2008, a survey conducted with 25 truck farmers, 564 households, and 33 market vendors enabled the collection of socio-economic and health data, which was encoded using Cs-Pro 2.6 (International Programs Center US Census Bureau 2005) and analyzed with SPSS (Statiscal Package for the Social Sciences http://www.spss.com/). The statistical study of the data has allowed to characterize the situation of the truck farmers, vendors and consumers, to identify the main forms of exploitation in marshy bottomlands and to diagnose the main met problems, to estimate income stemming from activities connected to the truck farming and finally to estimate prevalence and incidence of the hydric diseases.

Chemical, bacteriological, and parasitological analyses were run on six water samples taken during the dry and rainy seasons in different sampling points (Table 15.1). Ammonium (NH_4^+), nitrates (NO_3^-), phosphates (PO_4^{3-}), biological oxygen demand (BOD_5) and chemical oxygen demand (COD) were analyzed using the protocols described by Hach (1992). For the bacteriological indicators, faecal *Streptococci* (FS) and faecal Coliforms (FC) were analysed by membrane filtration (Rodier 1996). The parasitological analyses consisted of searching the samples for protozoan cysts and helminth eggs using a two-phase separation technique (Rodier 1996).

The parasitological analyses of the faecal samples were done using Bailenger's method (Bailenger 1979) on the samples of a cohort of fifty people of different social status, gender, and age groups. This phase was completed by consulting the health registries in order to determine the prevalence of waterborne diseases.

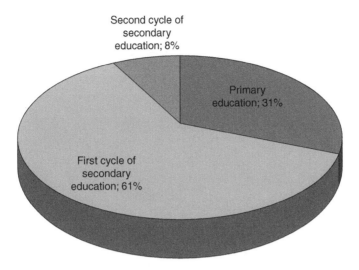

Figure 15.2 Educational levels of the truck farmers.

15.4 SOCIO-ECONOMICAL ASPECTS OF ACTOR COMMUNITIES

15.4.1 Truck farmers directly in contact with wastewater and soils

The breakdown of the people using wastewater in truck farming in Yaoundé is 73% men and 27% women, with ages ranging from 24 to 68 years. Overall, these people have rather low levels of schooling (Figure 15.2). Engaging in this activity is justified by the lack of employment (54%), the need for more income (38%), and family tradition (8%). Nearly 92% of truck farmers use large amounts of chicken droppings, pig slurry, and horse manure to fertilise their fields, while 8% use household refuses. They add chemical fertilizers (NPK, urea) to this process. They irrigate crops manually (96%) or by means of a power-driven pump (4%). Many unpleasant problems are observed in the course of handling the wastewater (Figure 15.3).

Truck farmers regularly use insecticides (cypermethrin 50 g/l, endosulfan 250 g/l, dimethoate 200 g/l) and fungicides (86% copper oxide, tridemorph 750 g/l, 80% maneb). Lettuce (*Lactuca sativa*), amaranth calula or "folon" (*Amaranthus viridis*), black nightshade or "zom" *(Solanum nigrum)*, Cayenne pepper *(Capsicum frutescens)*, Jew's mallow or "tegue" (*Corchorus olitorius*), okra *(Abelmoschus esculentus)*, and basil *(Basela alba)* are the main market garden crops.

The monthly income generated by this activity ranges from 7,500 to 90,000 CFA francs (12 to 137 Euros). The market gardeners mentioned many restrictions for their activity: real estate pressure, health risks, lack of supervision by the authorities, internal conflicts, flooding, and deseases. More than half of the market gardeners (61.5%) had intestinal amoebiasis and went to a health centre for treatment (68.3%) or medicated themselves (30.9%) in the three years prior to the survey. For this they spent between 5,000 and 25,000 FCFA (8 to 38 Euros) (Table 15.2) and lost from seven to thirty days of work in all. Some market gardeners also reported having serious diarrhoea (42%), typhoid fever (23%), cholera (4%), and dermatitis (31%) in this three-year period.

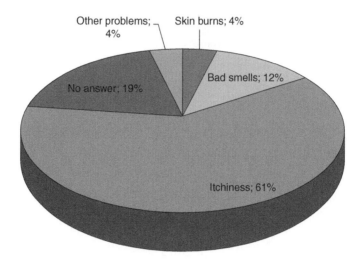

Figure 15.3 Problems reported by the truck farmers when using wastewate.

Table 15.2 Costs generated by sickness among workers of urban agriculture.

Cost of care in CFA francs (in Euros)	Truck farmers	Vendors	Households
5,000–15,000 (8 to 23 Euros)	68.3%	42.9%	63%
15,000–25,000 (23 to 38 Euros)	30.9%	35.6%	18.6%
25,000–50,000 (38 to 76 Euros)		35.6%	14.4%
more than 50,000 (more than 76 Euros)		14.3%	2.5%

15.4.2 Vendors: Vectors of transmission of the sanitary risks

Vendors' ages ranged from 20 to 45 years. They had completed primary school (38.5%), the first cycle of secondary school (42.3%), or the second cycle of secondary school (15.4%) or had no schooling (3.8%). Vendors wash their products with various types of water before selling it: tap water (30.8%), well water (11.5%), or river water (42.3%). The reasons for purchasing their vegetables in the marshy bottomlands where urban market gardening is done include the purchase prices, nearness of the place of purchase to the town, and the products' good quality. A large percentage of vendors (53.8%) had intestinal amoebiasis. The treatment methods are varied: attended a hospital (71.5%), self-medication (14.3%), and the use of medicinal plants (14.3%). The expenses that are incurred are shown in Table 15.2. Vendors also suffered from typhoid fever (30.8%), severe diarrhoea (26.9%), cholera (3.8%), and dermatitis (15.4%). As a result of these illnesses they lost 1 to 24 days of work.

15.4.3 Vegetables consumers and wastewater producers

The heads of households' ages ranged from 16 to 90 years; their levels of education varied considerably (Figure 15.4). The households' sizes ranged from 1 to 51 people

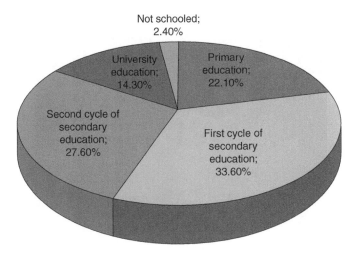

Figure 15.4 Educational level of household' heads.

Table 15.3 Ways for disposing domestic excrements.

Disposal method:	Pit latrines	Direct-discharge latrines	WC with septic tank	Improved latrines	Unspecified system	Total
Percentage	53.1%	6%	14.1%	25.5%	0.9%	99.6%

with a mean that revolved around six people per household. The ways in which they disposed of their excrement are described in Table 15.3.

Thirty-one percent of households had no idea what happened to the wastewater that they generated. They were also totally unaware of the fact that this wastewater was subsequently used in the city's bottomlands for truck farming. Before consuming market garden products, households washed the merchandize with the following tapwater (53.7%), spring water (5.4%), well water (25.1%), water with bleach (10.1%) or water with vinegar (4.3%). With respect to health, 59.5% of households had intestinal amoebiasis in the three years preceding the survey. In 76% of the cases they had been treated in hospital or by traditional practitioners (4%). Self-medication and application of medicinal plants was respectively reported in 14.7% and 4.7% of the cases. The expenses incurred to treat health conditions are shown in Table 15.2. Households lost up to 45 days of work due to these illnesses, which were aggravated by cases of typhoid fever (30%), cholera (12.2%), and severe diarrhoea (22.7%) in the three-year period that preceded the survey.

15.5 IRRIGATION WATER

The irrigation comes from wastewaters, used directly or diluted in River Abiergué. The water used in urban market gardening is degraded by pollutants coming almost totally from human activities. Wastewaters come from households, hospitals, markets, direct-discharge latrines set up along watercourses, pit latrines equipped with overflow

Figure 15.5 Latrine equipped with an overflow pipe.

Figure 15.6 Refuse dumped in ditches.

pipes that discharge their contents into the watercourses (Figure 15.5), solid waste dumped in the beds of the watercourses, makeshift drainage ditches (Figure 15.6), and industrial effluents. These pollution sources carry large amounts of contaminants into the watercourse identified be used to irrigate the market garden crops.

15.6 CHEMICAL, BACTERIOLOGICAL AND PARASITOLOGICAL PROFILE OF WATER

Chemical, bacteriological, and parasitological analyses were performed on 12 irrigation water samples in order to determine their quality. The results of these analyses are shown in Table 15.4 and Table 15.5.

Table 15.4 Chemical and bacteriological profile of irrigation water.

		Chemistry					Bacteriology	
Sample	Season	PO_4^{3-} (mg/l)	NO_{3-} (mg/l)	NH_4^+ (mg/l)	BOD_5 (mg/l)	COD (mg/l)	FC (CFU/ 100 ml)	FS (CFU/ 100 ml)
KAN_1	DS	1	13.2	0.5	25	54	21600	5550
	RS	0.1	25	0.5	300	100	1300	6000
KAN_2	DS	1.2	13.2	3.2	35	95	11400	2500
	RS	2	28	0.3	140	80	80000	1200
KAN_3	DS	1	13.2	7.1	10	15	5025	1960
	RS	1	124.5	0.6	300	170	222000	40000
KAO_4	DS	1	13.2	10.5	12	21	18200	8800
	RS	0.1	45	1.43	140	80	125000	11000
KAM_5	DS	12.3	52.8	4.6	56	105	190000	19600
	RS	0.1	9.6	1.45	200	140	780000	52000
$KEECV_6$	DS	31.7	70.4	37.2	360	601	3000000	138000
	RS	2.2	48.6	23.6	3770	1300	912000	22000

DS = Dry season; RS = Rainy season FC = Faecal Coliforms; FS = Faecal *Streptococci*

Table 15.5 Parasitological content of irrigation water.

		HELMINTH EGGS	PROTOZOAN CYSTS			
Sample	Season	(Eggs/L)	E. hystolitica (Cysts/L)	Giardia sp (Cysts/L)	Other (Cysts/L)	Total (Cysts/L)
KAN_1	DS	0	0	0	0	0
	RS	–	0	0	0	0
KAN_2	DS	0	0	0	2	2
	RS	–	0	0	0	0
KAN_3	DS	1	0	0	2	2
	RS	–	0	0	0	0
KAO_4	DS	7	2.2	0	0	2.2
	RS	–	0	0	0	0
KAM_5	DS	5	4.4	0	4	8.4
	RS	–	0	0	0	0
$KEECV_6$	DS	37	20.0	7	20	47
	RS	–	62	0	12	74

The high BOD_5 and COD values measured in some of the samples (KAM_5 and $KEECV_6$) confirm that these surface waters are polluted. The high ammonium concentrations indicate this pollution is recent. These wastewaters are relatively rich in phosphate ions and nitrates, which are useful for plant growth. The black water discharge pipe ($KEECV_6$) has a high load of pathogenic microorganisms and thus is a major pollution source.

On the whole, the various bacteriological and parasitological parameters exceed the guideline values set by the WHO (1989), which recommends fewer than 1000

CFU/100 ml for faecal coliforms and, less than one helminth egg per litre for water intended to irrigate food crops that are eaten raw.

15.7 POPULATIONS AFFECTED BY FAECAL PERIL

Parasitological analyses showed that *Entamoeba histolytica* eggs are present in the faeces of consumers, vendors, and market gardeners in fairly similar proportions, namely, in the faeces of three (15%) of the twenty consumers, four (20%) of the twenty market gardeners, and two (20%) of the ten vendors surveyed. After consulting health registries in health care units from the study area, it was found that 32% of people who consulted these units suffered from febrile gastroenteritis, a condition related to amoebiasis. Children five years old or younger are the most affected population group, while amongst adults, gastroenteritis affected more women than men.

15.8 SANITARY RISKS AMPLIFIED BY BAD PRACTICES

Urban truck farming in Yaoundé is a constant activity that concerns a large number of poor households, whether as gardeners or as vendors. This activity, which should improve the well-being of the groups that it concerns, is negatively affected by the location of cropland in marshy bottomlands, as well as the poor quality of the irrigation water. Indeed, the water used in this activity is constantly contaminated by many pollutants stemming from human activities. Many disseminated sources of pollution located up and down the watercourse degrade the quality of its water, which unfortunately is used untreated to cultivate the gardening market products. These repeated impacts imply serious health hazards for market gardeners, who regularly are in contact with wastewaters, and vendors and consumers who handle and eat the market gardeners' products.

Table 15.4 and Table 15.5 show the bacteriological and parasitological variations in the water River Abiergué. It is astonishing to inform that the black water from Cité-Verte's SIC housing division were found to contain even higher levels than the one reported in that Table: up to 3×10^6 faecal coliforms and 1.38×10^5 faecal *Streptococci* per 100 ml water, and 47 to 74 protozoan eggs and up to 37 helminth eggs per litre of water.

The pollution of water used in the market gardens explains the vulnerability to disease of the groups that use it on a daily basis. Nevertheless, while it is clear that all of the waterborne diseases that constantly plague the local population cannot be attributed to this practice, this source of contamination is a very important risk factor. The present findings for Yaoundé are very consistent many of authors who have documented the health risks linked to the use of wastewater in various cities in the developing world: Armar-Klemesu et al. (1998) for Accra (Ghana), Cifuentes et al. (2000) for Mexico City, Jacobi et al. (2000) for Dar es Salaam (Tanzania), Petterson et al. (2001) and Cissé (2002) for Ouagadougou (Burkina Faso), Niang (2002) for Dakar (Senegal), Keraita et al. (2002, 2008) for Kumasi (Ghana), Ensink et al. (2003) for Islamabad (Pakistan), Amoah et al. (2007) for Ghana, Mara et al. (2007) and Trang et al. (2007) for Hanoi (Vietnam).

The health care expenditures and days of work that are lost as direct consequences of these health impacts thwart the development of the households whom they affect. The main catalysts of the spread of these waterborne diseases are the risky behaviours and practices seen in the field, *i.e.*, the lack of appropriate equipment for the market gardeners, their constant handling of the wastewater (Figure 15.7), the cleaning of the gardeners' produce with wastewater before selling them (Figure 15.8), the practice of coating various parts of the body with used motor oil in order to prevent "the bacteria entering the body", the inclusion in the market gardening and sales workforce of very young children (under ten years of age) who do not observe the basic rules of hygiene, the fact that vendors in the various local markets place the market garden crops on the bare earth, and so on. The last two points are significant because they indicate that the main causes of bacterial contamination of fresh vegetables can be attributed to the distribution, handling,

Figure 15.7 Market gardener drawing river water.

Figure 15.8 Vendor washing lettuce with rive water.

and sales systems rather than their production (Armar-Klemesu et al. 1998; Bos et al. 2010).

15.9 HEALTH RISK MANAGEMENT

Urban truck farming in Yaoundé is an opportunity for many households that get the bulk of their income from this activity, despite the associated health risks. This practice, which was long banned by State authorities, has become a must in an environment marked by a persistent economic crisis, increasing degradation of water resources, and the farming of vacant lots. The phase of "punishment" or at least "prohibition" now belongs to the past and reality in the field calls for adopting a forward-looking approach to reduce the negative health impacts of this practice. Market gardening has been legitimised by the many advantages that it offers to workers and the general population. The complexity of the issue calls for synergistic actions involving various parties (the State, NGOs, market gardeners, vendors, and consumers) in a concerted process to cope with the health risks that are linked to the use of wastewater in urban market gardening. It is important for the various actors to agree on the approach to be implemented (Evans et al. 2010).

The State must legalise and supervise the practices that are linked to the use of wastewater in urban market gardening so as to control it most effectively. By opting for this approach, the State, which is currently unable to provide the population with effective water treatment services, will be able to carry out *in situ* work with the market gardeners to ensure that the wastewater is used "in a healthy manner". In this way, it will be possible to control the risky practices and behaviours that are contributing to the resurgence of waterborne diseases and to conduct a whole series of wide scale actions, for instance setting up fountains or digging settling ponds.

These initiatives should not make people to ignore the State's role in treating wastewater and managing solid waste appropriately. Koné et al. (2002) recommend the implementation of extensive treatment system at waste stabilization ponds. Jimenez et al. (2010) and Keraita et al. (2010) proposed water treatment methods (waste stabilization ponds, wastewater storage and treatment tanks, septic tanks) adapted to the developing countries with a double objective concerning reduction of the pathogenic bacteria and conservation of the nutrients. This strategy will allow to comply with the policies and operational programmes describes in the Hyderabad Declaration (IWMI 2002) with regard to wastewater use in agriculture as applied to urban truck farming, and guidelines for wastewater use in agriculture (WHO 2006). These actions can also make reference to the UNEP or FAO programmes concerning the reuse of waste water in urban truck farming (Jimenez and Asano 2008).

People involved in urban market gardening have none or incomplete schooling, and lack knowledge about how to use wastewater. NGOs should invest in "informing-educating-communicating" (IEC) the market gardeners and vendors about all of the facets of this practice, especially the hazards and risk associated to it. This long term education action constitutes a big and fundamental step in the strategy of reduction of the sanitary impacts specifically within the framework of the reuse of wastewater in urban truck farming (Jimenez et al. 2010 and UNHSP 2008). They can accompany the market gardeners on site in the various phases of production in order to weed out practices and behaviours that are conducive to the spread of waterborne disease. These campaigns could be extended to the population groups that are responsible for polluting the water sources upstream from the gardening. Such awareness-raising is all the more justified in that 31.5% of the households whom were unaware of what happened to their wastewater. In this context, iterative consultation frameworks must be set up in the various neighbourhoods to combat water pollution. Moreover, it is important that the people follow elementary rules of hygiene and wash the market garden produce thoroughly with drinking water before eating them.

The market gardeners must work with appropriate safety wear like boots and gloves, and avoid direct contact with wastewaters as much as possible. They could dig stabilisation ponds with special areas for drawing water (Figure 15.9). These settling ponds can reduce the concentrations of pathogens in water through the combined effects of the sun's rays and the settling process. Gardeners should irrigate their crops with well water during the pre-harvest fortnight. This proposal joins Keraita et al. (2008) which recommend stopping the irrigation with wastewater during the two last weeks preceding the harvest. Well water has lower microbial loads than surface waters, thus limiting the use of well water to this period is justified by its lack of organic matter and the hardship of drawing and transporting this liquid. Using well water could reduce microbial contamination of crops; hence gardeners and vendors would benefit by setting up fountains in the bottomlands where they wash the products before taking it to market. This simple wash with good quality water leads to a logarithmic reduction of pathogens (Xanthoulis 2008). These various actions join the global and integrated dynamics recommended by WHO (2006) within the framework of the reduction of the waterborne diseases related to the reuse of wastewater in urban truck farming. Moreover, all these actions should be based and strongly anchored on the local knowledge. About this, Keraita et al. (2010) specify that it is judicious to take the local knowledge into account in order to develop strategies aiming at reducing the health impacts related to the reuse of wastewater in urban truck farming.

Figure 15.9 Settling pond with a drawing area.

15.10 CONCLUSION

As in most of African big cities, urban truck farming becomes an increasingly popular common activity. The city of Yaoundé is not an exception to the rule. This activity is connected to a complex situation mainly related to hazards associated to truck farming, marshy bottomlands, faecal contamination of irrigation water, poor hygienic practices shown by truck farmers, vendors and consumers, and the sanitary risks which affect the health of the populations. The quality of the wastewater that is used in this activity is a corollary of defective water treatment and ineffectual refuse collection. The haphazard handling of this resource loaded with high microbial infection vectors, is responsible for waterborne diseases and intestinal amoebiasis in truck farmers, market vendors, and consumers. Reducing the health risks linked to the use of wastewater in urban agriculture in Yaoundé calls for a series of concerted actions involving actors from State institutions, NGOs and development associations.

The study, evaluation and assessment of this complex problem requires holistic trans-disciplinary approaches, consistent with the Ecohealth initiative to associate environmental, economic and social constituents, which ponders not only human health, but also the health of natural ecosystems. Linking representatives of producers, users and the general population with scientific experts could strengthen the empowerment and relevance of research results by the actors. A new partnership agenda shall pursue the effective application of the measures and recommendations herein proposed. The improvement of the living conditions of the populations, while benefiting from contributions of urban truck farming, will be possible as far as all the concerned parts (researchers and actors) are associated with fair representation and joint efforts.

ACKNOWLEDGEMENT

We thank the International Development Research Centre (IDRC-CRDI) in Ottawa, Canada, for the financial support that granted this community action research project. Photographic credit: © Guy Kouam-Kenmogne.

REFERENCES

Amoah P, Drechsel P, Abaidoo RC, Henseler M. 2007. Irrigated urban vegetable production in Ghana: Microbiological contamination in farms and markets and associated consumer risk groups. Journal of Water and Health 5(3): 455–466.

Armar-Klemesu M, Akpedonu P, Egbi G, Maxwell D. 1998. Food Contamination in Urban Agriculture: Vegetable production using waste water. In: Armar-Klemesu M and Maxwell D, editors. Urban Agriculture in the Greater Accra Metropolitan Area. Final report to IDRC. (project 0033149). University of Ghana. Noguchi Memorial Institute. Accra(Ghana).

Bailenger J. 1979. Mechanisms of parasitological concentration in coprology and their practical consequences. Journal of American Medical Technology 41:65–71.

Bos R, Carr R, Keraita B. 2010. Assessing and Mitigating Wastewater-Related Health Risks in Low-Income Countries: an Introduction. In: Dreschel P, Scott CA, Raschid-Sally L, Redwood M, Bahri A, editors. Earthscan/IWMI/IRDC. Wastewater irrigation and health. 432 p.

Broutin C, Commeat PG, Sokona K. 2005 [cited 2010 March 31]. Le maraîchage face aux contraintes et opportunités de l'expansion urbaine (Truck farming faced with pressures and opportunities of urban expansion). Le cas de Thiès/Fandène (Sénégal). Document de travail. [Internet].Gret. Enda Graf. Ecocité. Sénégal. Available from http://www.gret.org/ressource/pdf/07798.pdf

Cifuentes E, Blumenthal U, Ruiz-Palacios G, Bennett S, Quigley M. 2000. Health risk in agricultural villages practicing wastewater irrigation in Central Mexico: perspectives for protection. In: Chorus I, Ringelband U, Schlag G and Schmoll O, editors. Water Sanitation and Health. London: International Water Association Publishing. p 249–256.

Cissé G, Kientga M, Ouedraogo B, Tanner M. 2002. Développement du maraîchage autour des eaux de barrage à Ouagadougou : quels sont les risques sanitaires à prendre en compte ? (Development of truck farming around waters of dam in Ouagadougou: what are health risks to be taken into account?). Agricultures 11(1): 31–38.

[COPEH] Communities of Practice in Ecohealth. 2010 [cited 2010 March 31]. First African Meeting of Researchers and Actors in Ecosystem Approaches to Human Health. [Internet]. Cotonou. Benin. Available from http://www.copes-aoc.org/

Djeuda Tchapnga HB, Tanawa E, Ngnikam E. 2001. L'eau au Cameroun : Tome 1. Approvisionnement en eau potable (Water in Cameroon: Volume 1. Supply in drinking water). Presses Universitaires de Yaoundé editors. 359 p.

Ensink JHJ, Van Der Hoek W, Matsuno Y, Munir S, Aslam MR. 2003. The use of untreated wastewater in peri-urban agriculture in Pakistan: risks and opportunities. IWMI Research Report no. 64. International Water Management Institute. Colombo, Sri Lanka. 22p.

Evans AEV, Raschid-Sally L, Cofie OO. 2010. Multi-stakeholder processes for managing wastewater use in agriculture. In: Dreschsel P, Scott CA, Raschid-Sally L, Redwood M, Bahri A, editors. Earthscan/IWMI/IRDC. Wastewater irrigation and health. 432 p.

[FAO] Food and Agriculture Organization. 2007 [cited 2010 March 31]. L'agriculture biologique peut contribuer à la lutte contre la faim (Organic farming can contribute to the conflict against hunger). [Internet].Relations Medias. FAO. Available from www.fao.org/newsroom/fr/news/2007/1000726/index.html

Fayomi B. 2010. Mise en place de l'approche Ecosystème et Santé au Bénin (Ecosystem and health approach in the Benin). Communities of Practice in Ecohealth. First African Meeting of Researchers and Actors in Ecosystem Approaches to Human Health. Cotonou. Benin. CD of Communications.

Forget G, Lebel J. 2001. Une approche d'écosystème à la santé humaine (An approach to ecosystem and human health). Journal International de la Santé Professionnelle et Environnementale Supplément 7(2): 537–538.

Hach. 1992. Water analysis handbook. Colorado: Hach Company. p 220–746.

Houenou PV. 2006. Amélioration de la santé humaine et celle des écosystèmes dans la région de Buyo (Sud-Ouest de la Cote d'ivoire (Improvement of ecosystems and human health in the region of Buyo, southwest of the Ivory Coast). Côte d'ivore : CRDI. Rapport technique du projet de recherche CRDI N°100484–001. 29 p.

[IPC] International Programs Center Census Bureau. 2005 [cited 2010 April 19]. CSPro Data Entry. User's Guide Version 2.6. Washington(USA): 41p.Available from http://web.nso.go.th/poc/CSPro/start26.pdf

[IWMI] International Water Management Institute 2002 [cited 2010 March 31]. Reuse of wastewater for Agriculture. The Hyderabad declaration on wastewater use in agriculture. [Internet]. Hyderabad. India. Available from http://www.iwmi.cgiar.org/health/wastew/hyderabad_declaration.htm/

Jacobi P, Amend J, Kiango S. 2000. Urban agriculture in Dar es Salaam: providing an indispensable part of the diet. In: Bakker N, Dubbeling M, Gündel S, Sabel-Koschella U, Zeeuw H, editors. Growing cities, growing food: urban agriculture on the policy agenda, a reader on urban agriculture. Allemagne. Deutsche Stiftung fur Internationale Entwicklung. DSE. 257–283.

Jimenez B, Asano T. 2008. Water reclamation and reuse around the world. In: Jimenez B, Asano T. editors. Water Reuse: An International Survey of Current Practice. Issues and Needs. London: IWA Publishing. 648 p.

Jimenez B, Mara D, Carr R, Brissaud F. 2010. Wastewater treatment for pathogen removal and nutrient conservation: suitable systems for use in developing countries. In: Dreschsel P, Scott CA, Raschid-Sally L, Redwood M, Bahri A, editors. Earthscan/IWMI/IRDC. Wastewater irrigation and health. 432 p.

Keraita B, Drechsel P, Rashid L. 2002. Wastewater use uninformal irrigation in urban en peri urban areas of Kumasi. Ghana. Urban Agriculture Magazine 8:11–13.

Keraita B, Jimenez B, Drechsel P. 2008. Extent and implications of agricultural reuse of untreated, partly treated and diluted wastewater in developing countries. Agriculture. Veterinary Science. Nutrition and Natural Resources 3(58): 15 p.

Koné D, Cissé G, Seignez C, Holliger C. 2002. Le lagunage à laitue d'eau (Pistia stratiotes) à Ouagadougou : une alternative pour l'épuration des eaux usées destinées à l'irrigation. Cahiers d'études et de recherches francophones (Lettuce *Pistia stratiotes* watering in Ouagadougou: an alternative for the purgation of wastewater intended for irrigation. Agricultures 11(1): 39–43.

Keraita B, Konradsen F, Drechsel P. 2010. Farm-based measures for reducing microbiological health risks for consumers from informal wastewater-irrigated agriculture. In: Dreschsel P, Scott CA, Raschid-Sally L, Redwood M, Bahri A, editors. Earthscan/IWMI/IRDC. Wastewater irrigation and health. 432 p.

Kouam-Kenmogne GR, Rosillon F, Mpakam HG, Nono A. 2010. Enjeux sanitaires, socio-économiques et environnementaux liés à la réutilisation des eaux usées dans le maraîchage urbain à Yaoundé au Cameroun : cas du bassin versant de l'Abiergué (Health, socioeconomic and environmental stakes linked to the reuse of wastewater in urban truck farming in Yaoundé in Cameroon: case of the basin overturning of Abiergué). Revue électronique de Sciences de l'Environnement. 10(1). (Accepted).

[LESEAU] Laboratoire d'Environnement et des Sciences de l'Eau. 2002. Inventaire des déchets solides et liquides non ménagers de la ville de Yaoundé (Inventory of the not domestic solid and liquid waste of the city of Yaoundé). Yaoundé (Cameroun): Ecole Nationale Supérieure Polytechnique. Rapport Final. 153 p.

Mara DD, Sleigh PA, Blumenthal UJ, Carr RM. 2007. Health risks in wastewater irrigation : Comparing estimates from quantitative microbial risk analyses and epidemiological studies. Journal of Water and Health 5(1): 39–50.

Nguegang AP. 2008. L'agriculture urbaine et péri-urbaine à Yaoundé : analyse multifonctionnelle d'une activité montante en économie de survie (Urban and suburban agriculture in Yaoundé: multifunctional analysis of an activity going up in a survival economy). [dissertation]. [Bruxelles (Belgium)]: Université Libre de Bruxelles. 200 p.

Niang S. 2002. Utilisation des eaux usées dans l'agriculture urbaine au Sénégal : Cas de la ville de Dakar (Use of wastewater in urban agriculture in the Senegal : Case of the city of Dakar). Atelier international sur la réutilisation des eaux usées en agriculture urbaine: un défi pour les municipalités en Afrique de l'ouest et du centre. Ouagadougou. Burkina Faso.3-8 juin. Rapport final ETC/RUAF-CREPA-CTA. p 165–180.

Olanrewaju BS, Moustier P, Mougeot L, Abdou F. 2004. Développement durable de l'agriculture urbaine en Afrique francophone (Sustainable development of urban agriculture in French-speaking Africa). Enjeux, Concepts et méthodes. Ouagadougou (BF): CIRAD-CRDI.176 p.

Petterson SR, Ashbolt N, Sharma A. 2001. Microbial risks from wastewater irrigation of salad crops: A screening-level risk assessment. Water Environment Research 73(6): 667–672.

Raschid-Sally L, Jayakody P. 2008. Drivers and characteristics of wastewater agriculture in developing countries: results from a global assessment. Colombo (Sri Lanka): International Water Management Institute. Research Report 127

Rodier J. 1996. Analysis of water: natural waters, sedimentary waters, seawater. 8th edition. Paris: Dunod Editors. 1383 p.

Scott CA, Faruqui NI, Raschid-Sally L, editors. 2008. Wastewater use in irrigated agriculture: confronting the livelihood and environmental realities. Wallingford. (UK): CABI Publishing. 208 p.

Segalen P. 1994. Soil ferralitiques and their geographical sharing out. General introduction to soil ferralitiques: their identification and their immediate environment. Collection of Studies and Theses. Paris: ORSTOM Editors. 201 p.

Trang DT, Hien BTT, Molbak K, Cam PD, Dalsgaard A. 2007. Epidemiology and aetiology of diarrhoeal diseases in adults engaged in wastewater-fed agriculture and aquaculture in Hanoi, Vietnam. Tropical Medicine and International Health 12(2): 23–33.

[UNDP] United Nations Development Programme. 1996. Urban agriculture: food, jobs and sustainable cities. UNDP. 302 p.

[UNDP] United Nations Development Programme. 2000. Etudes socio-écologiques régionales au Cameroun (Socioecological studies in the Cameroun regions). Evaluation de la pauvreté: amélioration des conditions sociales. Province du Centre. MINPAT/Projet PNUD-OPS CMR/98/005/01/99.

[UNHSP] United Nations Human Settlements Programme UN-HABITAT and Greater Monetor Sewerage Commission. 2008. Atlas of Excreta, Wastewater Sludge, and Biosolids Management: Moving Forward the Sustainable and Welcome Uses of a Global Resource. LeBlanc R J, Matthews P and Richard R P editors. Nairobi (Kenya): UN-HABITAT. 632 p.

[WHO] World Health Organization. 1989. L'utilisation des eaux usées en agriculture et en aquaculture : recommandations à visées sanitaires (The use of wastewater in agriculture and in aquaculture: recommendations with health aims). Rapport d'un groupe scientifique de l'OMS. Série de Rapports Techniques 778. 74p.

[WHO] World Health Organization. 2006. Guidelines for the safe use of wastewater, excreta and greywater. Volume 2: Wastewater use in agriculture. 222 p.

Xanthoulis D. 2008. Low-Cost Wastewater Treatment. Development of Teaching and Training Modules for Higher Education. ASIALINK. EuropeAid. Contract VN/Asia-Link/012(113128) 2005-2008. 443 p.

Yongue-Fouateu R. 1986. Contribution à l'étude pétrologique de l'altération et des faciès du cuirassement ferrugineux des grains migmatitiques de la région de Yaoundé (Contribution to petrological studies of ferruginous alteration and curation phases of migmatitiques grains in the region of Yaoundé). [dissertation]. [Yaoundé (Cameroun)] : Faculté des Sciences Université de Yaoundé. 214 p.

Yonkeu S. 2005. Elaborations des stratégies de réduction des risques de maladies diarrhéiques pour les populations humaines dus aux petits barrages en Afrique de l'Ouest : cas du barrage de Yitenga (Elaborations of reductions strategies of diarrhoeic diseases risks for human populations in Western Africa: case of the dam of Yitenga). Ouagadougou (Burkina Faso): CRDI. Rapport Final de Recherche Groupe EIER-ETSHER/CRDI. 286 p.

Abandoned ship in the Aguirre Mangrove Forest in Guayama, Puerto Rico
(@Eddie N. Laboy-Nieves)

Chapter 16

Application of low-cost sorbents to remove chromium from industrial wastewater discharges

Iris N. Cosme-Colón, Evens Emmanuel and Eddie N. Laboy-Nieves

SUMMARY

Removal of toxic heavy metal contaminants from industrial wastewater discharges (IWD) is one of the most important environmental issues facing society today. Such pollutants do not degrade biologically, and they also accumulate in the food chain. Chromium contamination of water, especially Cr(VI), its most toxic, mutagenic and carcinogenic form, is a persistent and serious global threat to human health and natural water resources. There is an urgent need for effective low-cost techniques for chromium removal from IWD, in particular since operating costs are increasing, and as environmental laws become more stringent. Sorption technology appears to be the most promising, eco-friendly and economically suitable alternative to control this problem. This chapter highlights environmental issues of chromium in IWD, assesses the use of low-cost sorbents for removal of this pollutant, and analyses risk management with regards to pollutant removal and the protection of human and environmental health.

16.1 INTRODUCTION

Removal of toxic heavy metal contaminants from industrial wastewater discharges (IWD) is one of the most important ecological and human health research issues. Such discharges often contain significant amounts of heavy metals that would jeopardize public health and the environment, particularly marine and freshwater ecosystems (Laboy-Nieves 2009), if effluents are released without adequate treatment (Rengaraj et al. 2001). These pollutants do not degrade biologically like organics compounds, and they could bioaccumulate through food web (ATSDR 2000; Emmanuel 2007; Emmanuel 2009).

Chromium is one of the heavy metals considered a major pollutant. It has been widely used in industrial processes for leather tanning, dyes and paint preparation, textile manufacturing, paper mills, wood preservation, stainless steel production, and photography (El Nemr et al. 2008). Whereas, chromium concentrations in non-polluted fresh water vary from 0.1 to 0.5 ppm (De Filippis and Pallaghy

1994; Cervantes et al. 2001), the effluents of industries employing this heavy metal may contain concentrations ranging from tenths to hundreds of milligrams per liter (Rengaraj et al. 2001).

Chromium exists in several oxidation states. The most stable and common forms are trivalent chromium, Cr(III), and hexavalent chromium, Cr(VI), which exhibit contrasting biochemical properties and toxicokinetics (McGrath and Smith 1990; Cervantes et al. 2001). Cr(III) compounds occur naturally in the form of oxides, hydroxides or sulfates, and they are nutritionally necessary to humans for glucose, fat and protein metabolism (ATSDR 2000). In contrast, Cr(VI) compounds are mainly anthropogenic and highly toxic; its mutagenic and carcinogenic nature and high oxidation state enhances its ability to move into living cells (ATSDR 2000). Cr(III) and Cr(VI) interchangeability depends on their concentration in solution, pH, the redox potential (E_h) of the medium, and the presence or absence of a strong oxidant or reductant (Tadesse et al. 2006; El Nemr et al. 2008).

Chromium removal from IWD can be achieved by adsorption, reduction, precipitation, ion-exchange, reverse osmosis, and electrodialysis (Kurniawan et al. 2006). Currently, chromium sorption using natural materials or waste products has become the most economically feasible alternative (Bailey et al. 1999; Mohan et al. 2006). However, there is an urgent need for low-cost and effective removal techniques to comply with emergent and stringent environmental laws (Suksabye et al. 2007). This chapter aims to highlight environmental issues related to chromium in industrial wastewater discharges and to assess the use of low-cost sorbents for chromium removal. Risk management in relation to protecting human and environmental health is also discussed.

16.2 PHYSICAL AND CHEMICAL PROFILE OF CHROMIUM

Chromium is a relatively common element, naturally occurring in rocks, soil, plants, animals, fresh water and seawater, volcanic dust, and gases (ATSDR 2000; Zayed and Terry 2003). It is the 21st most abundant element in the Earth's rocks, especially in igneous rocks (Turekian and Wedepohl 1961). Its abundance in Earth's crust ranges from 100 to 300 µg/g (Cervantes et al. 2001). Chromium can form relatively stable, soluble and insoluble compounds when it combines with different nonmetals (e.g., oxygen and chlorine), nitrates and sulfates (ATSDR 2000; Motzer and Engineers 2005).

Chromium oxidation states range from −2 to +6, being Cr(III) and Cr(VI) the most common and stable forms (ATSDR 2000; Cervantes et al. 2001; Motzer and Engineers 2005). Cr(III), is the most thermodynamically stable form and occurs naturally associated to iron ores (ATSDR 2000; WHO 2009), as depicted in Figure 16.1. Cr(III) and Cr(VI) exhibit contrasting environmental mobility, chemical and biochemical behavior, bioavailability, and toxicological profiles (McGrath and Smith 1990; Cervantes et al. 2001; Stanin and Pirnie 2005). Cr(IV) and Cr(V) do not occur in nature, and both represent important intermediates that influence the reduction process of Cr(VI) (Zayed and Terry 2003). Cr(VI), the second most stable form, is primarily produced from anthropogenic sources, but it can be found naturally in the mineral crocoite ($PbCrO_4$) (Hurlbut 1971; ATSDR 2000).

Chromium speciation, and Cr(III) and Cr(VI) interchangeability in the environment, particularly in soils and groundwater, mainly depend on the E_h, and the

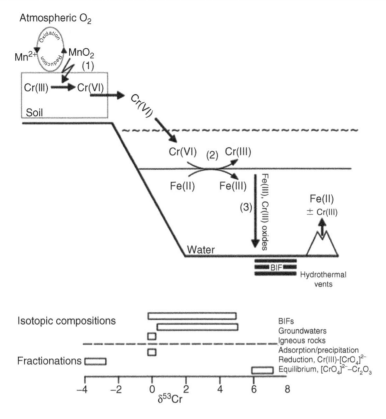

Figure 16.1 Schematic diagram of the surface chemistry of chromium (Frei et al. 2009).

predominant acidic or alkaline conditions of the medium (Mohan and Pittman 2006; Stanin and Pirnie 2005; Tadesse et al. 2006; El Nemr et al. 2008). Cr(III) is a hard acid that predominates as Cr^{3+} under reducing conditions and pH less than 3.5, which can form relatively strong complexes with oxygen and donor ligands (Cook et al. 2000; Motzer and Engineers 2005; Mohan and Pittman 2006). At higher pH values, Cr(III) hydrolysis produces several chromium hydroxy species: $CrOH^{2+}$, $Cr(OH)_2^+$, $Cr(OH)_3^0$, and $Cr(OH)_4^-$ (Rai et al. 1987; Palmer and Puls 1994). $Cr(OH)_3^0$ is the only solid species of Cr(III) that under slightly acidic to alkaline conditions can be precipitated as two different types of amorphous crystals [$Cr(OH)_3 \cdot 3H_2O$ or Cr_2O_3], depending on the suitable conditions (Palmer and Puls 1994; Cook et al. 2000). Cr(VI) is a strong oxidant, and exists primarily as the following species: salts of chromic acid ($H_2CrO_4^0$) at pH less than 1.0; bichromate or hydrogen chromate ion ($HCrO_4$) at pH between 1.0 and 6.0; and chromate ion (CrO_4^{2-}) at pH above 6.0 (Stanin and Pirnie 2005; Mohan and Pittman 2006).

16.3 ENVIRONMENTAL TOXICOLOGY OF CHROMIUM

The adverse effects of chromium as an environmental and human health noxious substance are the primary focus of its environmental toxicology (LeBlanc 2004).

Cr(III) is an essential micronutrient in the diet of human and other mammals (WHO 2009). The biologically active form of chromium is believed to be a complex of nicotinic acid, amino acids and Cr(III), which potentiates insulin termed as glucose tolerance factor (ATSDR 2000). However, Cr(III) can be toxic if it is ingested in large amounts (Kirpnick-Sobol et al. 2006; WHO 2009).

Chromium is not expected to be biomagnified in the aquatic food-chain. It has been found that aquatic organisms tend to bioaccumulate it at different rates and levels (Eisler 1986; WHO 2009; Dhir 2010; Rai 2010). Sensitive species of freshwater organisms have showed reduced growth, inhibited reproduction, and increased bioaccumulation of chromium at 10.0 μg/L of Cr(VI), and other adverse effects at 30.0 μg/L of Cr(III) (Eisler 1986). In the case of marine organisms it has been recorded accumulations in oysters and worms at 5.0 μg/L of Cr(VI), reduction in algal growth at 10.0 μg/L, and reproduction inhibition of polychaete annelid worms at 12.5 μg/L (Eisler 1986). In all circumstances, Cr(III) was less harmful than Cr(VI). The lethal concentration of Cr(VI) required to kill half of sensitive freshwater and marine species after 96 hours of exposure ranges from 445 and 2,000 ppb; for Cr(III) these values ranged from 2,000 to 3,200 ppb for sensitive freshwater species, and 3,300 to 7,500 ppb for marine organisms (Eisler 1986). For some freshwater green algae species, Cr(III) is five to ten times more toxic than Cr(VI) in circumneutral solutions, thus the actual toxicity of Cr(III) to aquatic organisms may has been underestimated (Vignati et al. 2010).

16.3.1 Behavior of chromium in aquatic systems

The percentage of soluble and insoluble chromium in the water column is very small, since the major part released into water will be deposited in sediments (ATSDR 2000). Cr(III) predominates under anaerobic and sub-anaerobic conditions, while Cr(VI) (highly soluble in water and thus very mobile in the aquatic environment) will prevails in aerobic conditions (Eisler 1986; Stanin and Pirnie 2005). The solubility of Cr(III) in an aqueous solution depends on the pH of the water: in neutral to basic pH conditions the metal tends to precipitate, while in acidic conditions it becomes soluble (Eisler 1986; Kimbrough et al. 1999). Oxygen concentration, pH, and the presence of reducing and complexing agents are factors that determine which chromium forms will predominate in the aquatic environment (Kimbrough et al. 1999).

Cr(VI) is reduced to Cr(III) by organic substances, hydrogen sulfide, sulfur, iron sulfide, ammonium and nitrite (Kimbrough et al. 1999). The rate at which the Cr(VI) reduction reaction occurs will depend on the pH, E_h, and the concentration of organic matter in water (Kimbrough et al. 1999; Lin 2002); the reaction is favored under acidic conditions (Kimbrough et al. 1999). Sunlight is another key factor that affects the redox reactions of this element, since it appears to degrade organically bound chromium, resulting in the release of inorganic chromium (Bartlett 1991; Kimbrough et al. 1999; Lin 2002). Zhang and Bartlett (1999) suggested that light-induced oxidation of aqueous Cr(III) in the presence of Fe(III) would be one potential pathway for the oxidation of Cr(III) to toxic Cr(VI) (more likely in acidic surface waters with low dissolved organic carbon content). Stanin and Pirnie (2005) reported that the transport and fate of chromium in surface waters is a function of temperature, depth, degree of mixing, oxidation conditions, and season (especially in the case of lakes with high biological activity and thermal stratification).

Chromium mobility in groundwater will depend on its solubility, its tendency to be sorbed by soil or aquifer materials, and on groundwater chemistry (Stanin and Pirnie 2005). Groundwater contamination by chromium can be more extensive in permeable aquifers (e.g., sand and gravel, and fractured rock) with a flow speed of 0.1 to 5 m/day, in contrast to aquifers of low permeability (e.g., clayey materials) which speeds few centimeters or less per year (Stanin and Pirnie 2005).

16.3.2 Human oral exposure through drinking water

The recommended standard concentration for chromium in drinking water is 0.05 mg/L (WHO 2008). It has been adopted by many environmental agencies or sanitary authorities through the world. Although modern methods of water treatment remove much of the naturally present chromium forms, chlorinated drinking water usually contains traces of toxic Cr(VI), possibly due to the presence of some available Cr(III) that becomes oxidized to Cr(VI) during chlorination process (WHO 1988; ATSDR 2000). As a core result of industrial pollution, it has been found that some populations worldwide are ingesting higher concentrations of Cr(VI) [e.g., 580 µg/L from a groundwater well in Hinkley, California, USA] in drinking water (Costa 2003). Insufficient epidemiologic studies have risen divided opinions as to if Cr(VI) ingestion (especially through drinking water) can cause an increased risk of cancer (Costa 2003; Kirpnick-Sobol et al. 2006; Stout et al. 2009).

There is consensus that Cr(III) is 500 to 1000 times less toxic to a living cell than Cr(VI), due to its poor ability to permeate the cell membrane. In contrast, Cr(VI) has the highest bioavailability, being able to be transported actively across both prokaryotes and eukaryotes cell membranes via their natural anion uptake systems (Kerger et al. 1996; De Flora 2000; Cervantes et al. 2001; Costa 2003). At physiological pH (6.8 to 7.4), Cr(III) does not resemble any biological nutrient, whereas Cr(VI) exists as an oxyanion (CrO_4^{2-}), which predominates over the dichromate anion, and resembles other oxyanions used extensively in many cellular biochemical processes, like sulphates and phosphates (Kerger et al. 1996; Costa 2003; Davidson et al. 2004).

Scientists disagree when considering the expected biotransformation that Cr(VI) undergoes as it encounters biological fluids, and when it enters the cells. Extracellularly, Cr(VI) is reduced to the less permeable and bioavailable Cr(III), and detoxified in most body fluids (including sweat, saliva, gastric juices, epithelial-lining fluid and blood) and in the intestinal lumen by the bacterial glutathione (De Flora 2000; James 2002; Stout et al. 2009). Cr(VI) reduction to Cr(III) is thought to occur primarily in the stomach, as a detoxification mechanism (De Flora 2000; Stout et al. 2009). The reduction capacity for human gastric juices is about 84 to 88 mg of Cr(VI) per day (De Flora et al. 1997; Stout et al. 2009). Cr(VI) is rapidly reduced to Cr(III) by ascorbate, glutathione, cysteine, hydrogen peroxide and riboflavin. This reduction leads either to oxidative DNA damage resulting from the formation of unstable and very reactive pentavalent and tetravalent chromium, and free radicals, or to DNA damage by Cr(III)-DNA direct interactions that may induce chromosomal alterations and mutational changes, and apoptosis (Cervantes et al. 2001;

Rengaraj et al. 2001; Kirpnick-Sobol et al. 2006; Stout et al. 2009). Both Cr(VI) as an environmental contaminant, and the nutritional supplement Cr(III), increase DNA deletions *in vitro* and *in vivo*, when ingested through drinking water, but surprisingly Cr (III) is a more potent inducer of DNA deletions than Cr(VI) once the first is absorbed (Kirpnick-Sobol et al. 2006).

In response to doubts about current water quality standards and their ability to protect the public from the possible health risks associated with chromium, the USA National Toxicology Program launched in 2007, a two-year animal study to evaluate Cr(VI) administered as sodium dichromate dihydrate in drinking water (Stout et al. 2009). It was reported that Cr(VI) is carcinogenic in rats and mice after chronic oral exposure. Furthermore, the toxicity of erythron, histiolytic infiltration, and uptake of Cr(VI) into tissues of rats and mice, suggested that: hexavalent chromium is not completely reduced in the stomach; unreduced Cr(VI) is transported into tissues; and the exposure of tissues to unreduced Cr(VI) by absorption and distribution is responsible for the observed effects (Stout et al. 2009). It can be inferred that if the reduction capacity of natural defenses is exceeded through oral exposure, then Cr(VI) toxicity, carcinogenicity, and genotoxicity may be expressed in human beings (De Flora 2000; James 2002; Costa 2003; Stout et al. 2009). Other important issues to be considered when assessing the possibility of human health risks of chromium ingested via drinking water are the individual genetic differences in uptake or in the efficiency of Cr(VI) to Cr(III) reduction in the stomach and at other organs (Costa 2003), and the possibility of Cr(VI) as a potent cocarcinogen for UV-induced skin tumors (Davidson et al. 2004; Salnikow and Zhitkovich 2008).

16.4 CHROMIUM REMOVAL FROM INDUSTRIAL WATER DISCHARGES

16.4.1 Conventional methods

Chromium removal from industrial water discharges (IWD) can by achieved by several methods, or a suitable combination between them, each one having advantages and disadvantages, as summarized in Table 16.1. The most used and simple method for chromium removal from IWD is chemical precipitation of Cr(III) at high pH (~9.0 to 10.0) using lime preceded by reduction of Cr(VI) to Cr(III) at low pH (~2.0) (Mohan and Pittman 2006), or using sodium metabisulfite, ferrous sulfate, zero-valent iron or dimethyldithio carbamate (Chang 2003). Ion exchange is considered a better choice for chromium removal from IWD than chemical precipitation (Rengaraj et al. 2001; Ahluwalia and Goyal 2007). The most common ion exchanger (i.e., a solid able of replace chromium cations in solution, with non-toxic ones) used is synthetic organic resins (Ahluwalia and Goyal 2007). In the reverse osmosis method, a portion of the wastewater is forced to pass through an appropriate semi-permeable membrane (i.e., permeable to water and impermeable to chromium salts) applying a higher pressure than the osmotic pressure, concentrating in the portion retained most of the dissolved chromium salts present in the original effluent (Pérez-Padilla and Tavani 1999). The electrodialysis process uses alternated selective semi-permeable membranes (i.e., cation exchange membrane and anion exchange membrane), which are fixed between the respective electrode in electrolytic cells. This induces the migration of chromium cations present in the wastewater towards the cathode through the cation exchange membrane under the influence of an electric field (Ahluwalia and Goyal 2007).

Among all the treatment technologies developed to remove chromium from IWD, sorption (adsorption and biosorption) has evolved as the most effective and versatile method (Kurniawan and Babel 2003; Kurniawan et al. 2006; Mohan and Pittman 2006; Mohan et al. 2006; Suksabye at al. 2007). Chromium biosorption is a property of certain types of living or death biological material (e.g., algae, fungi, bacteria, plants and plants-derived products) to bind and concentrate chromium forms from aqueous solution by active and passive uptake (living cells) or by passive uptake only in dead cells (Mohan and Pittman 2006).

16.4.2 Chromium adsorption kinetics

The main factors governing the adsorption mechanisms of chromium forms and the process influenced or affected by each parameter are summarized in Table 16.2. Chromium equilibrium measurements are used to determine the maximum or ultimate adsorption capacity (or loading) of the material tested, and equilibrium isotherm data are formulated into an adsorption isotherm model (Mohan and Pittman 2006). The most commonly used isotherm models are Freundlich (describes the equilibrium on heterogeneous surfaces and does not assume monolayer capacity), Langmiur (describes homogeneous surfaces assuming that all the adsorption sites have equal adsorbate affinity and adsorption at one site does not affect adsorption at an adjacent site), and Brunauner, Emmett, and Teller (popularly known as BET), which describes multi-layer surfaces assuming the partitioning of a compound between liquid and solid phases (Mohan and Pittman 2006; El Nemr et al. 2008; Altenor et al. 2009).

The adsorption of chromium is controlled by the: transport in the bulk solution; diffusion across the liquid film boundary surrounding the sorbent particles; intraparticle diffusion in the liquid contained in the sorbent pores and in the sorbate along the pore walls; and adsorption and desorption within the particle and on the external surface (McKay 1995; Mohan and Pittman 2006). The kinetics reactions occurring in the majority of sorbents proposed for chromium removal from aqueous solution are governed by first-order, second-order and two constant rate equations, which describe the adsorption rate based on adsorption capacity (Mohan and Pittman 2006; El Nemr et al. 2008).

16.4.3 Selecting a low-cost sorbent

The two key issues to consider for selecting the most economically feasible sorbent for chromium removal from IWD are technical applicability and cost-effectiveness (Kurniawan et al. 2006; Acharya et al. 2009). First, it is essential to characterize the IWD, since the choice of treatment will depend on parameters such as pH, concentration of chromium forms, temperature, flow volume, biological oxygen demand, economics involved, and standards set by regulatory agencies in a given country (Acharya et al. 2009; Lee and Jones-Lee 2010). Second, it is necessary to compare

Table 16.1 Conventional methods for chromium from industrial wastewater discharges.

Method	Advantages	References	Disadvantages	References
Chemical Precipitation	*Most simple and cost-effective *Remove up to ppm levels from the discharge	Mohan and Pittman (2006) Ahluwalia and Goyal (2007)	*Efficiency is strongly affected by pH and the presence of other ions *Requires addition of other chemicals that produce a high water content sludge (i.e., voluminous and toxic solid waste of expensive treatment and disposition) and *Lacks specificity and efficiency in removal at low concentration when lime, bisulfate or ion exchange processes are used with it	Ahluwalia and Goyal (2007) Chang (2003); Erdem and Tumen (2004); Mohan and Pittman (2006) Ahluwalia and Goyal (2007) Ahluwalia and Goyal (2007)
Ion exchange	*Remove up to ppb levels from the discharge while handling relatively large volumes *Use fewer chemical compounds and produce less sludge volume *Recovery of metal value and good surface area *High potential of regeneration with no adsorbent loss	Rengaraj et al. (2001); Ahluwalia and Goyal (2007) Rengaraj et al. (2001); Ahluwalia and Goyal (2007) Mohan and Pittman (2006); Ahluwalia and Goyal (2007); Mohan and Pittman (2006)	*High operational costs (i.e., it cannot handle concentrated metal solution as the membrane becomes fouled by organics or other solids in the effluent) *Efficiency depends on the pH of the solution and the type of resin used	Mohan and Pittman (2006); Ahluwalia and Goyal (2007) Mohan and Pittman (2006); Ahluwalia and Goyal (2007)
Reverse osmosis	*Produces water with a very low content of the metal salts in the portion that passes the membrane	Pérez-Padilla and Tavani (1999)	*Efficiency may depend on the integrity of the membrane and its capacity to retain in the concentrated portion the chromium salts dissolved in the original effluent	Pérez-Padilla and Tavani (1999)

Method	Advantages	References	Disadvantages	References
Electrodialysis	*High selectivity	Rengaraj et al. (2001)	*High cost of electrodes and ion exchange membranes *Short membrane life since they are exposed to a high density electrical field	Xu and Huang (2008) Xu and Huang (2008)
Sorption (Adsorption/Biosorption)	*Very effective and versatile, especially if combined with regeneration processes *Very high surface areas and fast kinetics *Eco-friendly and readily available natural and synthetic materials can be used	Kurniawan and Babel (2003); Crini (2005); Kurniawan et al. (2006); Mohan and Pittman (2006); Mohan et al. (2006); Suksabye et al. (2007) Mohan and Pittman (2006); Altenor et al. (2009) Bailey et al. (1999); Mohan and Pittman (2006); Joseph et al. (2009)	*Some of the materials used are still expensive and non-selective (e.g. activated carbon) *Biosorption: – Early saturation of sorption sites – Potential for biological process improvement is limited (i.e., cells are not metabolizing) – No potential for biologically altering the metal valence state	Mohan and Pittman (2006); Altenor et al. (2009)

Table 16.2 Main parameters governing adsorption mechanisms of chromium.

Parameter	Process or factor influenced or affected	References
pH	*Interchangeability between Cr(III) and Cr(VI) *Electrostatic attraction between chromium forms and the surface of the sorbent	Mohan and Pittman (2006); Tadesse et al. (2006); El Nemr et al. (2008)
Initial chromium concentration	*Number of active sites of the sorbent *Percentage of sorption of chromium	El Nemr et al. (2008); Acharya et al. (2009)
Temperature	*Transport of chromium forms to the sorbent surface or pores	Mohan et al. (2006)
Contact time	*Adsorption capacity of sorbent corresponding to equilibrium time of sorption	Acharya et al. (2009); El Nemr et al. (2008)
Sorbent dose	*Effective surface area or sorption sites	Rengaraj et al. (2001); El Nemr et al. (2008)
Particle size and surface area	*Rate of diffusion (pore, film or particle) and sorption	Acharya et al. (2009)
Degree of mixing or velocity gradient	*Residence time for the complete removal of chromium	Rengaraj et al. (2001)
Type of sorbent	*Efficiency and sorption capacity of material depending on the previous parameters	Rengaraj et al. (2001); Mohan and Pittman (2006)

of options of low-cost sorbents in terms of the parameters governing chromium adsorption kinetics, as depicted in Table 16.2, to rate optimum treatment conditions, technical viability, and the best adsorption capacities (Kurniawan et al. 2006). Unfortunately, cost information is seldom reported, but generally, a sorbent can be assumed as "low-cost" if it requires little processing, is locally available, is abundant in nature, or is a by-product or waste material from another industry (Bailey et al. 1999). A cost-benefit analysis of chemicals, electricity, labor, transportation and maintenance is needed to consider any expenditure (Kurniawan et al. 2006). Obviously, the cost-efficiency of a low-cost sorbent for chromium removal from IWD increases if it can be regenerated and reused without further waste by-products to treat, and its price may decrease, as more industries consider using it for treating their effluents.

16.5 RISK MANAGEMENT

Risk management identifies, evaluates, selects, and implements actions to reduce threat to human health and to ecosystems (CRARM 1997). The term risk has a relative interpretation, because it may address an exposure that: increases the probability of a specific result; increases the probability of occurrence of an infection or other outcomes; and a determinant that can be modified by an action, leading to a reduction of the probability of occurrence of results (Last 2010). Nevertheless, many countries have adopted standards to guide, assess and apply risk management strategies (Sutter II 2007).

As pointed out by Last (2010), the risk of an event, such as the death from a specified cause, is calculated from the incidence or death rate of the specified condition. Hence, the concept of relative risk is more complex, and it is not made easier by the fact that the term has more than one meaning, although its different usages are similar. The three common meanings are:

1. The ratio of the risk of a disease or death among those exposed to a specified risk to those not exposed to this risk. This is commonly called as the "risk ratio".
2. The ratio of the cumulative incidence rate in those exposed to a specified risk to the cumulative incidence rate in those not exposed to a specified risk. This is popularly known as the "cumulative incidence ratio".
3. Relative risk is probably most often equated with the "odds ratio" that is calculated from the results of analyzing the data obtained in a case-control study. Although the odds ratio is not a rate, if the condition or situation being studied is relatively rare, it approximates to what the rate would be if the samples examined were large enough. Thus, in published papers, relative risk is often used as a synonym for the odds ratio.

In relation to human health risk assessment of chromium in aquatic medium, there is evidence that extreme exposure to chromium causes renal and hepatic damage (Robson 2003). Problems of reduced growth, inhibited reproduction, and increased bioaccumulation have been identified in organisms living in water contaminated by chromium (Eisler 1986). No matter which of the three definitions presented by Last (2010), the goal of risk management is to procure scientifically sound, cost-effective, and integrated actions that reduce or prevent hazards while taking into account social, cultural, ethical, political, and legal considerations (CRARM 1997). For aquatic ecosystems polluted with chromium from industrial water discharges (IWD), risk management pursues employing low-cost sorbents to remove the contaminant, a strategy centered on prevention or to reduce the probability of contaminated discharges that endanger and inflict ecosystems and human health (Table 3). To attain this goal, it is of utmost importance to consider holistic approaches including stakeholders participation and using iterations, if new information is developed that could change the needs, or the nature of one part or the whole process been assessed. Any meaningful effort should recognize that humans are an integral part of the landscape, particularly in urban settings, and that natural resource baselines have permanently shifted (Laboy-Nieves 2009). Consequently, the human dimensions component of sustainability has become an integral part of ecological restoration and rehabilitation strategies (Laboy-Nieves 2009).

16.6 CONCLUDING REMARKS

Chromium contamination of water, especially due to Cr(VI), derived from industrial water discharges (IWD), is a persisting and serious global threat to human health

Table 16.3 Risk management relative to the use of low-cost sorbents to remove chromium from
          IWD (CRARM 1997).

Process	Discussion	
Problem definition in context to Cr contamination of water resources by IWD	The industrial use of chromium have inflicted and endangered aquatic ecosystems and human life, consequently risking water quality and availability. Nature and human health are connected, thus risk ought to be reduced or eliminated. It could help to use effluents emissions inventories, and environmental and biological monitoring.	
Analysis of risks	Hazard identification	Requires accurate determination of chromium oxidation states in nature and consideration of all the direct and indirect routes of exposure (especially, oral exposure), for understanding carcinogenesis, mutagenesis, genotoxicity, dermatosis or dermatitis.
	Dose-response assessment	Implies reviewing toxicological and epidemiological literature to relate responses to levels of exposure or dose of chromium.
	Exposure assessment	Needs completing a chromium template of the routes and media (especially, oral exposure and water), and establishing possible scenarios.
	Risk characterization	Requires careful combination of chromium dose-response information with exposure assessment to determine the level of risk to target populations.
Addressing risks	It is necessary to develop methods for identifying and exploring the available low-cost sorbents to remove chromium from IWD, in terms of their effectiveness, feasibility, costs, benefits, and possible consequences or social, cultural, economic, ethical, political, environmental, and legal impacts.	
Decision making about which option to implement	It is important to base the decision of which low-cost sorbent to use to remove chromium from IWD on the best available scientific, economic, and technical information. It must be given special attention to prevent more than to control the risks or remediate harms, and include incentives for innovation, evaluation, assessment, and research.	
Actions taking to implement decisions	The action taken to implement the use of the low-cost sorbent chosen to remove chromium from the IWD must be effective, expeditious, flexible, and supported by the stakeholders.	
Evaluation and assessment of actions	It is critical to determine if the use of low-cost sorbents is successful, needs modifications, present information gaps, and has unintended consequences or more benefits than expected. Assessing this stage is important to monitor any public policy or regulatory activity to address the actions taken.	

and natural water resources. Although, many treatment technologies have been proposed for chromium removal, it is clear that not all of them are applicable. Comparisons of the cost-effectiveness of low-cost sorbents evaluated are also difficult, because of a lack of information and inconsistencies in the data. Among all the IWD treatment technologies presented for chromium remediation, sorption appears to be the most promising, eco-friendly and economically feasible alternative to control the problem. Although commercial activated carbon has proven to be the most efficient adsorbent for chromium removal from IWD, due to its high surface area and fast kinetics, it is still very expensive and non-selective. However, there are several natural and synthetic materials accessible in large quantities from agricultural or industrial operations with a high potential as low-cost chromium sorbents. After conversion to activated carbon, they have demonstrated excellent removal capacity. Also, there are other raw materials such as sawdust, used tires, and red mud that, despite their adsorption capacities are not so extraordinary, they have the advantage of abundance and availability as solid wastes.

This chapter examined the assessment of sorbents' adsorptive capacities, but it is crucial that future work should take into account: the adsorption kinetics of materials in the presence of chromium and other industrial water discharges pollutants; applications of low-cost sorbents at a pilot-plant scale or at a variety of scenarios like tropical developing countries; the evaluation of economically suitable methods for the regeneration or recycling of sorbents once they are used; the possibility of mass production of sorbents and commercialization to encourage their use; and the impacts of the treated effluents in ecosystems and human health. Finally, the problem should not be limited to an economical spectrum, but for searching for integrated and multidisciplinary approaches consistent with each country' resources and potential to apply the different methods for chromium removal.

REFERENCES

Acharya J, Sahu JN, Sahoo BK, Mohanty CR, Meikap BC. 2009. Removal of cromium(VI) from wastewater by activated carbon developed from *Tamarind* wood activated with zinc chloride. Chemical Engineering Journal 150: 25–39.

Ahluwalia SS, Goyal D. 2007. Microbial and plant derived biomass for removal of heavy metals from wastewater, Bioresource Technology 98(12): 2243–2257.

Altenor S, Carene B, Emmanuel E, Lambert J, Ehrhardt JJ, Gaspard S. 2009. Adsorption studies of methylene blue and phenol onto vetiver roots activated carbon prepared by chemical activation. Journal of Hazardous Materials 165(1–3): 1029–1039.

[ATSDR] Agency for Toxic Substance and Disease Registry. 2000. Toxicological profile for chromium. Atlanta, Georgia (USA): U.S. Department of Health and Human Services. 461p.

Bailey SE, Olin TJ, Bricka RM, Adrian DD. 1999. A review of potentially low-cost sorbents for heavy metals. Water Research 33: 2469–2479.

Bartlett RJ. 1991. Chromium cycling in soils and water: Links, gaps, and methods. Environmental Health Perspectives 92: 17–24.

Cervantes C, Campos-García J, Devars S, Gutiérrez-Corona F, Loza-Tavera H, Torres-Guzmán JC, et al. 2001. Interactions of chromium with microorganisms and plants. Federation of European Microbiological Societies Microbiology Reviews 25: 335–347.

Chang, L. 2003. Alternative chromium reduction and heavy metal precipitation methods for industrial wastewater. Environmental Progress 22(3): 174–182.

Cook K, Sims R, Harten A, Pacetti J. 2000. *In situ* treatment of soil and groundwater contaminated with chromium. Technical Resource Guide, Report EPA 625/R-00/005 [Internet]. 2000 October; Cincinatti, Ohio (USA): United States Environmental Protection Agency. [cited 2009 Oct 20]. 97 p. Available from: http://www.epa.gov/nrmrl/pubs/625r00005/625r00005.pdf

Costa M. 2003. Potential hazards of hexavalent chromate in our drinking water. Toxicology and Applied Pharmacology 188: 1–5.

[CRARM] The Presidential/Congressional Commission on Risk Assessment and Risk Management. 1997. Framework for environmental health risk management, Final report, Vol. 1 [Internet]. Washington DC (USA): Commission on Risk Assessment and Risk Management. [cited 2009 Nov 26]; 70 p. Available from: http://www.riskworld.com/Nreports/1997/risk-rpt/pdf/EPAJAN.PDF

Crini G. 2005. Recent developments in polysaccharide-based materials used as adsorbents in wastewater treatment. Progress in Polymer Science 30: 38–70.

Davidson T, Kluz T, Burns F, Rossman T, Qunwei Z, Uddin A, et al. 2004. Exposure to chromium(VI) in the drinking water increases susceptibility to UV-induced skin tumors in hairless mice. Toxicology and Applied Pharmacology 196: 431–437.

De Filippis LF, Pallaghy CK. 1994. Heavy metals: Sources and biological effects. In: Rai LC, Gaur JP, Soeder CJ, editors. Advances in limnology series: Algae and water pollution. Stuttgart (Germany): E. Scheizerbartsche Press. p 37–77.

De Flora S. 2000. Threshold mechanisms and site specificity in chromium(VI) carcinogenesis. Carcinogenesis 21(4): 533–541.

De Flora S, Camoirano A, Bagnasco M, Bennicelli C, Corbett GE, Kerger BD. 1997. Estimates of the chromium(VI) reducing capacity in human body compartments as a mechanisms for attenuating its potential toxicity and carcinogenicity. Carcinogenesis 18(3): 531–537.

Dhir B. 2010. Use of aquatic plants in removing heavy metals from wastewater. International Journal of Environmental Engineering 2(1–3): 185–201.

Eisler R. 1986. Chromium hazards to fish, wildlife, invertebrates: A synoptic review. United States Fish and Wildlife Service Biological Report 85(1.6). 60 p.

El Nemr A, Khaled A, Abdelwahab O, El-Sikaily A. 2008. Treatment of wastewater containing toxic chromium using new activated carbon developed from date palm seed. Journal of Hazardous Materials 152: 263–275.

Emmanuel E, Angerville R, Joseph O, Perrodin Y. 2007. Human health risk assessment of lead in drinking water: a case study from Port-au-Prince, Haiti. International Journal of Environmental Pollution 31: (3/4): 280–291.

Emmanuel E, Pierre MG, Perrodin Y. 2009. Groundwater contamination by microbiological and chemical substances released from hospital wastewater: Health risk assessment for drinking water consumers. Environment International 35: 718–726.

Erdem M, Tumen F. 2004. Chromium removal from aqueous solution by ferrite process. Journal of Hazardous Materials B109: 71–77.

Frei R, Gaucher C, Poulton SW, Canfield DE. 2009. Fluctuations in Precambrian atmospheric oxygenation recorded by chromium isotopes. Nature 461: 250–253.

Hurlbut CS, editor. 1971. Dana's manual of mineralogy. 18th ed. New York (USA): John Wiley & Sons, Inc. p 346–347.

James BR. 2002. Chemical transformations of chromium in soils: Relevance of mobility, bio-availability and remediation. The Chromium File No. 8 [Internet]. 2002 February; International Chromium Development Association. [cited 2010 Feb 3];. 9 p. Available from: http://icdachromium.com/pdf/publications/crfile8feb02.htm

Joseph O, Rouez M, Métivier-Pignon H, Bayard R, Emmanuel E, Gourdon R. 2009. Adsorption of heavy metals on to sugar cane bagasse: Improvement of adsorption capacities due to anaerobic degradation of the biosorbent. Environmental Technology 30(13): 1371–1379.

Kerger BD, Paustenbach DJ, Corbett GE, Finley BL. 1996. Absorption and elimination of tri-valent and hexavalent chromium in humans following ingestion of a bolus dose in drinking water. Toxicology and Applied Pharmacology 141: 145–158.

Kimbrough DE, CohenY, Winer AM, Creelman L, Mabuni C. 1999. A critical assessment of chromium in the environment. Critical Reviews Environmental Science and Technology 29(1): 1–46.

Kirpnick-Sobol Z, Reliene R, Schiestl RH. 2006. Carcinogenic Cr(VI) and the nutritional supplement Cr(III) induce DNA deletions in yeast and mice. Cancer Research 66(7): 3480–3484.

Kurniawan TA, Babel S. 2003. A research study on Cr(VI) removal from contaminated wastewater using low-cost adsorbents and commercial activated carbon. In: Proceedings of the 2nd International Conference on Energy Technology towards a Clean Environment (RCETE), 2003 Feb 12-14; Phuket, Thailand, 2: 1110–1117.

Kurniawan TA, Chan GYS, Lo W, Babel B. 2006. Comparisons of low-cost adsorbents for treating wastewaters laden with heavy metals: A review. Science of the Total Environment 366: 409–426.

Laboy-Nieves EN. 2009. Environmental Management Issues in Jobos Bay, Puerto Rico. In: Laboy-Nieves EN, Schaffner F, Abdelhadi AH, Goosen MFA, editors. Environmental Management, Sustainable Development and Human Health. London: Taylor and Francis. p. 361–398.

LeBlanc GA. 2004. Basics of environmental toxicology. In: Hodgson E. ed. A textbook of modern toxicology. Hoboken, New Jersey (USA): John Wiley & Sons, Inc. p 463–477.

Last JM. 2010. Relative Risk [Internet]. Encyclopedia of Public Health. [cited 2010 May 09]. Available from: http://www.encyclopedia.com/doc/1G2-3404000729.html.

Lee GF, Jones-Lee A. 2010. Issues in monitoring hazardous chemicals in stormwater runoff/discharges from superfund and other hazardous chemical sites. Remediation Journal 20(2): 115–127.

Lin CJ. 2002. The chemical transformations of chromium in natural waters: A model study. Water, Air, and Soil Pollution 139:137–158.

McGrath SP, Smith S. 1990. Chromium and nickel. In: Alloway BJ, editor. Heavy metals in soils. New York (USA): John Wiley & Sons, Inc. p 125–150.

McKay G. 1995. Use of adsorbents for the removal of pollutants from wastewaters. Boca Ratón, Florida (USA): CRC Press. 186 p.

Mohan D, Singh KP, Singh VK. 2006. Trivalent chromium removal from wastewater using low cost activated carbon derived from agricultural waste material and activated carbon fabric cloth. Journal of Hazardous Materials B135: 280–295.

Mohan D, Pittman Jr. CU. 2006. Activated carbons and low cost adsorbents for remediation of tri- and hexavalent chromium from water. Journal of Hazardous Materials B137: 762–811.

Motzer WE, Engineers T. 2005. Chemistry, geochemistry, and geology of chromium and chromium compounds. In: Jacobs JA, Guertin J, Avakian CP, editors. Chromium(VI) Handbook. Boca Ratón, Florida (USA): CRC Press. p 23–88.

Palmer CD, Puls RW. 1994. Natural Attenuation of Hexavalent Chromium in Groundwater and Soils. Environmental Protection Agency Ground Water Issue. EPA/540/5-94/505 October 1994 [Internet]. [cited 2009 Nov 9]; 12 p. Available from: http://www.epa.gov/tio/tsp/download/natatt.pdf

Pérez-Padilla A, Tavani EL. 1999. Treatment of an industrial effluent by reverse osmosis. Desalination 126: 219–226.

Rai D, Sass BM, Moore DA. 1987. Chromium(III) hydrolysis constants and volubility of chromium(III) hydroxide. Inorganic Chemistry 26(3): 345–349.

Rai PK. 2010. Microcosm investigation on phytoremediation of Cr using *Azolla pinnata*. International Journal of Phytoremediation 12(1): 96–104.

Rengaraj S, Yeon KH, Moon SH. 2001. Removal of chromium from water and wastewater by ion exchange resins. Journal of Hazardous Materials B87: 273–87.

Robson M. 2003. Methodologies for assessing exposures to metals: Human host factors. Ecotoxicology and Environmental Safety 56:104–109.

Salnikow K, Zhitkovich A. 2008. Genetic and epigenetic mechanisms in metal carcinogenesis: Nickel, Arsenic and Chromium. Chemical Research in Toxicology 21(1): 28–44.

Stanin FT, Pirnie M. 2005. The transport and fate of Cr(VI) in the environment. In: Jacobs JA, Guertin J, Avakian CP, editors. Chromium(VI) handbook. Boca Ratón, Florida (USA): CRC Press. p 162–211.

Stout MD, Herbert RA, Kissling GE, Collins BJ, Travlos GS, Witt KL, et al. 2009. Hexavalent chromium is carcinogenic to F3441N rats and B6C3F1 mice after chronic oral exposure. Environmental Health Perspectives 117(5): 716–721.

Suksabye P, Thiravetyan P, Nakbanpote W, Chayabutra S. 2007. Chromium removal from electroplating wastewater by coir pith. Journal of Hazardous Materials 141: 637–44.

Sutter II GW. 2007. Ecological Risk Assessment. Boca Raton, Florida (USA): CRC Press. 538 p.

Tadesse I, Isoaho SA, Green FB, Puhakka JA. 2006. Lime enhanced chromium removal in advanced integrated wastewater pond system. Bioresource Technology 97: 529–34.

Turekian KK, Wedepohl KH. 1961. Distribution of the elements in some major units of the Earth's crust. Bulletin of Geological Society of America 72(1): 175–192.

Vignati DAL, Dominik J, Beye ML, Pettine M, Ferrari BJD. (2010). Chromium(VI) is more toxic than chromium(III) to freshwater algae: A paradigm to revise?. Ecotoxicology and Environmental Safety (in press).

[WHO] World Health Organization. 1988. Chromium. Environmental Health Criteria 61 [Internet]. Geneva (Switzerland): International Programme on Chemical Safety. [cited 2009 Jan 3]; 197 p. Available from: http://www.inchem.org/documents/ehc/ech61.htm

[WHO] World Health Organization. 2008. Guidelines for drinking-water quality [Internet]. 3rd ed. Recommendations, Vol. 1. Geneva (Switzerland): WHO Press. [cited 2010 Feb 25]; p 145–196. Available from: http://www.who.int/water_sanitation_health/dwq/fulltext.pdf

[WHO] World Health Organization. 2009. Inorganic chromium(III) compounds. Concise International Chemical Assessment Document 76. [Internet]. Geneva (Switzerland): WHO Press. [cited 2009 Dec 1]; 100 p. Available from: http://www.who.int/ipcs/publications/cicad/cicad76.pdf

Xu T, Huang C. 2008. Electrodialysis-based separation technologies: A critical review. American Institute of Chemical Engineers 54(12): 3147–3159.

Zayed AM, Terry N. 2003. Chromium in the environment: Factors affecting biological remediation. Plant and Soil 249: 139–156.

Zhang H, Bartlett RJ. 1999. Light and iron(III)-induced oxidation of chromium(III) in the presence of iron(III). Environmental Science and Technology 33: 588–594.

Chapter 17

Viability of the upflow anaerobic sludge process for risk management of wastewater treatment

Nancy Ma. Cáceres-Acosta, Miriam Salgado-Herrera, Eddie N. Laboy-Nieves and Evens Emmanuel

SUMMARY

In many countries, urban effluents are subjected to physiochemical and biological treatments. Anaerobic digestion continues to be the most widely used wastewater management process because it represents a sustainable system and an suitable method for developing countries. This chapter presents an evaluation of the up flow anaerobic sludge reactor (UASB) operation, its advantages, and use in water and environment conservation. The technique combines physical, and biological processes with anaerobic degradation such as hydrolysis, fermentation, acetogenesis and methanogenesis. Formation of granules is key step for the success of the UASB. Various groups of bacteria dominate the metabolic pathways in the granules and the wastewater decomposition. The advantages of the reactor are: conservation of the environment, low energy consumption, and biogas and low sludge production. Temperature, pH, flow rate, influent type and concentration, sludge retention time, nutrient availability and presence of xenobiotics may be limitations. The first section describes the role of anaerobic wastewater treatment, as a biological process, while the second explains its application and performance in domestic and industrial wastewater treatment. The third section presents advantages and disadvantages of the UASB reactor and assesses the effects of physical parameters on the operational process. The final part is devoted to studying the management of aquatic ecological risks.

17.1 INTRODUCTION

Interest in wastewater treatment has attracted the attention of the scientific community, governmental agencies and the general public because water pollution is associated with poor public health, degradation of water quality and diminution of aquatic resources (Mahmoud 2010). This problem has been exacerbated by the scarcity of potable water and a rapid increase in population worldwide. For the World Bank the utmost challenge in the water and sanitation sectors over the next decades will be low cost sewage treatment that parallels selective reuse of treated effluents for agricultural and industrial purposes (Leitão et al. 2006).

Human society produces wastes that can be employed as useful raw materials for the production of energy, and the recovery of byproducts and component water. However, sewage treatment systems demand a high cost of operation (Leitão et al. 2006). The level of wastewater treatment is a function of the load of pollution tolerated in natural waters and the pollution produced by municipal wastewater and industrial activities (Emmanuel et al. 2009). In the case of wastewater treatment, combinations of different methods can be used, such as physical, chemical and biological. While biological processes have been widely reported by the literature, the physical parameters and the physical-chemical mechanisms of solids removal have not (Mahmound et al. 2003). The treatment of wastewater using biological methods, such as anaerobic digestion has been broadly recognized as the core waste management process because it represents a sustainable and appropriate system for developing countries (Seghezzo et al. 1998; Hammes et al. 2000).

There are two basic biological processes for wastewater: the conventional which treats sludge from primary and secondary municipal plants, and the anaerobic treatment which is the microbiological processing of biodegradable pollutants. The later produces methane, carbon dioxide and biomass, and reduces the pollutant concentration through microbial coagulation and removal of non-settled organic colloidal solids. Organic matter is biologically stabilized so that no further oxygen demand is exerted. This process consists of a heated digestion tank containing waste and bacteria responsible for the anaerobic treatment. At the end of the process the mixed treated waste and microorganisms are usually removed together for final disposal. The biological treatment requires contact of the biomass with the substrate so that bacteria are not lost in the effluent, thus a digester is used (Bal and Dhagat 2001).

The upflow anaerobic sludge blanket popularly known as the UASB reactor is the most widely and successfully used high rate anaerobic system for sewage treatment (Lettinga 2001). Withing the spectrum of anaerobic treatment technologies, it offers great promise, especially in developing countries that are usually located in hot and moderate climatic zones (Tawfik 2010). Interest in anaerobic systems wastewater treatment was shown with the development of the UASB reactor in the early 70s (Lettinga et al. 1980; Seghezzo et al. 1998). The functioning of UASB systems depends on physical parameters and biological processes, which determine the final removal efficiency and conversion of organic compounds (Mahmoud et al. 2003; Sabry 2008). Several parameters affect particles removal in the sludge bed of a UASB system, for instance: 1) the reactor operational conditions (temperature, organic loading rate, hydraulic retention time, upflow velocity); 2) influent concentration, particle size distribution (PSD) and charges; 3) sludge bed PSD, exopolymeric substances, charges, and sludge hold up (Mahmoud et al. 2003), and reductions in biological oxygen demand (BOD) of 75 to 90 percent (Li et al. 2010).

17.2 ANAEROBIC WASTEWATER TREATMENT EVALUATED AS A BIOLOGICAL PROCESS

Domestic wastewater generally shows high flow variations due to the number of inhabitants and dwellings connected to the sewer system, specific characteristics of the sewerage (type, material, length, maintenance, infiltration, use of pump stations) as well as climate, topography, and commercial/industrial contributions (Leitão et al. 2006). Although domestic and industrial wastewaters are pollutants, after proper treatment they can be utilized as tap water, energy production, and as a fertilizer's source (Elmitwalli and Otterpohl 2007). Wastewater have been treated with anaerobic methods in alcohol distillation, pharmaceutical and chemical manufacturing, and landfill slaughterhouse as well as from meatpacking, fish and seafood processing and domestic and industrial sources (Oktem et al. 2007).

Many toxic and organic compounds can be degraded under anaerobic treatment. Anaerobic treatment is established as the best technology for the primary treatment of high Chemical Oxygen Demand (COD) mainly for soluble industrial effluents. MEF (2009) suggested the use of anaerobic treatment when the effluent has high or low total dissolved solids (TDS), high BOD and low difference between COD & BOD, and when the effluent is highly organic and fully biodegradable.

17.3 THE ANAEROBIC DIGESTION PROCESS

Anaerobic processes have been used for the treatment of concentrated domestic and industrial wastewater for well over a century (McCarty and Smith 1986). It is a natural process mediated by bacteria which degrade organic matter to its most reduced form, methane, in the absence of oxygen. The digestion process begins when insoluble organic polymers such as carbohydrates, cellulose, proteins and fats are broken down and liquefied by enzymes produced by hydrolytic bacteria. Carbohydrates, proteins and lipids are hydrolyzed to sugars which then decompose to form carbon dioxide, hydrogen, ammonia and organic acids. Proteins decompose to form ammonia, carboxylic acids and carbon dioxide. The organic acids formed in the hydrolysis and fermentation stage are converted to acetic acid by acetogenic micro-organisms. Methane and carbon dioxide are produced from the organic acids and their derivatives produced in the acidogenic phase (Alcina 2003).

The anaerobic degradation of organic material has been described as a multistep process (Gujer and Zehnder 1983) where stage products are used as substrates for the next stage. Gujer and Zehnder (1983) identified these stages as: extracellular hydrolysis of complex biomolecules; fermentation of sugars and amino acids; acetogenesis of alcohols and volatile fatty acids; anaerobic oxidation of fatty acids and; aceticlastic methanogenesis (Figure 17.1).

The hydrolysis process acts separately on three main groups of complex biomolecules: proteins, carbohydrates and lipids, which are hydrolyzed by extracellular enzymes to soluble products, small enough to allow their transport across the cell membrane. The separate hydrolysis processes produces amino acids, sugars and fatty acids, which are fermented (acidogenesis) or anaerobically oxidized to short chain polymers, CO_2, hydrogen and ammonia (Gujer and Zehnder 1983).

Proteins are a very important substrate in the anaerobic digestion process, because the amino acids derived from their hydrolysis have high nutritional value. Proteolytic enzymes, called proteases, hydrolyze proteins into peptides and amino acids. Some of these amino acids are used in the synthesis of new cellular material and the rest are degraded to volatile fatty acids, carbon dioxide, hydrogen, ammonia and sulfur in later stages of the process. Enzymes like cellulose, converts cellulose

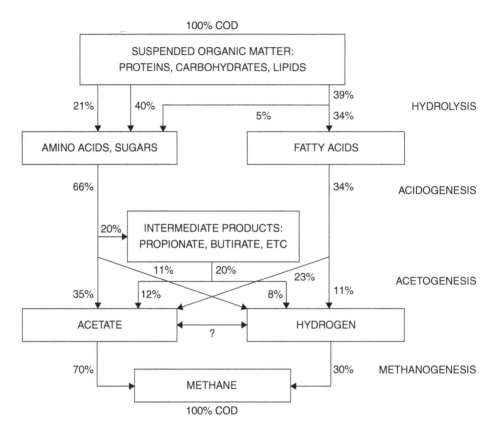

Figure 17.1 Anaerobic digestion process scheme (Van Haandel and Van Der Lubbe 2007).

into cellobiase and glucose, whereas other enzymes hydrolyze the hemicelluloses to produce pentose, hexose and uranic acids (Lettinga et al. 2001).

Acid fermentation (acidogenesis) is a constant BOD stage, because the organic molecules are only rearranged or stabilized. This step is essential for the second stage (methane fermentation) as it converts the organic material to a form usable by methane producing bacteria (Bal and Dhagat 2001). Acidogenesis is an energy yielding step, where soluble substrates such as amino acids and sugars, after being transported through the membrane into the cell, are degraded to organic acids and alcohols (Alcina 2003). Numerous microbial species are known to ferment dissolved organic compounds into organic acids, like acetic, propionic, butyric and lactic acid, and other fermentation products, such as alcohols, under anaerobic conditions. Acidogenic fermentation is an important step in anaerobic digestion of organic compounds to release methane and carbon dioxide (Bengtsson et al. 2008).

The acetogenesis process is obligately syntrophic with methanogens and is responsible of converting alcohols and volatile fatty acids into acetate and hydrogen to produce methane (Shin et al. 2001). Syntrophic acetogenic bacteria play an important role in consuming many of the short-chain acids that accumulate in the pathway, and the end products, predominately acetate and CO_2, become viable

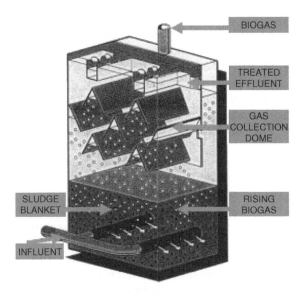

Figure 17.2 Schematic diagram of an UASB reactor (JAEE 2002).

substrates for methanogenesis (Parawira 2004). Several groups of fermentative and hydrogen-producing and consuming acetogenic bacteria, together with aceticlastic and CO_2 reducing methanogens, catalyze the reactions taking place during anaerobic digestion (Schmidt and Ahring 1996; Bal and Dhagat 2001).

17.4 APPLICATION AND PERFORMANCE OF THE UASB REACTOR IN DOMESTIC AND INDUSTRIAL WASTEWATER TREATMENT

Lettinga and Hulshoff Pol (1991) provided design considerations for the UASB reactor: wastewater characteristics in terms of composition and solids content; volumetric organic load; upflow velocity; reactor volume; and physical features including the influent distribution system and the gas collection system. Schmidt and Ahring (1996) divided this reactor into four compartments: the granular sludge bed, the fluidized zone, the gas-solids separator and the settling compartment (Figure 17.2). The granular sludge blanket is at the bottom where wastewater is pumped into and passes upward through the granular sludge bed where the organic compounds are biologically degraded and biogas is produced. Just above the granular sludge bed, a fluidized zone develops due to production of biogas separated from the liquid. Flocculated and dispersed bacteria are washed out of the reactor with the effluent, granules with good abilities return to the fluidized zone of the granular sludge bed.

The UASB reactor is an economical solution for the treatment of industrial effluent which has substantially dissolved pollutants. It combines modern high-rate treatment technology with simplicity of design. In the UASB reactor, special gas-solid-liquid separators are mounted which enable collection of biogas and recycle of anaerobic biomass. The avoidance of internal packing in the reactor greatly reduces the cost of reactor construction (Leitão et al. 2006).

The strong variations in flow and concentration may adversely affect the efficiency of an anaerobic treatment in the reactor. The effect of fluctuations in hydraulic and organic load generally depends on the applied hydraulic retention time, sludge properties, intensity and duration of the variations, and the reactor design of the three phase separator (Leitão et al. 2006). Wide variations in influent flow and organic loads can upset the balance between acid and methanogenesis in anaerobic processes. For soluble, easily degradable substrates, such as sugars and fermentation starches, the acidogenic reactions can be much faster at high loading and may increase the reactor volatile fatty acids and hydrogen concentration, and depress the pH (Tchobanoglous et al. 2003). High hydrogen concentrations can inhibit propionic and butyric acid conversion. Flow equalization or additional capacity must be provided to meet peak flow and loading conditions (Metcalf and Eddy 2004).

Reactor temperatures ranging from 25° to 35°C are generally preferred to support more optimal biological reaction rates for more stable treatment. Variations in temperature can dramatically affect the performance of anaerobic reactors because of the different responses of various methabolic groups of microorganism. A drop in the activity of methanogens occur at temperatures below 16°C, (Mahmoud 2002) which can lead to an accumulation of volatile fatty acids, VFA, and a drop in pH (Leitão et al. 2006). Generally, COD concentrations greater than 1500 to 2000 mg/L are needed to produce sufficient quantities of methane to heat the waster without an external source. At 1300 mg/L COD or less, aerobic treatment may be the preferred selection. Anaerobic treatment can be applied at lower temperature and has been sustained from 10° to 20°C in suspended and attached growth reactors. Lower temperatures imply slower reaction rates, and longer SRT, thus larger reactor volumes and lower organic COD loadings are needed (Leitão et al. 2006).

The alkalinity or buffering capacity of the wastewater is another important parameter, because it affects the pH. With a CO_2 content ranging from 30 to 50 percent, $CaCO_3$ (2000 to 4000 mb/L) is typically required to maintain the pH at or near neutral. The level of alkalinity needed is seldom available in the influent wastewater, but may be generated in some cases by the degradation of protein and amino acids (Metcalf and Eddy 2004).

In anaerobic treatments, a portion of the organic waste is converted to biological cells, while the remainder is stabilized by conversion to methane and carbon dioxide. During anaerobic processes less sludge means lower nitrogen and phosphorus for biomass growth. Depending on the characteristics of the substrate and the SRT value, typical nutrient requirements for nitrogen, phosphorus, and sulfur range from 10 to 13 mg per 100 mg of biomass, respectively. The value for nitrogen and phosphorus are consistent with the value for these constituents estimated on the basic of composition of the cell biomass (Buzzini and Pires 2002).

Inorganic and organic toxic compounds also affect the performance of the UASB. Proper analysis and treatability studies are needed to assure that a chronic toxicity does not exist for wastewater treated by anaerobic processes. The presence of a toxic substance does not mean the process cannot function, but rather that it could inhibit anaerobic methanogenic reaction rates. A high biomass inventory and low enough loading sustain the process. Pretreatment steps may be used to remove the toxic constituents and in some cases, phase separation can prevent toxicity problems by degrading toxic constituents in the acid phase before exposure to methanogenic bacteria (Lettinga and Hulsshoff Pol 1991).

The efficiency of anaerobic treatment is related to the solids retention time (SRT). The SRT is a fundamental design and operating parameter for all anaerobic processes. As the retention time is decreased, the percentage of microorganisms wasted from the digester each day is increased. At some minimum SRT, the microorganisms are washed from the system faster than they can reproduce themselves and failure of the process results. The formation of anaerobic granular sludge can be considered as the major reason of the successful introduction of the upflow sludge bed for the anaerobic treatment of industrial effluents. The organization of the various groups of bacteria in the granules depends on the dominating metabolic pathways in the granules and on the wastewater composition (Metcalf and Eddy 2004).

Suspended solids (SS) affect the anaerobic process. The formation of scum layers and foaming due to floating lipids retards or impede the formation and entrapment of granular sludge. These SS tends to fall apart or disintegrate, requiring a complete wash-out of the sludge in the reactor, with the consequent declination of the overall methanogenic activity (Anh 2009). SS effect on the anaerobic treatment of domestic sewage was studied in fed-batch one day recirculation experiments using raw sewage and sewage without SS (paper-filtered). The average removal efficiency for chemical oxygen demand (COD at treatment of raw and paper-filtered sewage under "steady-state" conditions was, respectively, 63 and 81%. The significantly higher removal efficiency for COD when treating paper-filtered sewage instead of raw sewage can apparently be attributed to the production of COD due to hydrolysis of coarse SS. This means that a properly designed two-step system would enable an improved removal of SS and colloids as well (Mu et al. 2007).

Chemical oxygen demand (COD) refers to the organic matter in the wastewater expressed as the weight of oxygen to combust it completely. The COD removal is limited in high rate anaerobic systems at low temperatures and, therefore a long hydraulic retention time (HRT) is needed for providing sufficient hydrolysis of particulate organic in domestic wastewater (Elmitwalli and Otterpohl 2007) found that a total COD removal of 52–64% was obtained at HRT between 6 and 16 h in a UASB reactor.

The preliminary examination of the seed material can give information about its ability to degrade wastewater (Schmidt and Ahring 1996). Biomass is retained as aggregated granules formed by the self immobilization of the bacteria (Shin et al. 2001), mainly from the AZ strain of *Methanosaeta concilii* or *Methanobacterium strain* growing under conditions of high H_2 pressures (Hulsshoff Pol et al. 2004). Plug flow or semi-plug reactor with a nearly neutral pH, an adequate source of nitrogen in the form of ammonium, and a limited amount of cystine, are other factors favorable for granulation (Hulsshoff Pol et al. 2004).

17.5 ADVANTAGES, DISADVANTAGES AND LIMITATIONS OF THE UASB REACTOR

UASB reactors are more suitable than traditional anaerobic treatment, because of their ability to retain high biomass concentrations despite the up flow velocity of the wastewater and the production of biogas (Lettinga 1995 2001; Schmidt and Ahring 1996; Seghezzo et al. 1998; Leitão et al. 2006). The costs of aeration and sludge handling, the two largest costs associated with aerobic sewage treatment, can be reduced dramatically, because no oxygen is needed in the process, and the production of sludge is 3–20 times smaller than in aerobic treatment. Moreover, the sludge produced in aerobic processes has to be stabilized in classic anaerobic sludge digesters before it can be safely disposed, given it is very resistant to anaerobic degradation (Sanders et al. 1996). The most frequent limitations found for the reactor performance are related to flow rate, influent type and concentration, sludge retention time, nutrient available, temperature (Mahmoud 2002), pH (Metcalf and Eddy 2004), post-treatment methods (Lettinga 1995), pathogen and nutrients removal (Seghezzo et al. 1998), and presence of xenobiotics (Leitão et al. 2006). Many companies may resist the adoption of this technology (van Haandel and Lettinga 1994). The advantages and disadvantages of anaerobic sewage treatment (with special emphasis on high-rate reactors) are described in Table 17.1.

17.6 MANAGEMENT OF AQUATIC ECOLOGICAL RISKS

The prospect of assessing human health risks from exposure to chemical mixtures looms as a nightmare for many scientists, especially toxicologists. Indeed, exposure at a variety of levels to large numbers of chemical compounds, either concurrently or sequentially via multiple pathways, is the environmental reality for just about everyone on the planet. As depicted from Figure 17.3, it is possible that adverse ecological effects will occur as a result of exposure to stress from various human activities, a statement consistent with Norton et al. (1992); EPA (1986, 1992), and Suter (1993). Risk management is a policy-based activity that defines end-points and questions from risk assessment so as to protect human health and ecosystems (SETAC 1994). For example: if the number of enteric bacteria in a beach is over an acceptable level, then, risk management is responsible to limit the risk to an acceptable level.

Chemicals mixtures are present in ground, surface, and drinking water, air, food, and as well as in soil surrounding leaking toxic waste disposal sites. Examples of environmentally prevalent chemical mixtures are cigarette smoke, diesel and automobile exhaust, disinfection by-products from chlorination, and dioxin-like compounds formed by incomplete combustion in hospital and municipal waste management's facilities. Despite the high adverse potential of these mixtures, the majority of exposure standards are for single compounds. Moreover, the bulk of toxicology studies examine the cancer and noncancerous effects of single chemicals (Lang 1995). Currently, more than 95% of the resources in toxicology are devoted to single-chemical studies. For most chemical mixtures and multiple chemical exposures, adequate data on exposure and toxicity are lacking (Lang 1995). Household detergents are the principal source of inorganic and organic chemical substances of domestic origin which are discharged into the environment, normally through wastewater drainage systems (Prats et al. 1997). This type of wastewater is readily treatable. Industrial wastewater may contain toxic substances, a high percentage of organic matter, or solids which can make treatment more difficult.

Anionic surfactants (AS) in sewage are found as a result of the use of domestic products like detergents, cleaning and disk washing agents, and personal care products (Mungray and Kumar 2008). The linear alkylbenzene sulfonate (LAS) is the largest group of AS (Holt and Bernstein 1992) which can be removed with the UASB

Table 17.1 Advantages and disadvantages of anaerobic wastewater treatment (Seghezzo et al. 1998).

Advantages	*High efficiency.* Good removal efficiency can be achieved in the system, even at high loading rates and low temperature.
	Simplicity. The construction and operation of these reactors is relatively simple.
	Flexibility. Anaerobic treatment can easily be applied on either a very large or a very small scale.
	Low space requirements. When high loading rates are accommodated, the area needed for the reactor is small.
	Low energy consumption. As far as no heating of the influent is needed to reach the working temperature and all plant operations can be done by gravity, the energy consumption of the reactor is almost negligible. Moreover, energy is produced during the process in the form of methane.
	Low sludge production. The sludge production is low, when compared to aerobic methods, due to the slow growth rates of anaerobic bacteria. The sludge is well stabilized for final disposal and has good dewatering characteristics. It can be preserved for long periods of time without a significant reduction of activity, allowing its use as inoculums for the start-up of new reactors.
	Low nutrients and chemicals requirement. Especially in the case of sewage, an adequate and stable pH can be maintained without the addition of chemicals. Macronutrients (nitrogen and phosphorous) and micronutrients are also available in sewage, while toxic compounds are absent.
Disadvantages	*Low pathogen and nutrient removal.* Low pathogen and nutrient removal. Pathogens are only partially removed, except helminthes eggs, which are effectively captured in the sludge bed. Nutrients removal is not complete and therefore a post-treatment is required.
	Long start-up. Due to the low growth rate of methanogenic organisms, the star-up takes longer as compared to aerobic processes, when no good inoculums is available.
	Possible bad odors. Hydrogen sulphide is produced during the anaerobic process, especially when there are high concentrations of sulphate in the influent. A proper handling of the biogas is required to avoid bad smell.
	Necessity of post-treatment. Post-treatment of the anaerobic effluent is generally required to reach the discharge standards for organic matter, nutrients and pathogens.

reactor. Mungray and Kumar (2008) argued that the risk generated by UASB effluents and sludge to both aquatic and terrestrial environments has not been assessed. Wastewater irrigation poses several threats to the environment via contamination by nutrients, heavy metals, and salts. Increased loads of nitrates in wastewater may increase the danger of groundwater contamination (Stagnitti et al. 1998). Risks can be reduced by matching plant production systems to effluent characteristics (Pant and Mittal 2007).

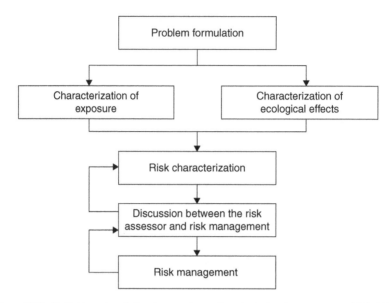

Figure 17.3 Definition sketch for the conduct of ecological risk assessment (EPA 1992).

The Clean Water Act (CWA) established the basic structure for regulating discharges of pollutants into the waters of the United States and regulating the quality standards for surface waters (EPA 1972). The basis of the CWA was enacted in 1948 and was called the Federal Water Pollution Control Act, but it was significantly reorganized and expanded in 1972. Under the CWA, EPA implemented pollution control programs such as setting wastewater standards for industry as well as for all contaminants in surface waters (EPA 1972). Individual homes connected to a municipal or septic system, or without a surface discharge are not required to have discharge permit, while industrial, municipal, and other facilities must obtain such permits if their discharges go to surface waters (EPA 1972).

Treated sewage discharged from UASB plants are likely to generated substantial risk to aquatic ecosystems. Post-treatment using 1–6 d detention, anaerobic, non-algal polishing pound was found ineffective. There is a need to utilize and aerobic method of post-treatment of UASB effluent in place of an anaerobic. Natural drying of UASB sludge on a drying bed (SDBs) under aerobic conditions results in a reduction of absorbed Anionic Surfactants (AS) by around 80%. Application of UASB sludge on SDBs was found simple, economical and effective. While disposal of treated UASB effluent may cause risk to aquatic ecosystem, use of dried UASB sludge is not likely to cause risk to terrestrial ecosystems (Mungray and Kumar 2008). In addition when using a UASB reactor in domestic sewage treatment, the ozonation can improve the effluent characteristics by removing organic matter, solids, surfactants, color and microorganisms (Gasi et al. 1990). The biodigester, however, produces an effluent still contaminated with pathogens. In order to protect public health disinfection is imperative (Gasi et al. 1990).

Al-Herrawy et al. (2005) tested a treatment system for domestic wastewater to remove *Cryptosporidium* oocysts. The process employed an up-flow anaerobic sludge blanket reactor; a Free Water Surface (FWS) unit; and sub-surface flow (SSF). They found that anaerobic treatment using UASB reactor could remove about 47.49% of *Cryptosporidium* oocysts present in raw wastewater, the FWS unit could remove about 94.35% of the oocysts that escaped from UASB treatment, and no oocysts were detected after SSF treatment. The physicochemical properties of the effluent were within the limits for wastewater reuse.

Some researchers see the slow sand filtration as a promising post treatment method for the UASB effluent. The slow sand filters principle of operation is that water seeps through a gelatinous layer found in the top few millimeters of the fine sand filter called hypogeal layer. As water passes through, particles are trapped in the viscous matrix and dissolved organic material is absorbed and metabolized by bacteria, fungi and protozoa (O'Neill 2009). Tyagi et al. (2009) found that slow sand filtration with 0.43 mm effective sand size is the most effective at a filtration rate of 0.14 m/h. It is capable of removing 91.6% of turbidity, 89.1% of suspended solids, 77% of chemical oxygen demand and 85% of bio-chemical oxygen demand, 99.95% of total and fecal coliforms and 99.99% of fecal streptococci. Slow sand filtration would be use for the pos-treatment of UASB reactor because effluent can be reused, as well as for safe discharge. This alternative reduces the risk and the environmental impact.

17.7 CONCLUDING REMARKS

The up flow anaerobic sludge blanket (UASB) reactor wastewater treatment system represents a proven technology for the treatment of raw sewage and a wide range of different industrial effluents. Whenever using the UASB it is important to make a preliminary examination of the seed material and wastewater both microbiologically and chemically. The operation of this reactor as a settling device is as important as its biological function especially in the treatment of domestic and industrial wastewater. The formation of anaerobic granular sludge is considered the major reason for its successful introduction. *Methanobacterium* growing under high H_2 pressures conditions is considered the key organism in granulation.

Conservation of the environment, low energy consumption, and biogas and low sludge production are the principal advantages of the UASB. The most frequent limitations in the reactor performance are flow rate, influent type and concentration, sludge retention time, nutrient availability, temperature, pH and presence of xenobiotics. The effluent needs post-treatment methods by which useful products like ammonia or sulfur can be recovered.

The world's freshwater resources are under severe stress. Although soil degradation and, salinisation, pollution of surface and groundwater, transport of harmful contaminants, impacts on vegetation and the transmission of diseases via the consumption of water-irrigated vegetables are some of the environmental risks associated with wastewater reuse, this strategy has been demonstrated to favor environmental conservation. Wastewater reuse is facing the challenge of minimize such risks and the UASB reactor is a promising technology to reach this end.

REFERENCES

Alcina-Alpoim M. 2003. Anaerobic biodegradation of long chain fatty Acids. Biomethanisation of biomass-associated LCFA as a challenge for the anaerobic treatment of effluents with high lipid/LCFA content. [dissertation]. Minho University.

Al-Herrawy AZ, Elowa SE, Morsy EA. 2005. Fate of Cryptosporidium during wastewater treatment via constructed wetland systems. International Journal of Environmental Studies 62(3): 293–300.

Anh NT. 2009. Methods for UASB reactor design, guest article. [internet] [cited 2010 Feb 18]. Available from: http://www.waterandwastewater.com/www_services/ask_tom_archive/methods_for_uasb_react or_design.htm.

Bal AS, Dhagat NN. 2001. Upflow anaerobic sludge blanket reactor- a review. Indian Journal of Environmental Health 43(2): 1–82.

Bengtsson S, Hallquist J, Werker A, Welander T. 2008. Acidogenic fermentation of industrial wastewaters: Effects of chemostat retention time and pH on volatile fatty acids production. Biochemical Engineering Journal 40: 492–499.

Buzzini AP, Pires EC. 2002. Cellulose pulp mill effluent treatment in an upflow anaerobic sludge blanked reactor. Process Biochemistry 38(5): 707–713.

Emmanuel E, Balthazard-Accou K, Joseph O. 2009. Impact of Urban Wastewater On Biodiversity of Aquatic Ecosystems, Environmental Management, Sustainable Development and Human Health. In: Laboy-Nieves EN, Schaffner FC, Abdelhadi AH and Goosen MFA, editors. Environmental Management, Sustainable Development and Human Health.Taylor and Francis Group, London UK; p. 399–422.

Elmitwalli TA, Otterpohl R. 2007. Anaerobic biodegradability and treatment of grey water in upflow anaerobic sludge blanket (UASB) reactor. Water Research 41 (6): 1379–387.

[EPA] Environmental Protection Agency (USA).1972. Summary of the Clean Water Act. [Internet] [cited 2010 Feb 16]. Available from: http://www.epa.gov/lawsregs/law/cwa.html

[EPA] Environmental Protection Agency (USA). 1986. Proposal Rules, part 797-Amended. Federal register, 51, 490–492 (Section 797.1310, Gammarid acute toxicity test).

[EPA] Environmental Protection Agency (USA). 1992. Framework for Ecological Risk Assessment. EPA/630-R-92/001, US, Environmental Protection Agency, Risk Assessment Forum, Washington DC. p. 41.

Gasi TMT, Amaral LAV, Pacheco CEM, Filho AG, Garcia Jr., AD, Vieira SM, Francisco Jr. R, Orth PD, Scoparo M de MSR, Días S and Magri ML. 1990. Ozone application for the improvement of UASB reactor effluent I. Physical-Chemical And Biological Appraisal, Ozone: Science & Engineering 13(2): 179–193.

Gujer W, Zehnder, A.J.B. 1983. Conversion process in anaerobic digestion. Water Science Technology 15: 127–167.

Hammes F, Kalogo Y, Verstraete W. 2000. Anaerobic digestion technologies for closing the domestic water, carbon and nutrient cycles. Water Science Technology. 41: 203–11.

Holt MS, Berstein SL. 1992. Linear alkylbenzenes in sewage sludge and sludge amended soils. Water Research 26: 613–624.

Hulsshoff Pol LW, de Castro-López SI, Lettinga G, Lens PNL. 2004. Anaerobic sludge granulation. Water Research 38: 1376–1389.

[JAEE]. Japanese Advanced Environmental Equipment (Japan). 2002. [Internet] [cited 2010 Apr 11]. IHI-UASB System. Available from http://www.gec.jp/JSIM_DATA/WATER/WATER_1/html/Doc_191.html

Lang L. 1995. Stranghe Brew: assessing risk of chemical mixtures. Environmental Health Perspect, Feb. 103(2):142–5. [Internet] [cited 2010 Apr 20] Available from: http://ehp.niehs.nih.gov/docs/1995/103-2/focus.html.

Leitão RC, van Haandel AC, Zeeman G, Lettinga G. 2006. The effects of operational and environmental variations on anaerobic wastewater treatment systems: A review. Bioresource Technology 97: 1105–1118.

Lettinga G, Rebac S, Zeeman G. 2001. Challenge of psychrophilic anaerobic wastewater treatment. Trends in Biotechnology 19 (9):363–370.

Lettinga G, van Velsen AFM, Hobma SW, De Zeeuw W, Klapwijk A. 1980. Use of the upflow sludge blanket (USB) reactor concept for biological wastewater treatment, especially for anaerobic treatment. Biotechnology and Bioengineering XXII: 699–734.

Lettinga G, Hulshoff Pol LWH, Zeeman G. 2000. Biological Wastewater treatment -Part I- Anaerobic Wastewater Treatment. Wageningen, The Netherlands, Wageningen University, p. 200.

Lettinga G. 2001. Digestion and degradation, air for life. Water Science Technology 44:157–76.

Lettinga G. 1995. Anaerobic digestion and wastewater treatment systems. Antonie van Leeuwenhoek 67: 3–28.

Lettinga G, Hulshoff Pol LW. 1991. UASB-process design for various types of wastewaters. Water Science and Technology 24(8): 87–107.

Li P, Wang Y, Wang Y, Liu K, Tong L. 2010. Bacterial community structure and diversity during establishment of an anaerobic bioreactor to treat swine wastewater. Water science and technology: a journal of the International Association on Water Pollution Research 61:234–52.

Mahmoud N. 2002. Anaerobic pre-treatment of sewage under low temperature (15°C) conditions in an integrated UASB-digester system. [dissertation].[Netherlands]: Wageningen University & Research Centre).[cited 2010 January 8]; Available from: http://library.wur.nl/WebQuery/wurpubs/317481

Mahmoud M. 2010. Development of novel naturally ventilated biotower for domestic wastewater treatment in developing countries. Science Topics.). [cited 2010 April 27]; Available from http://www.scitopics.com/Development_of_novel_naturally_ventilated_biotower_for_domestic_wastewater_treatment_in_developing_countries.html

Mahmoud N, Zeeman G, Gijzen H, Lettinga G. 2004. Anaerobic sewage treatment in a one-stage UASB reactor and a combined UASB-digester system. Water Research 38: 2348–2358.

Mahmoud N, Zeeman G, Gijzen H, Lettinga. 2003. Solids removal in upflow anaerobic reactors, a review. Bioresource Technology 90: 1–9.

McCarty PL and Smith DP. 1986. Anaerobic wastewater treatment. Environmental Science and Technology 20(12): 1200–1206.

MetCalf and Eddy 2004. Wastewater Engineering Treatment and Reuse. New York. International edition McGraw-Hill.

[MEF] Ministry of Environment and Forests (India). 2009. Technical EIA Guidance Manual for Common Effluent Treatment Plants. Final Draft by IL&FS Ecosmart Limited Hyderabad p. 3–15.

Mu SJ, Zeng Y, Tartakovsky B, Wu P.2007. Simulation and control of an Upflow anaerobic sludge blanket (UASB) reactor using an ADM1-Based Distributed Parameter Model American Chemical Society 46 (5): 1519–1526.

Mungray AK, Kumar P. 2008. Anionic surfactants in treated sewage and sludges: Risk assessment to aquatic and terrestrial environments. Bioresource Technology 99: 2919–2929.

Norton SB, Rodier DJ, Gentile van der Schalie WH, Wood WP, Slimak MW.1992. A framework for ecological risk assessment at the EPA. Environmental Toxicology and Chemistry 11: 1663–1672.

O'Neill S. 2009. Planning of laboratory scale grey water recycling systems. [dissertation]. TAMK University of Applied Sciences. p. 39.

Oktem YA, Ince O, Sallis P, Donnely T, Ince BK. 2007. Anaerobic treatment of a chemical synthesis pharmaceutical wastewater in a hybrid upflow anaerobic sludge blanket reactor. Bioresource Technology 99: 1089–1096.

Parawira W. 2004. Anaerobic Treatment of Agricultural Residues and Wastewater. Application of High-Rate Reactor. [dissertation].[Sweden] Lund University.

Pant A, and Mittal A K. 2007. Monitoring of Pathogenicity of Effluents from the UASB Based Sewage Treatment Plant. Environmental Monitoring and Assessment 133: 43–51.

Prats D, Ruiz F, Vázquez B, Rodríguez-Pastor M. 1997. Removal of anionic and nonanionic surfactants in a wastewater treatment plant with anaerobic digestion: A comparative study. Water Research 31(8):1925–1930.

Sabry T. 2008. Application of the UASB inoculated with flocculent and granular sludge in treating sewage at different hydraulic shock loads. Bioresource Technology 99: 4073–4077.

Sanders W, van Bergen D, Buijs S, Corstanje R, Gerrits M, Hoogerwerf T, Kanwar S, Zeeman G, van Groenestijn, Lettinga G. 1996. Treatment of waste activated sludge in an anaerobic hydrolysis upflow sludge bed reactor. EWPCA-Symposium 'Sludge treatment and reuse'. 7–9 May, München, Germany.

Seghezzo L, Zeeman G, van Lier JB, Hamelers HVM, Lettinga G. 1998. A review: The anaerobic treatment of sewage in UASB and EGSB reactors. Bioresource Technology 65:175–190.

[SETAC]. Society of Environmental Toxicology and Chemistry. 1994. Final report: Aquatic Risk Assessment Mitigation Dialogue Group. Society of Environmental Toxicology and Chemistry Press, Pensacola, FL. p. 220.

Schmidt JE, Ahring BK. 1996. Granular sludge formation in upflow anaerobic sludge blanket (UASB) reactors. Biotechnology and Bioengineering 49: 229–246.

Shin H S, Han SK, Song YC, Lee CY. 2001. Performance of UASB reactor treating leach-ate from acidogenic fermenter in the two-phase anaerobic digestion of food waste. Water Research 35: 441–3447.

Stagnitti F, Sherwood J, Allinson G. Evans L, Allinson M, Li L,Phillips I. 1998. Investigation of localized soil heterogeneities on solute transport using a multisegement percolation system NZ Journal of Agricultural Research 41: 603–612.

Suter, GW. 1993. Ecological Risk assessment. Boca Ratón, Fl., Lewis Publishers, II (ed.) p. 538.

Tawfik A, El-Gohary F, Temmink H. 2010. Treatment of domestic wastewater in an up-flow anaerobic sludge blanket reactor followed by moving bed biofilm reactor. Bioprocess Biosyst Engineering 33:267–276.

Tchobanoglous G, Burton FL, Stensel HD. 2003. Waste-water engineering: treatment and reuse. New Delhi, Tata McGraw-Hill Publishing Company Limited p. 1819.

Tyagi VK, Khan AA, Kazmi AA, Mehrotra I, Chopra AK. 2009. Slow sand filtration of UASB reactor effluent: A promising post treatment technique. Desalination 249: 571–576.

Van Haandel AC, Lettinga G. 1994. Anaerobic sewage treatment. A practical guide for regions with a hot climate. John Whiley and Sons. Great Britain.

Van Haandel A, Van Der Lubbe J. 2007 [cited 2010 May 03]. Handbook Biological Waste Water Treatment. [Internet]. Leidschendam – The Netherlands: Quist Publishing. Available from: http://www.wastewaterhandbook.com/webpg/th_sludge_83anaerobic_digestion.htm.

Water Treatment Plant. Sand filter. 2009. [Internet] [cited 2010 Apr 18]; Available from: http://www.thewatertreatmentplant.com/sand-filter.html.

Glossary

acahual: a secondary plant community derived from tropical rain forests; acahuales are the natural ecological restoration and production unit used by local communities

acceptable risk: risk that people could assume according with their cultural values, beliefs, education and experiences

acetogenic bacteria: microorganisms that generates acetate as a product of anaerobic respiration

acid deposition: precipitation of airborne dry acidic particles

activated sludge: the flocculent mass of microorganisms, mainly bacteria, that develops when sewage or liquid effluent is aerated

adsorption capacity: amount of an impurity adsorbed at equilibrium per weight of adsorbent used

adsorption isotherm: equation that describes the amount of a substance adsorbed onto a surface at a constant temperature; it depends on its pressure or concentration

advection: is a transport mechanism of a substance by a fluid, due to the fluid's bulk motion in a particular direction

aftershocks: series of tremors occurring after the main strike of an earthquake

albedo: reflectivity of the Earth's surface

amphipathic: molecules consisting of hydrophilic polar head group and a hydrophobic nonpolar tail group

anaerobic digestion: a series of processes in which microorganisms breakdown biodegradable material in the absence of oxygen

anthropogenic: derived from human activities

aquifer: an underground layer of water-bearing permeable rock or unconsolidated materials (gravel, sand, silt, or clay) that supplies water to wells and springs

aromatic compound: organic substances whose molecular structure includes one or more planar rings of carbon atoms

benefit-cost analysis: an organizational framework for identifying, quantifying, and comparing the cost and benefits (measured in dollars) of a proposed action

benefits: the gains or monetary values of direct and indirect uses of a resource or with improvements in the quality of a resource unit

bioaccumulation: accretion of substances in an organism at a rate greater than that at which the substance is lost

bioaugmentation: addition of microorganisms to enhance bioremediation where chemical release has occurred or indigenous bacteria are insufficient in number or capability to degrade the existing compounds

bioavailability: extent of absorption of a contaminant by living organism which can cause an adverse physiological or toxicological response.

biochemical oxygen demand (BOD):   the amount of oxygen (in mg) required by aerobic bacteria to decompose the biodegradable organic material in 1 liter of an effluent

biocultural resources: the natural and cultural heritage that exist in rural communities, related to resource management, ecosystem services or traditional knowledge

bioenhancing agents: compounds that improve the transport of immiscible wastes into solutions increasing remediation rates by stimulating bacterial growth

biogas: a mixture of methane (CH_4) and carbon dioxide (CO_2), produced by the anaerobic digestion of sludge or organic material

biomagnification: concentration increase of substance in the food web as a consequence of its persistence and low rate of degradation/excretion

biomonitor: any biological species or group of species whose function, population, or status can be used as an indicator of ecosystem or environmental integrity

bioremediation: the process where microorganisms transform xenobiotics compounds into a less toxic form, or completely mineralizes them into CO_2, inorganic compounds and water

biotreat: an aqueous-based nutrient mixture that shows surfactant properties, usually marketed for the cleanup of contaminants and for wastewater treatment

capability approach: a strategy that provides the intellectual foundation for human development, via participation, human well-being and freedom

capital basis of society: the set of human, man-made, and natural capital

carbolina: an industrial or domestic disinfectant

cataclysmic: a violent disturbance that causes destruction or brings about fundamental changes

cfu: unit of bacteria counting after growth on solid culture media in laboratory

chemical oxygen demand: an analytical method used to determine the content of biodegradable and non biodegradable organic material in wastewater

coliform: rod-shaped gram-negative bacteria abundant in the feces of warm-blooded animals, found in aquatic and terrestrial environments and on vegetation; their presence is used to indicate that other pathogenic organisms of fecal origin may be present

cometabolic: a process which relies on another compound for a carbon energy source to enhance biodegradation

community: group of people with different backgrounds who share the same interest

competitive advantages: when a company exhibits high performance levels than its contenders

complex: a molecular unit where a central metal ion is surrounded by nonmetal atoms or molecules through co-ordinate or dative-covalent bonds

condensation nuclei: small particles (typically 0.2 μm) about which cloud droplets coalesce; also known as cloud seeds

conductivity: measurement of the ability of an aqueous solution to carry an electrical current; it is directly proportional to the concentration of dissolved constituents

contingent valuation: a survey technique using direct questioning of people to estimate individuals' willingness to pay

continuous improvement: continuously perfect processes/activities of organizations in order to seek excellence

convection: the upward movement of an air mass due to heat

cost-benefit analysis: economic method for assessing the gain and prices of achieving alternative fitness-based standards with different levels of health protection.

culture: information patterns learned, shared and used by a group, including roles, values, and beliefs

defensive methods: cost of abatement practices

design for disassembly: development of products so that their restructuring and recycling can is facilitated, thus the re-usage and increase of the life cycle of the product and its components can also be achieved

design for environment: development of a product or process to reduce its environmental impact

developing country: a state with low per capita incomes relative to world standards

disinfection: chemical or physical methods aimed at reducing or eliminating pathogens (bacteria, viruses) in treated effluent

domestic wastewater: combination of liquid wastes which may include chemicals, household wastes, human excreta, animal or vegetable matter in suspension or solution

downwind: the direction in which the wind blows

drainage: facilities used to collect and carry off excess water

Ecohealth: an approach to support research on the relationships between all components of an ecosystem to define and assess priority problems that affect the health of people and the sustainability of the ecosystems they depend upon (http://www.irdc.ca/ecohealth/)

ecological risk assessment: the process that evaluates the likelihood that adverse ecological effects may occur or are occurring as a result of exposure to one or more stressors

eluate: the solution that results from the elution process

emulsions: suspensions of fine droplets of one liquid on another substance

endemism: the occurrence of a species in a particular limited locality or region

environmental justice: when all people is treatment equal, enjoy the same protection and health hazards regardless of race, color, income, nationality, or religion

environmental management practices: a set of strategies designed to reduce environmental impacts and explore competitive opportunities in organizations

envits[2]:   a biodegradable degreaser commercially available to homogenize and degrade organic compounds in the water matrix

ethnoecological restoration: management strategy focused on the recovery of the main environmental functions and based on traditional biocultural resources

ethnoecology: an interdisciplinary approach that studies the ways in which nature is and has been understood by different cultures

eutrophication: an increase in the concentration of nitrogen and phosphorus to an extent that there is an raise in the primary productivity of the ecosystem.

exposure: proximity a process or agent that can inflict the health of humans and the environment

externality: a side-effect on others following from the actions of an individual or group

fatty acid: a group of carboxylic acids which impart a foul, soapy flavor to beer, and contributes to its staling

fault: a crack in the Earth's crust resulting from the displacement of one side with respect to the other

flood: temporary inundation onto normally dry land

genuine savings: a weak sustainability indicator system developed by the World Bank

hazard: a condition that increase the possibility of injury

health risk assessment: the process of quantifying the probability of a harmful effect to individuals or populations from certain human activities

heavy metal: metallic chemical element that has a relatively high density and generally, its trace concentration can become toxic at higher levels

hedonic methods: form of values measurement that relies on natural experiments to determine economic values of environmental impact

hydraulic retention time (hrt):   the measurement of the average length of time that a soluble compound remains in a constructed reactor

hydrodynamic dispersion: the process in which a localized pulse of a dynamically neutral tracer disperses in a flow field under the combined action of convection and molecular diffusion

hydrolysis: chemical reaction, molecules of water is split into hydrogen (H^+) and hydroxide (OH^-)

hygroscopic nuclei: small particles which tend to attract and condense atmospheric water

imposed risk: danger that people could not control

inipol eap22:   microemulsion of a saturated solution of urea and oleic acid, containing tri(laureth-4)-phosphate and butoxy-ethanol

landscape management: an instrument for local policy development and social inclusion

life cycle analysis: assessing the environmental load associated with the product or process after the identification and quantification of energy and materials associated to the extraction of raw-material, manufacturing, transport, distribution, usage/re-usage and recycling

ligands: atoms or molecules that surround or bond a central metal ion in a complex compound.

methanogenesis: generation of methane (60%) and carbon dioxide (40%) from organic acids and their derivatives in the acidogenic phase

model: formal or informal framework for analysis that highlights some areas of the problem in order to better understand complex relationships

monitoring: procedures and techniques used to systematically analyze, inspect, and collect data on operational parameters of a facility or the environment

mutation: a change of the DNA sequence within a gene or chromosome resulting in the creation of a new character or trait not found in the parental type

negative gravity: a phenomena that occurs when the actual gravitational force is smaller than the calculated force in a given place, thus it's difference is negative

non-potable: water that may contain objectionable pollution or infective agents, and is considered unsafe and/or unpalatable for drinking

npk: a nitrogen (N), phosphorus (P), and potassium (K) chemical agent considered important to force crop production

nucleation: the condensation of gaseous matter of low vapor pressures

organic loading rate: the amount and concentration of organic matter processed by a bio-reactor and expressed in COD g/l/day

outrage: an act associated with anger, frustration and anxiety

pathogenic: organisms capable of causing diseases in a person

pollutant: any substance of strictly anthropogenic origin introduced into a biotope in sufficient amounts to adversely affect the quality of the resource

polycyclic aromatic hydrocarbon: organic molecules that consist of three or more benzene rings and are commonly produced by fossil fuel combustion

potable water: water that does not contain objectionable pollution, minerals, or infective agents, thus it is considered safe for drinking

pyroclastic flows: a hot, fast-moving and high-density mixture of gases and unsorted volcanic fragments, crystals, ash, pumice, and glass shards

radiative forcing: the difference between incoming and outgoing radiation energy in a given climate system

rainwater: water that has fallen as rain and contains little dissolved mineral matter

remediation techniques: procedures developed to eliminate or attenuate hazardous waste impacts due to the danger they represents to the environment and human health

riparian: the interface between land and a stream

risk: the possibility that something negative could happen as a result of an exposition

risk analysis: examination of the potential hazards and vulnerability of conditions that could pose a potential threat or harm to people, property, livelihoods and the environment

risk communication: interactive process of sharing information and opinions between individuals, groups and institutions

risk management: process of identifying, evaluating, selecting, and implementing actions to reduce risk to human health and to ecosystems.

risk perception: community interpretation of hazards

runoff: the portion of precipitation that drains from an area through stream channels

seismogenic: capable of generating earthquakes

smog: air pollution resulting from the combination of smoke and fog

social capital: the collective value of all social networks that is considered a key component to building and maintaining democracy

sonication: is the act of applying sound (usually ultrasound) energy to agitate particles in a sample

stormwater: rainfall that is collected after it has runoff urban surfaces

stressor: any physical, chemical, or biological entity that can induce an adverse effect

strong sustainability: the entire capital stock of natural, human and man-made capital that can be handed over to the next generation

subduction: the sliding down of one crustal plate below another crustal plate as the two converge

surfactant: amphipathic molecules consisting of a hydrophilic polar head group and a hydrophobic nonpolar tail group

sustainability measurement: a method to minimize risks and reveal the true welfare state of a country

sustainable communities: ecologically literate communities that understand that all members of an ecological community are interconnected in the web of life

sustainable development: since the Brundtland Report in 1987, the model for social and political processes that complies with the current demands and ensures that the future generations can also meet theirs

sustainable organizations: organizations that combine economic performance with environmental and social performance

synergy: interaction among different components within a system

technology transfer: shifting or sharing knowledge, skills, processes, and technologies through education and training

tectonic: the force that triggers the deformation of the Earth's crust, or the structures or features produced by such deformation

thermal inversions: the settlement of a warm air layer over a layer of cooler air that lies over the ground

toluene: a clear water-insoluble aromatic hydrocarbon widely used as an industrial feedstock and as a solvent.

toxicokinetics: studies the sequential interaction phases (exposure, absorption, distribution, biotransformation, accumulation, and excretion) between a chemical substance and a living organism

toxicology: the interpretation of the health risk associated to the adverse effects of chemical and physical on organisms

trace gases: substances that make up less than 1% by volume of Earth's atmosphere

transcurrent fault: a large-scale strike-slip fault where in which the surface is steeply inclined

troposphere: the lower layer of the atmosphere which contains approximately 75 percent of its mass, and 99 percent of its water vapor and aerosols

tsunamigenic: capable of generating tsunamis

unacceptable risk: risk that people could not assume according with their cultural values, beliefs, education and experiences

up-flow anaerobic sludge blanket (UASB): anaerobic digester used in the treatment of wastewater, forming a blanket of granular sludge

urban hydrology: the study of the water dynamic processes occurring within the urban environment

value: what one is willing to give up in order to obtain a good, service, experience, or state of nature

volatile organic compounds (VOC's): organic chemical compounds that have high enough vapor pressures under normal conditions to significantly vaporize and enter the atmosphere

voluntary risk: a risk that people agree to be expose (ex. smoking cigarettes)

wastewater: water discarded after been used in domestic, commercial, industrial and/or agricultural activities

water stress: a situation when the demand for water exceeds the available amount during a certain period or when poor quality restricts its use

watershed: an extent of land where water drains downhill into a river, lake, reservoir, estuary, wetland, or the sea

weak sustainability: the preservation of the entire capital stock and not to the preservation of its parts

weathered crude oil: petroleum that, owing to evaporation and other natural causes during storage and handling, has lost an appreciable quantity of its more volatile component

ξ potential: characteristic of the repulsive forces between particles, after measuring the speed of motion of the particle subjected to an electric field

CAPECO facilities after a massive explosion in Cataño, Puerto Rico (@Eddie N. Laboy-Nieves)

Subject index